CONTENTS

Contents

*Exercises with an asterisk * include Quick Response (QR) Codes that link print manual to online imagery or media. Exercises with a plus sign + include optional parts that make use of Google Earth™.*

INTRODUCTION

NEW TO THIS EDITION

All of the exercises found in the previous edition of this Lab Manual remain, along with one new exercise and many new features:

- A new exercise on *Groundwater* (Exercise 25) has been added.

- A new exercise on *Geographic Information Systems and Remote Sensing* (Exercise 8) replaces the last edition's GIS exercise. It has been written by Ryan Jensen of Brigham Young University, a recognized GIS expert.

- Sixteen new pages of color maps and images are included in the back of the Lab Manual, making a total of 32 pages of color maps. In previous editions of the Lab Manual, many of these new color maps appeared in grayscale.

- Many of the exercises underwent major revisions or expansions, including *Weather Variability and Climate Change* (Exercise 24), *Biomes and Ecological Land Units* (Exercise 26), *Aerial Photographs and Stereograms* (Exercise 32), and *Mass Wasting* (Exercise 38).

- The art has been improved and updated throughout the Lab Manual, including additional diagrams by illustrator Dennis Tasa from the twelfth edition of *McKnight's Physical Geography: A Landscape Appreciation*.

- The order of exercises has been rearranged slightly. At the suggestions of instructors, the exercises introducing contour line and topographic map skills now appear in the middle of the Lab Manual, just before the exercises on geomorphology that require these skills.

- Additional online resources, such as color maps, photographs, satellite images, and Google Earth™ "videos," have been added for some exercises. Students may immediately access nearly all of these materials by scanning QR (Quick Response) codes using a smartphone or other mobile device, or by visiting the MasteringGeography website.

- The new edition includes **MasteringGeography™ with Pearson eText**, specifically designed for this Lab Manual. The Mastering system helps teachers maximize lab time with media-rich, customizable, easy-to-assign, automatically graded assessments that motivate students to learn outside of class and arrive prepared for lab.

TO THE STUDENT

The exercises in this Lab Manual give you the opportunity to apply many of the concepts studied in your physical geography course.

Lab Manual Organization: The Lab Manual begins with some basics, such as metric conversions, latitude and longitude, and time zones. Next, you'll gain proficiency with mapping skills, such as using map scales, map projections, online mapping programs such as Google Earth™, and the fundamentals of GIS. The next section of the manual focuses on weather patterns and processes, the interpretation of weather maps, weather satellite images and climate data, as well as groundwater, biomes and soil. In the final section of the manual you will study the development of Earth's surface features, including the interpretation of topographic maps and aerial photographs, plate tectonics, volcanoes, faulting, the work of streams, underground water, wind, glaciers, and coastal waves.

Background Material: Each exercise begins with a brief introductory section that reviews key concepts and provides background information for the exercise problems. A reference to relevant pages in the textbook, *McKnight's Physical Geography: A Landscape Appreciation*, Twelfth Edition, by Darrel Hess, is provided at the beginning of most exercises. Key terms are marked in bold type, and a glossary is found at the back of the Lab Manual in Appendix I. It is likely that your instructor will assign several exercise problem sets to you each week. The length and relative difficulty of the problems vary from exercise to exercise.

Online Resources: Many of the exercises in the Lab Manual have problems using Google Earth™, or are based on photographs, satellite images or weather data you can access over the Internet. The Lab Manual website (**www.MasteringGeography.com**) provides easy links to all the Internet sites and images you'll need to complete these exercises, as well as various study tools and media to help you master course concepts.

Using Your Smartphone: You can access nearly all of the Internet content for the Lab Manual, including color maps, photographs, satellite images, and Google Earth™ "videos," by scanning the QR (Quick Response) code for an exercise with your smartphone or other mobile device. (You may need to download QR scanning software for your phone before using these codes.) This lets you view this enhanced Lab Manual content anywhere you can use your phone.

Color Topographic Maps: Thirty-two pages of topographic maps and aerial images are reproduced in color in the back of the Lab Manual. These maps are referred to in the exercises as "Map T-1," "Map T-2," and so on. A series of graphic map scales and a color map showing global precipitation is found inside the front cover of the Lab Manual. Charts showing standard symbols used on topographic maps and metric system conversion formulas are found on the inside of the back cover.

Stereo Aerial Photographs: Several of the exercises include stereo aerial photographs ("stereograms"). To view the aerial photographs in stereo, you will use a lens stereoscope supplied by your instructor. It is possible, however, to complete the exercises that include stereo aerial photographs even if you don't have a stereoscope or if you have difficulty using one.

Math Skills Practice Worksheet: You will find a "Math Skills Practice Worksheet" at the back of the Lab Manual in Appendix II. This ungraded worksheet will help you practice the kinds of math and charting skills you'll need to complete the exercises in the Lab Manual. Hints are provided to help you with the worksheet.

Supplies Needed: Few supplies are needed to complete the exercises. A ruler (about 6 inches long; graduated to at least 1/16 inch; scaled in both inches and centimeters), a 3-foot length of string, and blue, green, and red pencils will be useful. An inexpensive magnifying glass (about 5x) may be helpful in some map reading exercises. For a few of the exercises you will need access to a 25-centimeter (10-inch) diameter or larger globe and a small atlas with an index.

Unless otherwise directed by your instructor or the exercise problems, you should round off numbers in your answers to one decimal place (for example, round off 12.437 to 12.4).

MasteringGeography

The MasteringGeography website for the Lab Manual includes color maps, images, photographs, satellite movie loops, Google Earth™ KMZ files, and links needed to complete the Internet-based problem sets, along with various study tools and media to help you master course concepts. Go to **www.MasteringGeography.com**. The URLs of the recommended Internet sites needed to complete an exercise are also printed in the Lab Manual (instructors may recommend additional or different Internet sites for students).

TO THE TEACHER

This new edition of the Lab Manual retains all of the exercises found in the previous edition, along with a new exercise on *Groundwater* (Exercise 25) and a new exercise on *Geographic Information Systems and Remote Sensing* (Exercise 8), replacing the last edition's GIS exercise. The order of exercises has been adjusted slightly, grouping the topographic map interpretation skills in the middle of the Lab Manual before the exercises on geomorphology.

Short Exercises: For greater flexibility, some major topics are covered over several exercises, and the problems for most exercises are divided into two or more parts. The length and difficulty of the problem sets vary greatly from exercise to exercise. Although most exercises are designed to stand alone, in a few cases one exercise builds upon the previous one. For example, the exercise on the adiabatic processes assumes an understanding of relative humidity, and before assigning exercises using topographic maps or Google Earth™, those skills should be reviewed.

Students will be called upon to use several key skills repeatedly. For example, the interpretation of various kinds of isolincs will be required throughout the weather and climate exercises, and the interpretation of topographic maps is needed throughout the geomorphology exercises. These skills should be emphasized early on.

QR Codes: The addition of QR (Quick Response) codes with many exercises gives students access to Lab Manual Internet content wherever they can use a smartphone or other mobile device. By scanning the QR codes, they can view color maps, satellite movie loops, color photographs, and Google Earth™ "videos" needed to complete nearly all of the Internet-based problems in the Lab Manual. The **www.MasteringGeography.com** website also provides all of the content accessed with the QR codes, along with KMZ files for the problems using Google Earth™.

S.I. and English Units: Although either English units or S.I. units may be used to complete many of the exercises, the emphasis in the weather and climate exercises is on S.I. units; because the topographic maps used in the geomorphology exercises are based on English units, most of the topographic map interpretation exercises retain their emphasis on English units.

Stereograms: All of the exercises that include stereo aerial photographs have matching topographic maps. If your school doesn't have lens stereoscopes for classroom use, you can visit online sites such as **www.forestry-suppliers.com** or **www.amazon.com** to purchase inexpensive "student model pocket stereoscopes." As in earlier editions of the Lab Manual, if students are unable to see the images in stereo, it is still possible to complete the exercise problems. Further, all topographic map- and stereogram-based problem sets also include the latitude and longitude of key features so that these areas may be studied with Google Earth™ or the USGS National Map.

Math Skills Practice Worksheet: A Math Skills Practice Worksheet is found in Appendix II. This ungraded worksheet helps students practice the kinds of math and charting skills they will need to complete exercises in the Lab Manual. Hints (but not the answers) are provided to help students with the worksheet.

Answer Key: An answer key for the exercises in the Lab Manual is available. The Answer Key also includes a sample course syllabus, and suggestions on supplementing Lab Manual exercises. Contact your local Pearson representative to receive a copy of the Answer Key.

MasteringGeography: The **Mastering** platform is the most widely used and effective online homework, tutorial, and assessment system for the sciences. It delivers self-paced coaching activities that provide individualized coaching, focus on course objectives, and are responsive to each student's progress. The Mastering system helps teachers maximize lab time with customizable, easy-to-assign, and automatically graded assessments that motivate students to learn outside of class and arrive prepared for lab.

MasteringGeography (**www.MasteringGeography.com**) is now available with this Lab Manual, as well as with *McKnight's Physical Geography: A Landscape Appreciation Twelfth Edition,* and offers:

- **Assignable activities** that include Pre- and Post-Lab assessments for each Lab Exercise, Geoscience Animation activities, *Encounter Physical Geography* Google Earth™ Explorations, Video activities, GIS-inspired MapMaster™ Interactive Map activities, Map Projection activities, GeoTutor coaching activities on the most challenging topics in geography, end-of-chapter questions and exercises, reading quizzes, Test Bank questions, and more.
- **Student Study Area** with Geoscience Animations, Videos, GIS-inspired MapMaster™ interactive maps, weblinks, glossary flashcards, *In the News* RSS feeds, chapter quizzes, an optional Pearson eText (including versions for iPad and Android devices), and more.

ACKNOWLEDGMENTS

I would like to thank the many instructors who offered suggestions for improving this and earlier editions of the Lab Manual, especially Maura Abrahamson, Morton College; Edward Aguado, San Diego State University; Mary Bates, College of the Canyons; Barbara Batterson, San Diego Mesa College; Peter Combs, Hunter College-CUNY; John Conley, Saddleback College, Orange Coast College, and Fullerton College; the late James E. Court, City College of San Francisco; Purba Fernandez, DeAnza College; Justin Fuller, Central New Mexico Community College; George Gaston, University of North Alabama; Dafna Golden, Mount San Antonio College; Megan Harlow, Saddleback College; Ann Harris, Eastern Kentucky University; Tim Krantz, University of Redlands; Wayne Kukuk, Kellogg Community College; Dean Lambert, San Antonio College; Jen Lipton, Central Washington University; W. Franklin Long, Coastal Carolina Community College; Michael Madsen, Brigham Young University, Idaho; Robert Manlove, City College of San Francisco; Anya-Kristina Marquis, Moreno Valley College; Blake Mayberry, Red Rocks Community College; Armando Mendoza, Cypress College; Victor Mesev, Florida State University; Peter Mortenson, Richland Community College; Michael Pool, Austin Community College; Bradley Rundquist, University of North Dakota; Deborah A. Salazar, Oklahoma State University; Rich Smith, Harford Community College; Alice Luthy Tym, University of Tennessee at Chattanooga; Janet Valenza, Austin Community College; Brad Watkins, University of Central Oklahoma; Elizabeth Weaver, U.S. Military Academy; Shawn Willsey, College of Southern Idaho; and Craig ZumBrunnen, University of Washington. Of course, any errors remaining are my responsibility alone.

Many thanks to Ryan Jensen of Brigham Young University for developing and contributing the lab exercise on *Geographic Information Systems and Remote Sensing*, and to Armando Mendoza of Cypress College for a thoughtful critique of the entire Lab Manual.

The staff of the U.S. Geological Survey Earth Sciences Information Center in Menlo Park, California, provided valuable assistance in obtaining many of the maps and photographs used in this manual, as did the USGS/EROS Data Center in Sioux Falls, South Dakota. Thanks to Brendan Kelly and Greg Durocher of the USGS in Anchorage, Alaska, for their help selecting aerial photographs from that part of the country; to the staff at the Alaska Climate Research Center, University of Alaska, Fairbanks, for assistance obtaining climate data; and to Christopher Bailey and David Novak of NOAA for their help obtaining archived weather maps. Ed Shelden, Jr. provided much appreciated help with the processing of some of the digital images used in the Lab Manual.

Many thanks to all at Pearson Education, especially to Executive Geography Editor Christian Botting and Content Producer Anton Yakovlev for their helpful suggestions and steadfast support throughout this project. Thanks also to SPi Global Project Manager Christian Arsenault, International Mapping Senior Project Manager Kevin Lear, Media Producer Tim Hainley, and Executive Marketing Manager Neena Bali, for helping me through the production process.

Thanks go to all of my City College of San Francisco Earth Sciences Department colleagues, Ian Duncan, Carlos Jennings, James Kuwabara, Joyce Lucas-Clark, Suzanne Maher, Russell McArthur, Kirstie Stramler, Katryn Wiese, and Gordon Ye, for helping cultivate the most stimulating and supportive of academic work environments. My special thanks go to Department Chair Chris Lewis for his help and suggestions on a number of exercises in this Lab Manual.

Most of all, I wish to thank my students who, often without knowing it, helped me develop and improve the exercises for this Lab Manual.

If students or instructors have any comments, please address them to:

Darrel Hess
Earth Sciences Department
City College of San Francisco
50 Phelan Avenue
San Francisco, CA 94112
dhess@ccsf.edu

ABOUT OUR SUSTAINABILITY INITIATIVES

Pearson recognizes the environmental challenges facing this planet, as well as acknowledges our responsibility in making a difference. This book is carefully crafted to minimize environmental impact. The binding, cover, and paper come from facilities that minimize waste, energy consumption, and the use of harmful chemicals. Pearson closes the loop by recycling every out-of-date text returned to our warehouse.

Along with developing and exploring digital solutions to our market's needs, Pearson has a strong commitment to achieving carbon-neutrality. As of 2009, Pearson became the first carbon- and climate-neutral publishing company, having reduced our absolute carbon footprint by 22% since then. Pearson has protected over 1,000 hectares of land in Columbia, Costa Rica, the United States, the UK and Canada. In 2015, Pearson formally adopted The Global Goals for Sustainable Development, sponsoring an event at the United Nations General Assembly and other ongoing initiatives. Pearson sources 100% of the electricity we use from green power and invests in renewable energy resources in multiple cities where we have operations, helping make them more sustainable and limiting our environmental impact for local communities.

The future holds great promise for reducing our impact on Earth's environment, and Pearson is proud to be leading the way. We strive to publish the best books with the most up-to-date and accurate content, and to do so in ways that minimize our impact on Earth. To learn more about our initiatives, please visit **www.pearson.com/social-impact/sustainability/environment.html**.

PEARSON

EXERCISE **1**
Metric Conversions

Objective:	To practice making unit conversions between the English system and the metric (S.I.) system of measurement.
Reference:	Hess, Darrel. *McKnight's Physical Geography,* 12th ed., pp. 6–8 and A1–A2.

METRIC CONVERSIONS

Although the general public in the United States still uses the so-called English system of measurement (e.g., feet, miles), most of the rest of the world—and the entire scientific community—uses the metric system (e.g., meters, kilometers). Today, the metric system has been incorporated into what is formally known as the *Système International* or *S.I.* system of measurement. In this Lab Manual you will encounter both English units and S.I. units of measure. It is useful for you to be comfortable using both systems, and for you to be able to convert units from one system into the other.

There are two levels of conversion precision that may be useful to you. First, it is helpful to have a rough idea of the equivalents—the kind of conversion you can do quickly in your head without a calculator. For example, it is useful to know that 1 kilometer is about two-thirds of a mile. The second kind of conversion is a precise equivalent—for example, 1 kilometer = 0.621 mile. These exact conversions are necessary if a precise measurement in one system must be duplicated in the other system. Some commonly used conversions are given in Figures 1-1 and 1-2 (to an accuracy of 3 decimal places). Additional conversions factors are found on the back cover of the Lab Manual.

Conversions: *S.I. to English*

	Exact Conversions	**Approximate Conversions**
Distance:	cm $\times$ 0.394 = inches	1 centimeter = a little less than $^{1}/_{2}$ inch
	m $\times$ 3.281 = feet	1 meter = a little more than 3 feet
	km $\times$ 0.621 = miles	1 kilometer = about $^{2}/_{3}$ mile

1 cm (centimeter) = 10 mm (millimeters)
1 m (meter) = 100 cm
1 km (kilometer) = 1000 m

Volume:	liters $\times$ 1.057 = quarts	1 liter = about 1 quart
Mass (Weight):	g $\times$ 0.035 = ounces	1 gram = about $^{1}/_{30}$ ounce
	kg $\times$ 2.205 = pounds	1 kilogram = about 2 pounds

1 kg (kilogram) = 1000 g (grams)

Temperature:	($^{\circ}$C $\times$ 1.8) + 32 = $^{\circ}$F	1°C change = 1.8°F change

---

Figure 1-1: S.I. to English system conversions.

1

Conversions: *English to S.I.*

	Exact Conversions		Approximate Conversions	
Distance:	inches $\times$ 2.540	= centimeters	1 inch	= about $2\frac{1}{2}$ cm
	feet $\times$ 0.305	= meters	1 foot	= about $\frac{1}{3}$ m
	yards $\times$ 0.914	= meters	1 yard	= about 1 m
	miles $\times$ 1.609	= kilometers	1 mile	= about $1\frac{1}{2}$ km

1' (foot) = 12" (inches)
1 yard = 3'
1 statute mile = 5280'

Volume:	quarts $\times$ 0.946	= liters	1 quart	= about 1 liter
	gallons $\times$ 3.785	= liters	1 gallon	= about 4 liters

1 gallon = 4 quarts

Mass (Weight):	ounces $\times$ 28.350	= g	1 ounce	= about 30 g
	pounds $\times$ 0.454	= kg	1 pound	= about $\frac{1}{2}$ kg

1 lb. (pound) = 16 oz (ounces)

Temperature:	(°F − 32) ÷ 1.8	= °C	1°F change	= about 0.6°C change

Figure 1-2: English system to S.I. conversions.

ROUNDING

In scientific work, many of the numbers used are measured quantities and so are not exact—they are limited by the precision of the instrument used in the measurement. Further, calculations based on measured quantities can be no more precise than the original measurements themselves. Therefore, measurements and the results of calculations should be recorded in a way that shows the degree of measurement precision. For example, if you use an electronic calculator to divide the following two measured quantities, you would get:

$$5.7 \text{ centimeters} \div 1.75 \text{ minutes} = 3.2571429 \text{ centimeters/minute}$$

But is 3.2571429 centimeters/minute a truly correct answer? Not really. In general, the greater the number of digits in a measurement or calculation answer, the greater the implied precision of measurement. A mathematical operation cannot make your measurements more precise. In the previous example, our distance measurement is only accurate to tenths of centimeters (perhaps limited by the measuring device we used), and our final answer can be no more precise than this. So:

$$5.7 \text{ centimeters} \div 1.75 \text{ minutes} = 3.3 \text{ centimeters/minute}$$

When rounding off numbers, if the first digit to be dropped is less than 5, leave the preceding digit unchanged; if the first digit to be dropped is 5 or greater, increase the preceding digit by one. So, 6.74 becomes 6.7, whereas 6.75 becomes 6.8.

Your instructor may introduce the concept of *significant digits* to you. This will further extend your understanding of the proper rounding of measured quantities.

Name_____   Section _____

EXERCISE 1 PROBLEMS/SOLUTIONS—PART

1. Complete the following conversions using exact conversion factors (round your answers to 1 decimal place).

S.I. Units		English System Units
(a)	14 centimeters	_____ inches
(b)	29 meters	_____ feet
(c)	175 kilometers	_____ miles
(d)	42 liters	_____ quarts
(e)	57 grams	_____ ounces
(f)	65 kilograms	_____ pounds
(g)	37°C	_____ °F

2. Complete the following conversions using exact conversion factors (round your answers to 1 decimal place).

English System Units		S.I. Units
(a)	3 inches	_____ centimeters
(b)	4.3 feet	_____ meters
(c)	18 yards	_____ meters
(d)	73 miles	_____ kilometers
(e)	6.2 quarts	_____ liters
(f)	10 gallons	_____ liters
(g)	14 ounces	_____ grams
(h)	155 pounds	_____ kilograms
(i)	47°F	_____ °C

Name_____   Section _____

EXERCISE 1 PROBLEMS—PART II

1. Complete the following conversions using exact conversion factors (round your answers to 1 decimal place).

	S.I. Units	English System Units
(a)	72 centimeters	_____ inches
(b)	24 meters	_____ feet
(c)	1300 kilometers	_____ miles
(d)	4.5 liters	_____ quarts
(e)	144 grams	_____ ounces
(f)	228 kilograms	_____ pounds
(g)	12°C	_____°F

2. Complete the following conversions using exact conversion factors (round your answers to 1 decimal place).

	English System Units	S.I. Units
(a)	55 inches	_____ centimeters
(b)	1774 feet	_____ meters
(c)	220 yards	_____ meters
(d)	23,900 miles	_____ kilometers
(e)	24 quarts	_____ liters
(f)	300 gallons	_____ liters
(g)	26 ounces	_____ grams
(h)	4500 pounds	_____ kilograms
(i)	88°F	_____°C

Location

Objective:	To review the system of latitude and longitude and provide experience using atlases and globes.
Materials:	25 cm (10 inch) or larger diameter globe. World atlas (with index). Internet access (optional).
Reference:	Hess, Darrel. *McKnight's Physical Geography,* 12th ed., pp. 12–17.

LATITUDE AND LONGITUDE

Any location on Earth can be described using the grid system, or **graticule**, of **latitude** and **longitude**. Latitudes and longitudes are angular measures, with latitude describing north–south location, and longitude describing east–west location.

Lines of latitude on a map or globe are called **parallels** because they are all parallel to each other (Figure 2-1a). Latitude ranges from 0° at the equator to 90° north latitude at the North Pole and 90° south latitude at the South Pole.

Lines of longitude are known as **meridians** (Figure 2-1b). The meridians are farthest apart at the equator and converge at the poles.

The starting point for measuring longitude is the **prime meridian**, which runs through the Royal Observatory at Greenwich, England (just outside central London). Locations east of the prime meridian are described in degrees east longitude and locations west of the prime meridian in degrees west longitude. Longitude ranges from 0° at the prime meridian to 180° (on the opposite side of the Earth from the prime meridian). The complete grid system is shown in Figure 2-1c.

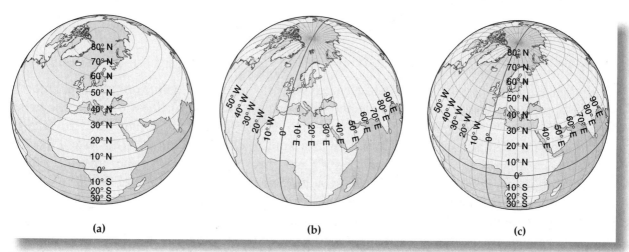

(a)   (b)   (c)

Figure 2-1: The geographic grid: (a) Parallels of latitude. (b) Meridians of longitude. (c) Complete grid system. (From McKnight and Hess, *Physical Geography,* 9th ed.)

When more exact descriptions of location are required (as when using detailed maps of a region), fractions of degrees of latitude and longitude are used. One degree is divided into 60 "minutes" (often written $60'$) and each minute can be further divided into 60 "seconds" ($60''$). Therefore, $1° = 60'$, and $1' = 60''$. When describing angular measures such as latitude and longitude, minutes and seconds are *not* referring to time, but to fractions of degrees.

As an example, the precise location of the crater of Mount St. Helens in Washington is $46°11'55''$ North, $122°11'15''$ West.

Decimal Degrees: With the increasing use of **Global Positioning System (GPS)** satellite technology to determine location, it has become common in some circumstances to indicate fractions of degrees in decimal units. For example, $45°35'$ N can be written $45.583°$ N, while $32°23'55''$ N can be written $32°23.917'$ N.

In some applications, north, south, east, and west can be omitted when specifying latitude and longitude. In these circumstances, north latitudes and east longitudes are shown with positive values, whereas south latitudes and west longitudes are shown with negative values. So, $38.5611°$ N, $110.0544°$ W would be designated 38.5611, −110.0544. *Note: With some applications, longitude is given before latitude.*

GLOBES AND ATLASES

Parallels and meridians are typically marked on globes in $10°$ or $15°$ increments. If the parallels and meridians are not marked on the globe, latitude and longitude are determined by using the degree markings on the arms or rings supporting the globe.

When searching for a location in an atlas, take advantage of the atlas index. The index will often comprise more than one-third of the pages in an atlas. In the index, cities, rivers, mountains, and other features are listed alphabetically. For each location, the index will typically provide the page number of the best map to use, the country, and often its latitude and longitude. Some atlases provide a pronunciation guide as well.

Some atlases do not refer to places in the index by latitude and longitude. Instead they provide a coordinate (such as "F7") that refers to a simplified grid system marked along the margins of each map in the atlas.

ONLINE GLOBE AND MAP PROGRAMS

In addition to printed atlases and physical globes, a number of online programs and mobile device apps allow you to access latitude and longitude information for the world. For example, a simple Google™ search will provide you with the precise latitude and longitude of any city in the world. You can also use Google Earth™ as a kind of electronic globe. When using Google Earth™ or Google Earth Pro™, go to "View" and check "Grid" to have the graticule appear around Earth—but don't zoom in so far that you lose a sense of the curvature of Earth.

Although it is easy to find latitude and longitude information online, the purpose of this lab exercise is to help you become familiar with the grid system as a whole—getting the correct numbers isn't as important as visualizing the grid and understanding how it works!

Name _____   Section _____

EXERCISE 2 PROBLEMS—PART I

1. Using a globe, determine the latitude and longitude of the following cities. Be sure to indicate if the location is north or south latitude, and east or west longitude. Indicate latitude and longitude to the nearest whole degree (round down if less than 30′; round up if 30′ or more).

	City	Latitude	Longitude
(a)	Chicago, Illinois	_____	_____
(b)	Tokyo, Japan	_____	_____
(c)	Sydney, Australia	_____	_____
(d)	Singapore	_____	_____
(e)	Buenos Aires, Argentina	_____	_____

2. Using a globe, determine which major city is located at the following coordinates:

	Latitude	Longitude	City
(a)	14° N	100° E	_____
(b)	56° N	38° E	_____
(c)	19° N	99° W	_____
(d)	1° S	37° E	_____
(e)	37° S	175° E	_____

3. (a) What is the latitude and longitude of your school? (Estimate to the nearest minute of latitude and longitude; be sure to indicate if the location is north or south latitude, and east or west longitude.)

    (b) What resource did you use to determine this?

Name _____   Section _____

EXERCISE 2 PROBLEMS—PART II

1. On the diagram shown, plot the following coordinates with a dot. Then label each dot with its corresponding letter.

    (a)   10° N, 40° W

    (b)   50° N, 40° E

    (c)   40° N, 25° W

    (d)   5° S, 10° W

    (e)   65° N, 70° E

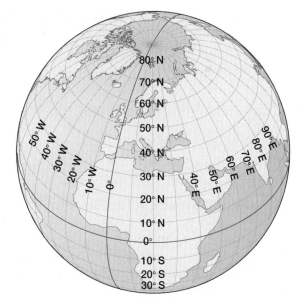

2. Use the index of an atlas to find the following places. Determine the latitude and longitude to the nearest degree.

Place	Latitude	Longitude
(a) Pusan (Busan)	_____	_____
(b) Reykjavik (Reikjavik)	_____	_____
(c) Walvis Bay	_____	_____
(d) Tuvalu (Ellice Islands)	_____	_____

3. If you start at the equator and travel to 10° N, approximately how many kilometers (or miles) north of the equator will you be? Take the circumference of Earth to be 40,000 kilometers (24,900 miles). Show your calculations.

4. If you travel west through 10° of longitude along the equator, the distance traveled will be very different from the distance traveled through 10° of longitude at 60° N. Why?

EXERCISE 3

Time

Objective:	To learn to calculate time and day differences around the world.
Reference:	Hess, Darrel. *McKnight's Physical Geography*, 12th ed., pp. 22–25.

LOCAL SUN TIME

Although few people today are concerned with the local **Sun time**, it is a useful starting point for a discussion of time. Local Sun time is based on the position of the Sun in the sky. The local Sun time "noon" for a given location is the moment in the day when the Sun reaches its highest point in the sky. However, at the same moment that it is local Sun time noon at our location, at locations east or west of us the local Sun time is different.

Earth rotates from west to east (looking down at the North Pole on a globe, Earth would appear to be spinning counterclockwise). This means that at the same moment the Sun is low in the morning sky in Honolulu, it is high in the sky at noon in Denver, and low in the afternoon sky in New York. In other words, as we travel to the east, the time becomes progressively later.

STANDARD TIME

Rather than having people continually adjusting clocks to local Sun time when moving east or west, 24 **standard time zones** have been established by international agreement. Each time zone is a band of longitude, within which it is the same standard time (although, of course, the local Sun time varies slightly within the time zone). When moving from one time zone to the next, we adjust our watches by 1 hour.

The time zones are based on **central meridians** spaced 15° of longitude apart. Earth rotates through 360° of longitude in 24 hours, and so rotates through 15° of longitude in 1 hour ($360° \div 24 = 15°$). Although a standard time zone is 15° of longitude wide, the actual time zone boundaries have been adjusted over most inhabited areas of Earth (Figure 3-1).

TIME ZONE CALCULATIONS

The map in Figure 3-1 shows standard time zones around the world. If we remember that it is always later in New York than in San Francisco, it is easy to calculate time differences. It becomes 1 hour later for each time zone we cross moving from west to east (from San Francisco toward New York), and 1 hour earlier for each time zone we cross moving from east to west. New York is three time zones to the east of San Francisco, so New York time is 3 hours later than San Francisco time.

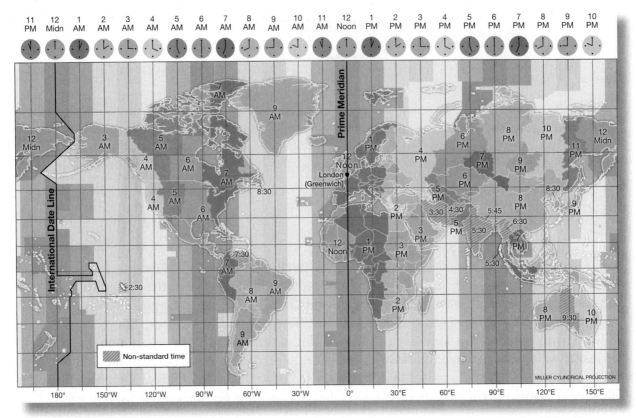

Figure 3-1: Standard time zones of the world. (From Hess, *McKnight's Physical Geography*, 12th ed.)

To avoid confusion, it is usually best to refer to "12:00 midnight" and "12:00 noon" rather than to 12:00 A.M. and 12:00 P.M., respectively.

Notice that a few time zones are based on the half-hour (such as for Newfoundland and India), but the same logic applies. For example, India is $5\frac{1}{2}$ hours later than Greenwich, England.

In 1884, **Greenwich Mean Time (GMT)** was established as the world reference for standard time (the Greenwich time zone is based on the prime meridian). Today, Greenwich time is known as **Universal Time Coordinated (UTC)** or **Zulu** time (Zulu time uses a 24-hour clock, so that "1530Z" would be 3:30 P.M. Greenwich time).

If you know the central meridian of a city's time zone, it is also possible to calculate time differences mathematically by determining the number of degrees of longitude between two locations. For example, Tokyo time is based on the 135° E central meridian and Rome is based on the 15° E meridian, a difference in longitude of 120°:

$$135° - 15° = 120° \text{ difference between Tokyo and Rome}$$

Fifteen degrees of longitude represents 1 hour of time, so:

$$120° \div 15° = 8 \text{ hours difference between Tokyo and Rome.}$$

Because Tokyo is east of Rome, the time will be 8 hours later in Tokyo than in Rome.

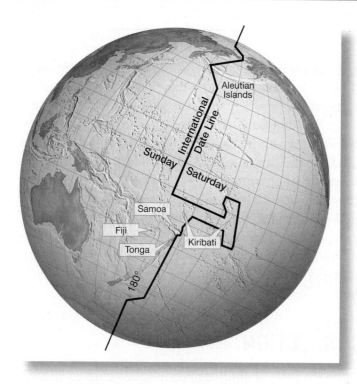

Figure 3-2: The International Date Line generally follows the 180° meridian, but deviates around various island groups, most notably Kiribati. (From Hess, *McKnight's Physical Geography*, 12th ed.)

INTERNATIONAL DATE LINE

When determining time differences between two places around the world, remember that the day may also be different. The day changes under two circumstances. First, the day changes at midnight. When traveling from west to east, when we cross into the time zone where it is midnight, it becomes the next day. Conversely, when traveling from east to west, when we cross into the 11 P.M. time zone, it becomes the previous day.

The day also changes at the **International Date Line** (IDL), which generally follows the 180° meridian down the middle of the Pacific Ocean (Figure 3-2). When crossing the IDL going from west to east (from China toward Hawai'i), it becomes the previous day. When crossing the IDL going from east to west (from Hawai'i toward China), it becomes the next day.

The International Date Line runs down the middle of a time zone. When first entering into this time zone, the hour changes, but the day remains the same until crossing the IDL (at which point only the day changes, not the time). Figure 3-3 shows the IDL and the bordering time zone boundaries. Sample times and days are shown in the diagram.

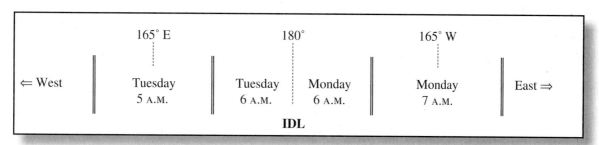

Figure 3-3: International Date Line (IDL) and bordering time zones, shown here with their central meridians along with sample days and times.

DAYLIGHT-SAVING TIME

A variation of standard time is **daylight-saving time**. Daylight-saving time is used throughout most of the United States during part of the year. In the summer, the days are longer than in the winter, and so there is a period of daylight that is "wasted" in the morning before people go to work. By shifting time ahead by 1 hour, there is, in effect, an "extra" hour of daylight in the afternoon after people come home from work. For most of the United States, daylight-saving time begins on the second Sunday in March and ends on the first Sunday in November.

Daylight-saving time calculations are easy. Simply remember the saying: "spring forward, and fall back." In other words, in the spring when going on daylight-saving time, we "spring forward" by adding 1 hour. When returning to standard time in the fall, we "fall back" by subtracting 1 hour.

When calculating time differences around the world, if both cities are observing daylight-saving time, you need not change your calculation procedure. However, if only one of the cities is observing daylight-saving time, convert that city back to standard time (by subtracting 1 hour), then proceed with your calculations as before.

SUNRISE AND SUNSET TIME CORRECTION

Local news media often provide us with the time of sunrise and sunset calculated specifically for our city. Because the local Sun time varies if we move east or west, the actual time of sunrise and sunset at locations east or west will vary slightly from that stated for our city. In locations to the east of our city, the exact time of sunrise and sunset will be earlier, while in locations to the west, the exact time will be later.

The sunrise/sunset time correction for different longitudes is easy to calculate. Earth rotates through 15° of longitude in 1 hour. Therefore, Earth rotates through 1° of longitude in 4 minutes (60 minutes ÷ 15° = 4 minutes per 1°). Locations to the east of us will experience sunrise/sunset 4 minutes earlier for each degree of longitude. Locations to the west will experience sunrise/sunset 4 minutes later for each degree of longitude. Note: This ignores differences in latitude that may also affect the sunrise/sunset time.

For example, if the stated sunset time for 75° W is 6:15 P.M., at 73° W, sunset will occur 8 minutes earlier (at 6:07 P.M.), while at 78° W, sunset will occur 12 minutes later (at 6:27 P.M.).

Name _____   Section _____

EXERCISE 3 PROBLEMS—PART I

Using the longitude of a time zone's central meridian (which has been provided for you), answer the following questions. Be sure to indicate if the time is A.M. or P.M.; however, refer to "noon" or "midnight" rather than to 12:00 P.M. or 12:00 A.M. It may be helpful to draw a simple diagram showing the central meridian of each time zone, such as Figure 3-3, when making your calculations.

1.   If it is 10:00 A.M. Monday in Denver (based on 105° W), what time and day is it in New York City (75° W)?

2.   If it is 11:00 A.M. Thursday in Seattle (120° W), what time and day is it in Seoul, South Korea (135° E)?

3.   A satellite image of the United States was taken at "0900Z." What was the local standard time in Chicago (90° W)?

4.   If it is Friday at 3:00 P.M. daylight-saving time in Kansas City (90° W), what is the day and time in Quito, Ecuador (75° W), where daylight-saving time is not being observed?

Name _____   Section _____

EXERCISE 3 PROBLEMS—PART II

1.   (a)   Your plane leaves Boston (75° W) at 7:00 A.M. on Saturday, bound for Los Angeles (120° W). The flight takes 5 hours. What is the time and day when you arrive in Los Angeles?

    (b)   Your connecting flight to Taipei (120° E) leaves Los Angeles at 1:00 P.M. on that same day. The flight takes 11 hours. What is the time and day when you arrive in Taipei?

EXERCISE 3 PROBLEMS—PART III

1.   For a given latitude, if the stated time of sunset is 6:45 P.M. at 90° W, what is the time of sunset at 91° W?

2.   For a given latitude, if the stated time of sunrise is 6:10 A.M. at 120°00′ W, what is the time of sunrise at 117°30′ W?

EXERCISE **4**

Map Scale

Objective:	To review the concept of map scale and to practice determining distances on a map using graphic and fractional scales.
Reference:	Hess, Darrel. *McKnight's Physical Geography*, 12th ed., pp. 30–32.

MAP SCALE

The **scale** of a map indicates how much Earth has been reduced for reproduction on that map. In practical terms, scale is the relationship between the distance shown on a map and the actual distance that this represents on Earth. There are three common ways to indicate the scale of a map.

Graphic Scales: The graphic scale for a map is a bar graph, graduated by distance. For example, Figure 4-1 shows the graphic map scales from a U.S. Geological Survey topographic map. To use a graphic scale, simply measure a distance on the map (or mark off the distance along the edge of a piece of paper), then compare the measured distance to the bar graph to determine the actual distance represented. On some graphic scales, "zero" is not at the far left. Graphic scales are useful because they remain accurate even if the map is enlarged or reduced in size.

In some cases, one graphic scale may not be accurate for all parts of the map. For example, some maps have several different graphic scales that are to be used for specified latitudes.

Fractional Scales: The fractional scale (also called the **representative fraction**) expresses the scale of a map as a fraction or ratio: 1/24,000 or 1:24,000.

This scale (read "one to twenty-four thousand") says that 1 unit of measurement on the map represents 24,000 units of measurement on Earth. At this scale, 1 centimeter on the map represents an actual distance of 24,000 centimeters on Earth, whereas 1 inch on the map represents an actual distance of 24,000 inches on Earth. Note that the units of measurement must be the same in both the numerator and the denominator.

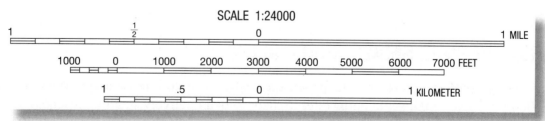

Figure 4-1: Graphic scales from a map with a fractional scale of 1:24,000. (From U.S. Geological Survey)

15

Verbal Scales: Map scale can also be expressed with words. Such **verbal map scales** are simply mathematical manipulations of the fractional scale. For example, there are 63,360 inches in 1 mile, so a map with a fractional scale of 1/63,360 can be expressed verbally as "one inch represents one mile."

COMPUTING DISTANCES WITH FRACTIONAL SCALES

In addition to using the graphic scale, it is possible to determine distances represented on a map by using the fractional scale:

1. Use a ruler to measure the distance on the map in centimeters (or inches). This is the *measured distance*.

2. Multiply the *measured distance* by the map's fractional scale denominator. This will give you the *actual distance* in centimeters (or inches).

3. To convert your *actual distances* in centimeters (or inches) to other units, use the following formulas:

To determine the distance in *meters*:   *Actual Distance* in centimeters ÷ 100
To determine the distance in *kilometers*:   *Actual Distance* in centimeters ÷ 100,000

To determine the distance in *feet*:   *Actual Distance* in inches ÷ 12
To determine the distance in *miles*:   *Actual Distance* in inches ÷ 63,360

For example, if we have a map with a scale of 1/50,000, a measured distance of 22 centimeters on the map represents an actual distance of 1,100,000 centimeters:

$$22 \text{ cm} \times 50,000 = 1,100,000 \text{ cm}$$

To calculate the actual distance in meters and kilometers:

$$1,100,000 \text{ cm} \div 100 \quad = 11,000 \text{ meters}$$
$$1,100,000 \text{ cm} \div 100,000 = 11 \text{ kilometers}$$

If we have a map with a scale of 1/24,000, a measured distance of 8.25 inches on the map represents an actual distance of 198,000 inches (8.25 inches × 24,000 = 198,000 inches). So:

$$198,000 \text{ inches} \div 12 \quad = 16,500 \text{ feet}$$
$$198,000 \text{ inches} \div 63,360 = 3.1 \text{ miles}$$

LARGE- VERSUS SMALL-SCALE MAPS

Large-scale maps refer to maps with a relatively large representative fraction (such as 1/10,000), whereas **small-scale maps** refer to maps with a relatively small representative fraction (such as 1/1,000,000). Large-scale maps show a small area of Earth in great detail, whereas small-scale maps show large areas in less detail.

Name _____   Section _____

EXERCISE 4 PROBLEMS—PART I

For questions 1 through 4, calculate the following distances using the fractional map scale. (Your instructor may ask you to show your work in the space provided.)

1. On a map with a scale of 1:24,000, a measured distance of
    1 inch represents an actual distance of: _____ feet

2. On a map with a scale of 1:62,500, a measured distance of
    4.5 inches represents an actual distance of: _____ miles

3. On a map with a scale of 1:250,000, a measured distance of
    4.5 inches represents an actual distance of: _____ miles

4. On a map with a scale of 1:50,000, a measured distance of
    7.5 centimeters represents an actual distance of: _____ kilometers

5. Map T-1 (in the back of the Lab Manual) shows part of the island of Hawai'i at a scale of 1:250,000. Using the appropriate graphic scale on Map T-1, determine the distance from the "Patrol Cabin" (at the summit of Mauna Loa) to the "Rest House" (northeast of the summit of Mauna Loa):

    _____ miles (statute miles)

    _____ kilometers

6. Map T-9 shows an area near Park City, Kentucky, at a scale of 1:24,000. Using the appropriate graphic scale, determine the distance from the "Fairview Church" (in the southwest corner of the map) to the "X" marked "BM 585" along the Louisville and Nashville railway:

    _____ feet

    _____ miles

    _____ kilometers

Name _____   Section _____

EXERCISE 4 PROBLEMS—PART II

1.   If a measured distance of 10 inches on a map represents an actual distance of 5 miles, what is the fractional scale of the map?

2.   Express a fractional scale of 1:100,000 as a verbal scale: One centimeter represents _____ kilometer(s).

Questions 3 through 4 are based on this set of graphic scales for a map with a fractional scale of 1:50,000:

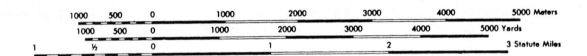

3.   Why isn't "0" at the far left of the scales?

4.   If a map with these graphic scales is enlarged along with the scales (such as by using a photocopy machine):

   (a)   Will the fractional scale of the map change? Why?

   (b)   Will the graphic scales (as shown) still be usable? Why?

EXERCISE **5**

Map Projections

Objective:	To examine the characteristics of different map projections.
Materials:	25 centimeter (10 inch) or larger diameter globe.
Reference:	Hess, Darrel. *McKnight's Physical Geography*, 12th ed., pp. 32–36.

CONFORMAL VERSUS EQUIVALENT MAPS

Only a globe can show the true area, shape, direction, and distance relationships of the spherical surface of Earth. It is impossible to show all of these relationships on a map without distortion.

Of the many different properties of maps, **equivalence** and **conformality** are perhaps the most important. An **equivalent map** (also called an **equal area map**) shows correct area relationships over the entire map. With an equivalent map, the area of one region on the map can be directly compared with the area of any other region. In contrast, a **conformal map** shows the correct angular relationships over the entire map. In practical terms, a conformal map shows the correct shapes of features in a limited area, although the true shapes of the continents can only be shown with a globe.

Figure 5-1 compares a conformal map (a) with an equivalent map (b). When compared with a globe, you will notice that the conformal map maintains the basic shapes of the continents, but the areas of the continents are severely distorted near the poles. On the other hand, the equivalent map shows the areas of the continents accurately, but the shapes are severely distorted in the high latitudes. It is

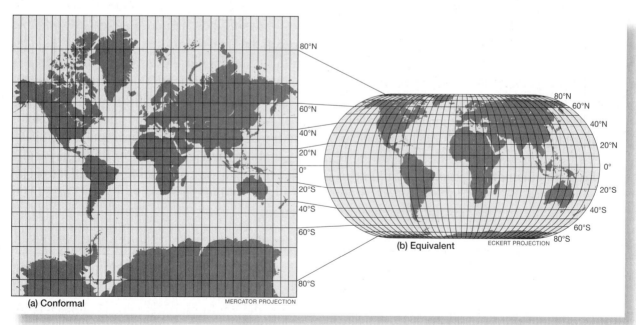

Figure 5-1: (a) Conformal projection—the Mercator. (b) Equivalent projection—the Eckert. (Adapted from McKnight and Hess, *Physical Geography*, 9th ed.)

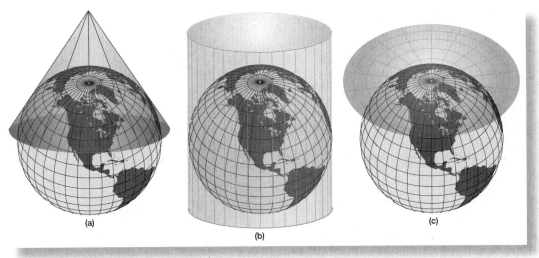

Figure 5-2: Three common families of map projections: (a) conic, (b) cylindrical, and (c) planar. (Adapted from McKnight and Hess, *Physical Geography*, 9th ed.)

impossible for a map to be both equivalent and conformal, and many maps are neither, but are instead a "compromise" map. (Note that these distortions are most pronounced on world maps—on large-scale maps showing limited areas, these distortions may not be a serious problem.)

Properties other than equivalence and conformality may also be maintained on a map. For example, true direction can be retained in some projections, and true distances can be shown on *equidistant maps,* but only from the center of the projection or along a specific set of lines.

MAP PROJECTIONS

Cartographers transfer the surface features of Earth to a map by mathematically "projecting" the **graticule** (the grid of latitude and longitude) out from the sphere onto a flat surface. Three common families of map projections are shown in Figure 5-2. In each case, there is one latitude or one point at which the map is tangent to ("touches") Earth. These latitudes are called *standard parallels* and represent the location of least distortion on the final map.

In the example shown, the **planar projection** (also called a *plane* or *azimuthal projection*) is tangent to the North Pole, and so would be suitable for maps of polar regions. The **cylindrical projection** is tangent to the equator, and would produce a map with low distortion in the equatorial regions. The **conic projection** is tangent to a parallel in the midlatitudes, making it a good choice for the midlatitude regions. Some cylindrical and conic projections are based on more than one standard parallel.

A fourth family of map projection is called **pseudocylindrical**. Pseudocylindrical projections are mathematically based on a cylinder, tangent to the equator, but the cylinder "curves back" down toward the poles so that the projection gives a sense of the curvature of Earth. The Eckert (Figure 5-1b) is based on a pseudocylindrical projection.

CHARACTERISTICS OF MAP PROJECTIONS

No single map projection is ideal for all purposes. Different kinds of projections produce maps that are suitable for different uses. For example, the conformal Mercator (Figure 5-1a) is based on a cylindrical projection. On the Mercator, any straight line is a **rhumb line** (a *loxodrome* or line of constant direction), making these maps useful for navigation.

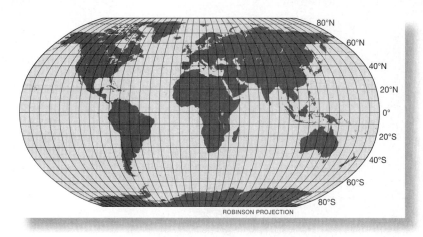

Figure 5-3: The Robinson is a compromise projection. (Adapted from U.S. Geological Survey *Map Projections* poster)

The Robinson (Figure 5-3) is a pseudocylindrical compromise projection that is widely used for maps of the world. It is neither conformal nor equivalent, but offers a good balance between correct shape and correct area.

The Lambert Conformal Conic projection (Figure 5-4) uses two standard parallels, and is often used by the U.S. Geological Survey for large-scale topographic maps.

Orthographic projections (Figure 5-5a) are known as perspective maps. They make the Earth appear as it would from space. On **gnomonic maps** (Figure 5-5b), a straight line represents a path along a **great circle** (the largest circle that can be drawn on a sphere) and shows the shortest path between two points. Both orthographic and gnomonic maps are based on planar projections.

An interesting type of cylindrical projection is the "Transverse Mercator" (Figure 5-5c). Instead of being tangent to the equator, the Transverse Mercator is tangent to a standard meridian (the 90° W/90° E meridian in the example shown), but notice that unlike a normal Mercator, most parallels and meridians are shown as curved lines. The Transverse Mercator is conformal, and is used on many U.S. Geological Survey topographic maps.

Figure 5-4: Lambert Conformal Conic. (Adapted from U.S. Geological Survey *Map Projections* poster)

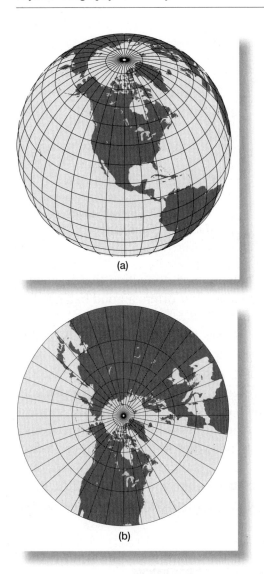

(a)

(b)

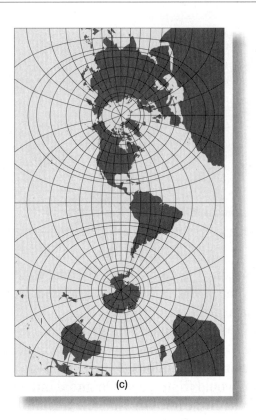

(c)

Figure 5-5: (a) Orthographic. (b) Gnomonic. (c) Transverse Mercator. (Adapted from U.S. Geological Survey *Map Projections* poster)

Goode's Interrupted Homolosine Equal Area projection (Figure 5-6) is widely used to show the distribution of phenomena on the continents. The Goode's Interrupted projection is equivalent, yet the shapes of the land masses are also very well maintained.

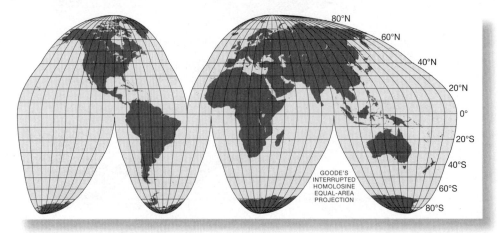

Figure 5-6: Goode's Interrupted Homolosine Equal Area Projection. (Adapted from McKnight, *Physical Geography*, 4th ed.)

Name _____   Section _____

EXERCISE 5 PROBLEMS—PART I

1. Compare the Mercator projection (Figure 5-1a) to a globe.

    (a) Are all of the lines of latitude parallel to each other on both the globe and the Mercator projection?

    (b) Do all of the parallels and meridians cross each other at right angles on both the globe and the Mercator? (Hint: Look carefully at just the immediate intersection of a parallel and a meridian on the globe.)

    (c) On a globe, the meridians converge toward the poles. Describe the pattern of meridians on the Mercator.

    (d) Is north always straight toward the top of the Mercator projection?

    (e) How would the North Pole be represented on the Mercator?

    (f) Could a single graphic scale be used to measure distances on a Mercator projection? Explain.

2. Study the Eckert projection (Figure 5-1b).

    (a) Do all of the parallels and meridians cross each other at right angles?

    (b) How does the Eckert maintain equivalence in the high latitudes (what happens to the meridians)?

    (c) What happens to the shape of Greenland?

    (d) Is north always straight toward the top of the Eckert? Explain.

Name _____   Section _____

EXERCISE 5 PROBLEMS—PART II

1.   Study the Goode's Interrupted projection (Figure 5-6):

   (a)   Are ocean areas "left off" this map? Explain.

   (b)   The Goode's is based on two different projections, one for the low latitudes and one
          for the high latitudes. At approximately what latitude does the projection change?
          (Hint: Look for the change in the shape of the map margins in the North Pacific.)

2.   On a globe, use a piece of string to find the shortest path between Yokohama, Japan (near
      Tokyo) and San Francisco. This path is a "great circle" path. Two maps are shown here, a
      Gnomonic (i) and a Mercator (ii).

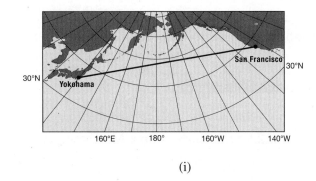

(i)

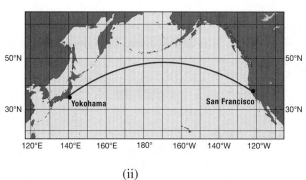

(ii)

   (a)   Is the path of the string on your globe the same as the heavy line shown on just one
          of these maps, or on both of these maps? (Hint: Look carefully at the string on the
          globe in relation to the Aleutian Islands of Alaska [at about 50° N, 175° W].)

   (b)   In terms of a navigator trying to maintain a constant compass heading, why would
          the great circle path shown be difficult to follow exactly?

   (c)   How would both a Mercator and a Gnomonic map be used together in navigation?

Isolines

Objective:	To practice interpreting and drawing isolines.
Reference:	Hess, Darrel. *McKnight's Physical Geography*, 12th ed., pp. 37–39.

ISOLINES

Often in geography we are interested in mapping particular characteristics of an area, such as the elevation, the amount of rainfall, or the temperature. A common and very useful method of showing varying levels or concentrations of some phenomenon is with **isolines**. An isoline is a line on a map that connects points of equal value.

For example, **contour lines** on topographic maps are isolines that show elevation (contour lines are discussed in Exercise 28). In our study of weather and climate we will use several kinds of isolines, such as **isotherms**, to show temperature, and **isobars** to show atmospheric pressure.

There are just a few basic rules pertaining to all isolines:

(a) An isoline connects points on a map where the value of some phenomenon is the same.

(b) Isolines are drawn at regular intervals (e.g., for every 5° of temperature difference).

(c) Isolines are always closed lines, although they often close beyond the margins of a map.

(d) Isolines never cross each other.

(e) Where isolines are close together, they show an abrupt horizontal change in the phenomenon; where they are far apart, they show a gradual horizontal change.

(f) Values inside a closed isoline are either higher or lower than those outside the closed isoline (it is usually clear which is the case based on the pattern of adjacent isolines).

The following example will help illustrate how isotherms are drawn. Figure 6-1 shows a simple map with temperatures plotted for 17 different cities.

We will draw isotherms at 5° intervals (15°, 20°, 25°, etc.). An isotherm will pass through any point with the same value as the isotherm, but between higher and lower values. On one side of the line, the temperatures will be higher than the value of the isotherm, while on the other side, temperatures will be lower.

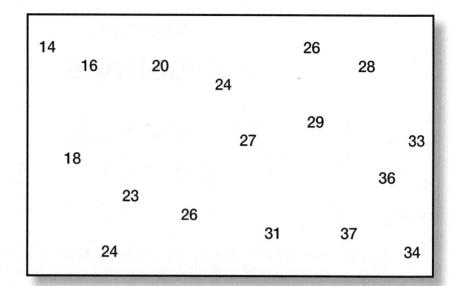

Figure 6-1: Map showing the temperatures in 17 cities.

Drawing isolines involves *interpolation*. For example, the 15° isotherm passes between the 14° and 16° locations, whereas the 27° location is about halfway between the 25° and 30° isotherms. Figure 6-2 shows the completed isotherm map. Notice that isotherms show the spatial pattern of temperature more clearly than the temperatures of the cities alone.

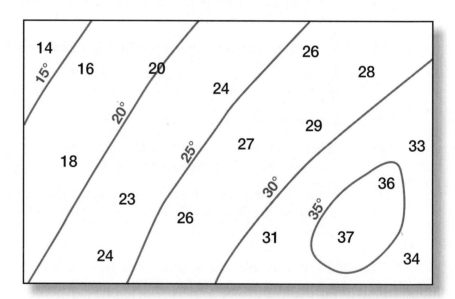

Figure 6-2: Temperature map with isotherms drawn.

Name _____   Section _____

EXERCISE 6 PROBLEMS—PART I

The following questions are based on the isotherm map in Figure 6-3, showing average January sea-level temperatures in °C and °F. Eight lettered points (labeled "A" to "H") are shown on the map.

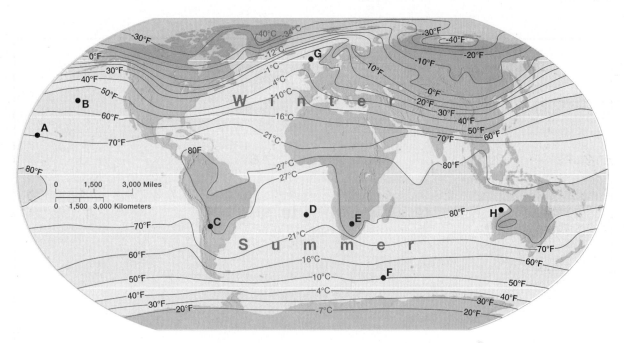

Figure 6-3: Average January sea-level temperatures (°C and °F). (Adapted from McKnight and Hess, *Physical Geography*, 9th ed.)

Determine the average January sea-level temperature at the following eight lettered points. Indicate if your answers are in °C or °F.

°C or °F (circle scale used)

A _____   E _____

B _____   F _____

C _____   G _____

D _____   H _____

Name _____   Section _____

EXERCISE 6 PROBLEMS—PART II

This map shows the location of 52 cities. The temperature of each city is given in degrees. Draw isotherms at 5° intervals, beginning with the 0° isotherm in the upper right corner. Label each isotherm.

Landscape Analysis with Google Earth™ & The National Map

Objective:	To learn how to use the online mapping services of *Google Earth™* and *The National Map* to analyze landscapes.
Resources:	Internet access.
Reference:	Hess, Darrel. *McKnight's Physical Geography*, 12th ed., pp. 39–40 and 42.

GOOGLE EARTH™ AND THE NATIONAL MAP

In this exercise we introduce two of the most popular and useful online mapping services for landscape analysis: *Google Earth™* and the U.S. Geological Survey's *The National Map*. In order to use Google Earth (GE), you must download and install free software on your computer (**http://earth.google.com**). You may use either Google Earth™ or Google Earth Pro™ for the exercises in this Lab Manual. Google Earth Pro™ is now free and has additional useful features not available with the original Google Earth™. The National Map (TNM) is accessed through a website (**http://nationalmap.gov**) and does not require you to install any software. In some regards, both applications let you do similar things: easily move to different parts of the United States (or the world, in the case of GE), then zoom in on a location, viewing roads, topography, and detailed aerial imagery. However, the applications differ in some important ways as well.

Google Earth™ contains an ever-growing amount of location information that is easily accessed, placemarked, and shared by individuals and businesses. With GE you can zoom in, rotate, and tilt your view of the surface below (aerial imagery "draped over" a three-dimensional digital elevation model of the terrain) and virtually "fly" over the landscape—a remarkable tool for landscape study in physical geography.

The National Map currently lacks the visual "bells and whistles" of GE, but it makes up for it by providing a remarkable number of high-quality map layers: consistently high-quality remotely sensed imagery, place names, hydrology, geology, contour lines, and many others. The purpose of TNM is to provide a seamless map of the entire United States that can be viewed at many different scales. Nearly all of the content of The National Map is in the public domain, free for all to use and download, whereas the Google Earth™ base map and much of its imagery may not be reproduced without permission.

In subsequent exercises in this Lab Manual, the latitudes and longitudes of many features are provided, enabling you to study these landscapes using Google Earth™ or The National Map. In addition, many exercises include specific questions based on Google Earth™, with predetermined locations you can visit with GE by opening files available on the Hess *Physical Geography Laboratory Manual*, 12th edition, website (**www.MasteringGeography.com**) or by scanning a QR ("quick response") code with a mobile device such as a smartphone to open a Google Earth™ "video."

We begin with a quick introduction to Google Earth™ and The National Map. Only the most basic functions of each are described, and with further exploration you will uncover much greater capabilities in both applications. Be sure to take advantage of the "Help" and tutorial functions offered by both applications to learn more about specific details. *Note: Google Earth™ and The National Map are constantly evolving applications—their functions and procedures may vary from that described here.*

GETTING STARTED WITH GOOGLE EARTH™

After installing the latest version of the free Google Earth™ or Google Earth Pro™ software on your computer, you may want to go through some of the introductory tutorials (click "Help" along the top of the screen). GE opens with a map of the world on the right and a sidebar with three menu panels on the left: *Search*, *Places*, and *Layers* (Figure 7-1). Any of these panels may be hidden from view by clicking the triangle next to the panel name.

Search Panel: The "Search" feature lets you find locations by entering a place name or address. You may also type in latitude and longitude coordinates and instantly "fly to" a specific location—this is one way you can navigate to locations when completing exercises in this Lab Manual. Enter latitude and longitude values in the same order as coordinates are given in this Lab Manual, but you needn't enter the symbols for degrees (°), minutes (′), or seconds (″)—just leave a blank space between each number and N or S for latitude, and E or W for longitude. So, the location of Mount St. Helens in Washington (46°11′55″ North, 122°11′15″ West) would be entered in the "Search" box as: 46 11 55 n 122 11 15 w (you may also enter latitude and longitude in decimal degree format: 46.1986 −122.1875; see Exercise 2). As soon as you click the search button, you'll fly to and zoom in on that location.

When searching for locations by street address, you may only need to enter the street address and zip code. For example, to find City College of San Francisco with an address of 50 Phelan Avenue, San Francisco, California 94112, you can simply enter "50 Phelan 94112" (this won't work for all addresses). When you search for locations by place name, a list of commercial establishments that have your search terms included may also appear.

Places Panel: The "Places" panel lets you keep track of locations you've "placemarked," as well as files opened from other sources (such as the Lab Manual website). (The procedures for creating placemarks and opening files are discussed later in this exercise.)

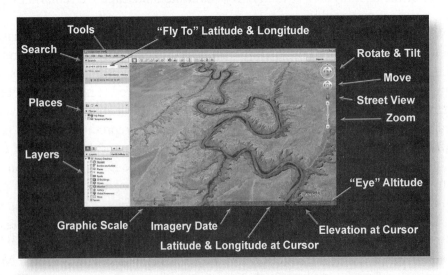

Figure 7-1: Google Earth Pro™ screen showing panels and navigation controls. (Google Earth™ is a trademark of Google, Inc.)

Layers Panel: The "Layers" panel lets you click on and off the many map layers of information available in GE. In this Lab Manual we're mostly interested in the terrain, so start off by checking "Terrain" and unchecking most (or all) of the other layers. You can then hide the Layers panel. If a "clock" icon appears next to the Imagery Date, you can select to view imagery from earlier years.

Navigating in Google Earth™: Once you have arrived at your map destination, use the controls along the right side of the screen to move around and change your perspective. The *Zoom* control changes the scale of the map—and so the apparent altitude of your "eye." The *Rotate and Tilt* control lets you rotate your vantage point and tilt your view up or down. The *Move* control moves you over the landscape in the direction you choose; if you've set the tilt control to a low angle, you'll have the sensation of flying over the virtual landscape below. You may also change location by clicking and dragging on the map with your mouse.

As you change the location of your cursor, notice that the latitude, longitude, and elevation values (shown along the bottom of the screen) also change. (If you don't see the latitude and longitude at the bottom of the screen, click "View" on the top toolbar and select "Status Bar.")

You can keep track of the direction you're viewing in GE by looking at the position of the "N" on the ring around the Rotate and Tilt control—when the N is in the "12:00 position" (top of the circle), the view is facing north; if the N is in the 6:00 position, you're facing south, and so on. To quickly reorient the map with north at the top of the screen, just click on the N.

Although the default settings of GE generally work well, you can adjust characteristics such as the measurement units shown, the format of latitude and longitude, and the amount of **vertical exaggeration** in the landscape. Select "Tools" from the top toolbar, and then "Options."

Ruler: You can measure distances on the GE map by using the "Ruler" function. From the top toolbar, select "Tools," then "Ruler" from the dropdown menu, or simply click on the Ruler button (Figure 7-2). Select the units of measurement (meters, miles, etc.). Use "Line" to measure distances along a single straight line, or "Path" to measure along a curved path by using a series of mouse clicks. With GE Pro you can also measure the area and perimeter of polygons you designate with mouse clicks. The ruler function measures straight line ("as the crow flies") distances, even though the ruler line you see on the screen may drape over the three-dimensional landscape and appear as if it is measuring the distance uphill or downhill.

Street View: To see the street-level images that are available in many locations, drag and hold the *Street View* icon over an area of interest. Release the icon and the screen will show ground-level photography (the quality and usefulness of the street view imagery varies somewhat from place to place).

Placemarks: You can mark locations in GE by adding *Placemarks*. Click the Placemark button near the top of the screen and an icon will appear on the map. Drag the icon to the location you desire and then give it a name in the dialogue box that appears. Your placemark then becomes an entry in the "My Places" folder on the "Places" panel (you may also add "Folders" to organize your placemarks). You can fly back to the location of any placemark on your list by double clicking on the placemark name.

Opening Lab Exercise Files in Google Earth™: For parts of some lab exercises you will be directed to the Hess *Physical Geography Laboratory Manual*, 12th edition, website at **www.MasteringGeography.com**, where you will select the appropriate Google Earth™ exercise from the menu. GE will open, and a KMZ file (compressed "Keyhole Markup Language" file) will

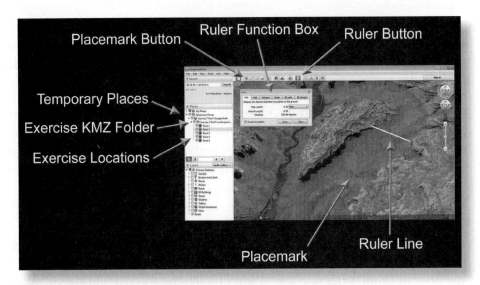

Figure 7-2: Google Earth Pro™ screen showing Temporary Places folder, placemarks, and ruler function. (Google Earth™ is a trademark of Google, Inc.)

appear in your "Temporary Places" folder on the Places panel. You may need to click on the "+" icon next to the folder to expose the one or more locations (*Point 1*, *Point 2*, and so on) you'll visit when completing the exercise. Be sure that the tiny box next to each placemark name is checked so that the placemark icons appear on the map.

Simply double click on a location name on the Places panel and GE will fly you to the correct location. Once there, you can further zoom in or change your view in GE; to return to the original view, just double click again on the location name on the Places panel.

Google Earth™ Videos: If you scan a QR code with a mobile device to view a Google Earth™ video for one of the lab exercises, GE software does not open. Instead, you'll see a screen-shot video showing the landscape and placemarks noted in the exercise. You can pause or replay these videos as needed to study the landscape more carefully.

The Three-Dimensional Landscape in Google Earth™: It is important to keep in mind that the landscape you're viewing in Google Earth™ is a virtual one. In most cases, vertical (or near vertical) aerial imagery is "draped" over a digital elevation model (DEM) of the landscape by the GE software. In order to view such an image-over-DEM landscape from a low angle, the vertical aerial imagery must necessarily be "stretched" down over the topography. Although such views can be extremely useful when studying landforms, keep in mind that this computer-generated image does not provide the exact same view you would have were you *actually* flying over the landscape—interpret what you see with a critical eye.

GETTING STARTED WITH THE NATIONAL MAP

The National Map (TNM) can be accessed in two ways: with a simple "Online Viewer" (**http://viewer.nationalmap.gov/viewer/**), or with a more sophisticated "Download Platform" (**http://viewer.nationalmap.gov/basic/**). Both allow you to view maps, imagery, and data, although the Viewer is easier to use when just viewing maps and imagery (as you'll do in this exercise), whereas the Download Platform must be used to obtain imagery and data. For the problems in this exercise, using the Viewer is recommended. The USGS is in the process of allowing the acquisition and viewing of their maps, imagery, and data through a number of additional platforms—over time, the interface to TNM may vary from that described here.

Find a Place: You can search for locations by place name, or by typing in latitude and longitude coordinates in the "Search" box (Figure 7-3). As with GE, you can enter latitude and longitude without degree, minute, or second symbols, but you should leave space between each value. So, 38°33′40″ North, 110°03′16″ West, would be entered as: 38 33 40 n 110 03 16 w. Also as with GE, you may enter latitude and longitude in decimal degree format: 38.5611 −110.0544 (when using the Download Platform you must use decimal degree format with longitude first: −110.0544 38.5611). The latitude and longitude of the cursor position on the map is shown along the bottom of the screen (from the "Options" icon at the top right of the screen, you can change the display format of latitude and longitude). To hide the "Results" panel that appears at the left of the screen after a search, click the triangle on the upper right side of the panel.

Navigating in The National Map: You can point and click on the map to zoom in to any location. You may also change locations by clicking and dragging on the map with your mouse. To center the map on a particular point, hold down the shift key and click on that spot. The zoom control lets you quickly change scales of the map (the map scale is shown along the bottom of the screen). You may also zoom in by simply double clicking on the map.

Base Maps and Imagery: You can toggle between a shaded-relief topographic map ("Topo"), satellite/aerial imagery ("Imagery"), a combination of both ("Imagery Topo"), as well as hydrologic data and a simple shaded relief map, with the buttons in the upper right of the screen. The many available map layers are shown along the left side of the map by selecting "Overlays"— you can hide or open this left panel by clicking the triangle along the top right of the panel.

Tools: A number of measurement tools and other functions are available by opening the tool functions along the top of the map. For example, a "Ruler" function in the "Advanced" tab

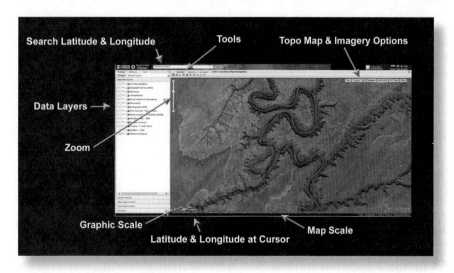

Figure 7-3: The National Map Online Viewer screen showing layers panel and navigation controls. (Adapted from U.S. Geological Survey)

lets you measure distances on the map or image—click once to start your measurement line; double click at the end of your measurement line and the distance will appear in a pop-up window (to clear your measurement lines off the screen, you may need to go to the "Standard" tab and click the "Clear Graphics" icon).

To determine elevations, from the "Standard" tab click on the "Spot Elevation" icon; click on the map, and the elevation of that point appears; click the Clear Graphics icon to clear elevations off the screen.

Downloading Maps: You can download maps and imagery from The National Map by using the Download Platform. However, if you simply want PDF files of topographic maps or orthophoto imagery maps, it is easier to use *U.S. Topo* (**http://nationalmap.gov/ustopo/**) described in Exercise 29.

GRADIENT

In landscape analysis, one way to describe the steepness of a slope is with **gradient**. Gradient describes the elevation change of a slope over a given distance. For example, if working with English units of measure, gradient is usually stated in feet of elevation change per mile. For example, if a slope changes elevation by 3200 feet over a distance of 2.5 miles, the gradient is:

$$\text{Gradient} = \frac{\text{Elevation change}}{\text{Number of miles}} = \frac{3200 \text{ ft.}}{2.5 \text{ miles}} = 1280 \text{ feet/mile}$$

In subsequent lab exercises, you will learn how to calculate gradients using topographic maps. Gradients may also be determined in Google Earth™ and The National Map by using their ruler and elevation functions. For example, to determine gradients in GE, open the ruler function (with distances set to miles); click your mouse to anchor the ruler—note the elevation at your starting point. Extend the ruler along the slope to your stopping point and click again, noting the elevation at the stopping point. Divide the difference in elevation by the distance measured by the ruler to determine the gradient of the slope. You can measure the distance along a curved path (such as a stream course) by setting the ruler to "Path" and using multiple mouse clicks along the way.

Because gradient describes the ratio of elevation change to distance, if you are measuring a uniform slope (in other words, a slope that has a constant gradient over a long distance), your calculations should yield the same gradient for that slope whether you make your measurements over a short distance or over a long distance. In most cases, however, even if the slope is uniform, you'll end up with more accurate results if you take your measurements over a longer distance.

WHICH TO USE—GOOGLE EARTH™ OR THE NATIONAL MAP?

An obvious question to ask is which is better for landscape analysis, Google Earth™ or The National Map? The short answer is that both are useful. The National Map has a large—and growing—number of consistently high-quality public-domain data layers that may be viewed and easily downloaded. As with many other free Internet applications, Google Earth™ has become a way for businesses to reach potential customers, but it also provides the geographer with easy ways to save and share location information, and to manipulate the view of the landscape in remarkable ways.

Name _____   Section _____

EXERCISE 7 PROBLEMS—PART I—GOOGLE EARTH™

Answer the following questions after you have installed the free Google Earth™ software from **http://earth.google.com** and opened the program on your computer.

1. In the "Search" box, enter the latitude and longitude of Crater Lake in Oregon; remember, you don't need to enter the symbols for degrees, minutes, or seconds, but you do need to leave a space between each value: 42°56′20″ North, 122°06′25″ West.

    (a) What is the apparent altitude of your "eye" in this opening view?

    _____ feet

    (b) Zoom out by clicking on the "−" symbol at the bottom of the zoom control until you can see the entire lake. What is the apparent altitude of the "eye" now?

    _____ feet

    (c) Move your cursor around the rim of the caldera. What is the highest elevation you find around the rim of Crater Lake?

    _____ feet

2. In the "Search" box, enter the address of your house or school. Zoom in as much as you can on your house or school building before image quality begins to break down.

    (a) Location: Latitude _____ Longitude _____

    (b) Elevation: _____

    (c) What time of day was this image taken? (circle) Morning / Midday / Afternoon

    (d) How can you tell?

    (e) What is the date of the imagery? _____

To answer problems 3 to 6, go to the Hess *Physical Geography Laboratory Manual*, 12th edition, website at **www.MasteringGeography.com**, then Exercise 7. Select "Exercise 7 Part I Google Earth" to open a KMZ file in Google Earth™. The opening view is the same region north of Canyonlands National Park and the Green River in Utah shown in Figure 32-3, Figure 32-4, and Map T-7.

3. In the Places panel, double click on "Point 1" to fly to this
    location. What is the elevation of Point 1?   _____ feet

4. (a) Double click on Point 2 to change your vantage
         point. What is the elevation of Point 2?   _____ feet

    (b) What is the relief (the difference in elevation)
         between Point 1 and Point 2?   _____ feet

5. Fly to Point 3. Click the "+" button on the Zoom control to move in closer to this sharp
    turn called "Bowknot Bend" along the Green River. Using the forward arrow on the Move
    control, fly over the landscape until you have a clear view of Point 3 along the river bot-
    tom. From this angle, it may appear that clouds are draped over the ridge between the
    tight loops in the river's course. The historical imagery you see was taken in 2003 from a
    vantage point nearly directly overhead of this area and the clouds were not really this low
    to the ground, so why do the clouds appear this way in Google Earth™?

6. Fly to Point 4 a short distance away. You are going to measure the gradient of this slope.
    Double click on Point 5 to move in closer and look directly down on this slope. Determine
    the gradient of the slope between Point 4 and Point 5.

    _____ ÷ _____ = _____ feet/mile gradient
    (Elevation change)   (Distance in miles)

EXERCISE 7 PROBLEMS—PART II—INTERNET

Go to The National Map (**http://viewer.nationalmap.gov/viewer/**). Once the viewer has opened,
in the "Search" box, enter the latitude and longitude of the area north of Canyonlands National
Park shown in Figure 32-3, Figure 32-4, and Maps T-7 and T-20: 38°33′40″ North, 110°03′16″
West. Select "Imagery" to see aerial imagery of this area. Zoom in until you have a close view of
this half-circle-shaped hill encircled by a dry gorge.

1. Use the Spot Elevation function (from the "Standard" tab) to determine the relief of this hill:

    _____ − _____ = _____ feet
    (Elevation at top   (Elevation in bottom   (Relief)
     of hill)   of dry gorge to north)

2. (a) Was this aerial image taken in the morning or in the afternoon? _____

    (b) How can you tell?

EXERCISE 8

Geographic Information Systems and Remote Sensing

by Ryan Jensen, *Brigham Young University*

Objective:	To learn the basics of GIS and remote sensing, become familiar with ArcGIS Online by displaying and investigating online datasets, and perform a proximity analysis using buffers and overlay in ArcGIS Online.
Resources:	Computer (laptop or desktop) or tablet; Internet access.
Reference:	Hess, Darrel. *McKnight's Physical Geography*, 12th ed., pp. 41–50. Jensen, J. and R. Jensen. *Introductory Geographic Information Systems* (2013).

INTRODUCTION

Geographic information systems (GIS) are important tools that geographers and others use to map, analyze, and query spatial data. Remote sensing data, usually acquired from either an orbiting satellite or plane/drone/helicopter, are commonly used in GIS analysis. This lab will introduce you to the basics of GIS and remote sensing by stepping you through exercises using an online mapping program: Esri's *ArcGIS Online*.

First, we explain the basics of GIS and remote sensing. Then, you will create a free account in ArcGIS Online and begin to view and analyze data. You will finish the exercise by completing a GIS proximity and overlay analysis.

GIS AND REMOTE SENSING

Geographic information systems are computer software programs specifically designed to view, analyze, map, query, and explore spatial data. Professionals from many different disciplines regularly use GIS to examine spatial relationships. For example, police officers use GIS to map crime and determine where to maximize police presence. A tax assessor may use GIS to examine market values of homes in different neighborhoods and determine if there are any outliers above or below the mean housing value in each neighborhood. Wildlife biologists use GIS to identify habitat areas suitable for mule deer in the western United States. Epidemiologists use GIS to map disease and disease vectors to better understand disease dynamics and transmission. Wastewater managers use GIS to map the wastewater network. GIS allows researchers to examine a problem spatially using multiple data layers.

You have probably used both GIS data and analysis with a personal navigation device to get from one point to another. In fact, if you have a smartphone with you, it probably has the ability to first locate places of interest near you (e.g., restaurants, ATMs) and then identify the best route to get you there. Your phone combines very accurate GIS data layers and global positioning system (GPS) locations to determine the best routes.

Remote sensing refers to collecting information about an object without actually touching that object. All of you have probably used a remote sensing device, such as a camera. A camera measures reflected light from objects or phenomena within its view. It does so without ever touching the objects. A camera may measure reflected sunlight on a sunny day or reflectance from its own flash. Most cameras that people regularly use only measure information in visible light. Conversely, most scientific remote sensing instruments measure visible light and information in other electromagnetic regions (e.g., near infrared, infrared, **radar**). Remote sensing data collection usually occurs via orbiting sensors, aerial cameras, and active sensors, such as radar and *lidar*. Remote sensing images are viewed every day on local television during the weather segment when Doppler radar (see Exercise 21) and visible satellite images (via GOES; see Exercise 20) are displayed to demonstrate weather and rainfall patterns. Remote sensing data are used by the National Aeronautics and Space Administration (NASA) to measure global vegetation patterns (Figure 8-1; shown in color in Map T-31a in back of Lab Manual) and to monitor forest fires (Figure 8-2; shown in color as Map T-31b).

GIS and remote sensing are often used together. For example, remote sensing data can derive very accurate land cover maps using spectral information that can then be analyzed in a GIS. You have probably used an app where both GIS and remote sensing data are utilized. For example, when a navigation app routes you from one point to another it relies on many different datasets including the network and traffic data that may have been derived from traffic sensors (e.g., cameras), individual smartphones (where users have enabled certain information to be used), and fleet cars such as taxis. All of these data are utilized dynamically in real time to enable you to travel on the fastest route. In fact, if traffic congestion occurs along your route, the navigation app may attempt to reroute you to a faster route that may have been longer in distance.

Figure 8-3 (shown in color as Map T-31c in the back of the Lab Manual) highlights a route from Los Angeles to Malibu in California, delimited in several colors along the route: Blue means

Figure 8-1: A vegetation map of the world (November 1 – December 1, 2007) derived using remote sensing data. This map is shown in color as Map T-31a in the back of the Lab Manual. (From NASA)

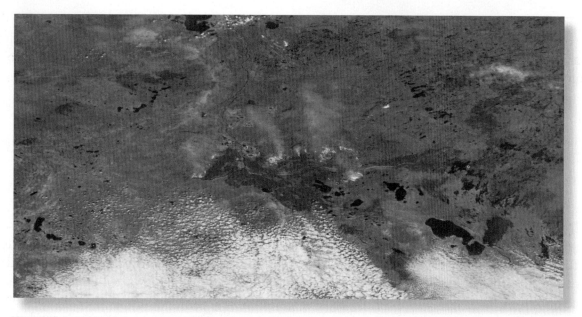

Figure 8-2: This image was acquired of the Fort McMurray, Canada, wildfire in May 2016. This map is shown in color as Map T-31b in the back of the Lab Manual. (From NASA)

no traffic and normal speeds; yellow signifies slow traffic; red indicates very slow or stopped traffic. Note the optional route in gray and the option for public transit, and that many cultural (city names, etc.) and physical geography (mountain ranges, ocean bathymetry, etc.) are described. Also, the land area displays a remote sensing image to give context to the route's landscape. Although this routing ability may seem simple, many different layers of GIS and remote sensing data and analysis are incorporated into the final solution. Routing apps, software, and websites integrate these different datasets and arrive at a solution seamlessly so that end users see only the results.

Figure 8-3: Route from Los Angeles, California to Malibu, California in derived in Google Maps. This map is shown in color as Map T-31c in the back of the Lab Manual. (From Google Maps™)

GIS is particularly useful for combining (or overlaying) different maps to determine the relationships between different variables. In addition, finding proximity and enclosure of multiple variables is also useful. This is done by combining different map layers into a single map and joining their attributes. Examples of this kind of analysis include the following:

- A biologist may combine a vegetation layer with an elevation layer to see if elevation has any effect on vegetation.
- A developer may want to determine what parcels of land for sale are within the 100-year floodplain.
- A police officer may be interested in what crimes occur in what neighborhoods.

There are two ways to obtain GIS and remote sensing datasets: (1) create them yourself, or (2) use what others have already created. Usually spatial data analysts will first search for necessary data from reputable online data repositories, because many spatial datasets have already been collected. Most governmental entities collect and serve a huge amount of spatial data layers. Usually, these data are available for free or at cost of reproduction based on governmental laws and policies that regulate data ownership and distribution. Further, data obtained and served by the government must often adhere to certain accuracy standards before they can be served. These accuracy standards are dependent on the agency/government collecting the data and how the data were originally designed to be used.

Before using any spatial data, it is important to consider the following data characteristics:

- Who collected the data?
- How were the data collected?
- When were the data collected?
- How accurate are the data?
- What projection and coordinate system do the data use?
- Are there any restrictions to using the data?
- Do you have other questions about the data?

The answers to each of these questions are often found in a "metadata" file that describes the characteristics of each dataset. Always make sure that you read and understand the metadata before you use any data that you did not collect yourself. Further, if you collect data yourself using GPS or some other method, make sure that you create a metadata file that answers all of the previous questions.

GIS and remote sensing data can be found in many different locations. Figure 8-4 is a table that contains a list of Internet sites where data can be found.

Name	Owner	Address
Earth Explorer	U.S. Geological Survey	http://earthexplorer.usgs.gov
U.S. National Wetlands Inventory	U.S. Fish and Wildlife Service	https://www.fws.gov/wetlands/data/data-download.html
Water Resources of the United States	U.S. Geological Survey	http://water.usgs.gov/maps.html
Data from NASA's Missions, Research, and Activities	National Aeronautics and Space Administration	http://www.nasa.gov/open/data.html
Datasets and Images	National Aeronautics and Space Administration	http://data.giss.nasa.gov
U.S. Census Data	U.S. Census Bureau	http://www.census.gov/data.html

Figure 8-4: Useful Internet sites to find GIS and remote sensing data.

Name _____   Section _____

EXERCISE 8 PROBLEMS—PART I—INTERNET

ARCGIS ONLINE

In this section you view data layers in ArcGIS Online. ArcGIS Online provides users with multiple datasets and the ability to analyze data.

You will need to create a free 60-day account to use ArcGIS Online. Go to the website **https://www.arcgis.com/home/index.html** and click on "Sign-up now" to create your account. On the next page click on "TRY ARCGIS" and follow the steps of setting up your account, including clicking on a link in an email. After you have set up your account, sign in to ArcGIS online and click on "Map." This will take you to the main Map page where you will be able to analyze spatial data and create maps online. You should see a screen that resembles Figure 8-5.

A GIS map of the United States should appear. Note that there are three tutorials along the left side of the screen that you could use to acquaint yourself with some of the basic abilities of ArcGIS Online.

You will now add a remote sensing data layer for a basemap. Click on "Basemap" and then on "Imagery." The map should now show an image of the United States.

Note in the upper right corner of the map a search box. Type in your address as completely as you can (e.g., 620 East 800 North, Provo, Utah 84606) and press the Return key.

1. How close is the map location to the actual location of your home? If it is not close, what are some reasons why?

Figure 8-5: Initial map screen in ArcGIS Online. Map image is the intellectual property of Esri and is used herein under license. Copyright © 2014 Esri and its licensors. All rights reserved.

Highest Points

Find the Terrain dataset to your map by clicking on "Add," then "Browse Living Atlas Layers." In the dropdown list, select "Landscape" then "Elevation" and "Terrain." Hover your mouse over the layer's tile to see a brief description of the dataset.

2.   What is the description of the dataset?

Add the layer to your map by clicking on "Add layer to map (as layer)." You can determine the characteristics of any spot in the map by clicking your cursor on the point of interest in the map.

3.   Mount Marcy is the highest point in New York state. Using the search tool and the terrain layer, determine the elevation of Mt. Marcy in meters (do this by clicking your cursor as close as you can to Mount Marcy).

_____ meters

Perform a web search to determine the highest point in your state. Then, use the map to determine its elevation in meters.

4.   What is the highest point of your home state?   _____

5.   What is its elevation?   _____ meters

6.   What is the highest spot in the conterminous United States? What is its elevation?

7.   What is the elevation of your home address?

Earthquake Risk

Earthquakes are one of the biggest natural disaster risks in some parts of the world. In this exercise you will add the USA Earthquake Risk layer to your map and assess the earthquake risk in several areas.

Find the USA Earthquake Risk layer by selecting "Add" then "Browse Living Atlas Layers." Next select "Landscape" then "Ecology" and "USA Earthquake Risk."

Add the layer to your map by clicking "Add layer to map (as layer)."

In the Content window on the left side of the screen, click on the ellipses below the layer name and then select "Show item details." A new screen should appear in your web browser that describes the dataset in the Description and Dataset Summary sections. If more information about the dataset is desired, users can click on "Link to source metadata" toward the bottom of the page. Answer the following questions about the dataset.

8.   Describe the characteristics of the "3.9 – 9.2%g (% of the acceleration of gravity)" Peak Acceleration class.

9.   What can you do with this layer?

10.   What areas do the data cover?

11.   How is ground acceleration different from the Richter scale?

12.   Who owns the dataset?

13.   When were the data created?

Zoom out to the entire United States to examine the earthquake risk trends throughout the continental United States.

14.   What are the general trends of earthquake risk in the continental United States (i.e., where earthquake risk is high and where risk is low)?

Sometimes it is useful to make a layer transparent so that the basemap (in this case the imagery) can be seen below the layer. Click on the ellipses below the USA Earthquake Risk layer name and select "Transparency." Set transparency to 50%.

15.   What happened to your map?

16.   Did this make it easier or more difficult to interpret earthquake risk in the United States?

Name _____   Section _____

EXERCISE 8 PROBLEMS—PART II—INTERNET

Assume you are an avid hiker who is interested in natural springs in the National Forest in the State of Utah. In particular, you want to identify natural springs that are within 500 meters of existing trails in part of the National Forest. In this part of the exercise you will use GIS analysis to solve this problem.

Go to the Hess *Physical Geography Laboratory Manual*, 12th edition, website at **www.MasteringGeography.com** and download the "springs.zip" and "trails.zip" datasets. These are partial datasets of the original data downloaded from the Uinta-Wasatch-Cache National Forest website (http://www.fs.usda.gov/detail/uwcnf/landmanagement/gis/?cid=stelprdb5434510). More information about these data—including their metadata—can be found on the website.

Open the datasets by clicking "Add" then "Add Layer from File." Select the zipped files where you saved them. (Note: do not uncompress, or unzip, the files. ArcGIS Online can read directly from the zipped files.) Do this for both the springs and trails datasets. After adding both layers your map should resemble the one in Figure 8-6, shown in color in the back of the Lab Manual as Map T-32a.

Figure 8-6: Map after adding the springs and trails layers. This map is shown in color as Map T-32a in the back of the Lab Manual. (Map image is the intellectual property of Esri and is used herein under license. Copyright © 2014 Esri and its licensors. All rights reserved. Springs and Trails data courtesy United States Forest Service.)

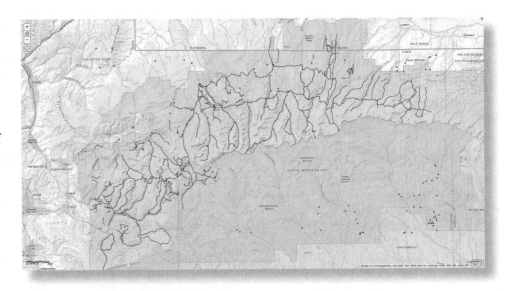

The next step is to buffer the trails layer to 500 meters. This will create a 500-meter area around each of the trails. Go to "Analysis" then "Use Proximity" and "Create Buffers." Fill out the parameters as shown in Figure 8-7.

After running this analysis your map should look like that in Figure 8-8 and Map T-32b (in the back of Lab Manual).

You will now overlay the Spring points with the 500-meter buffered trails. To do so, click on "Analysis" then "Manage Data" and "Overlay Layers." Fill out the parameters as shown in Figure 8-9.

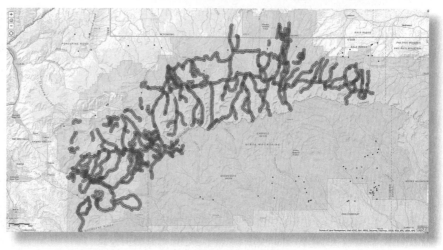

After completing this step, turn off all layers except for the trails layer and the layer you just created. (You turn off layers by clicking on and removing the check box next to the layers' names.) Your map should now look like the one in Figure 8-10 (shown in color as Map T-32c), which has the springs within 500 meters of a trail.

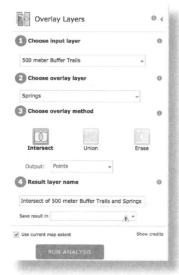

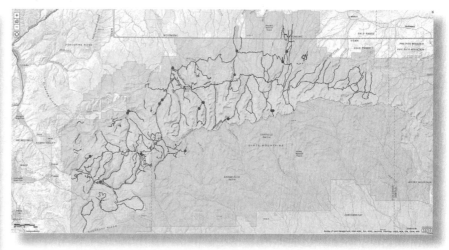

Next, determine the number of springs that are within 500 meters of trails. In the Details window click on the name of the newly created layer and click on the table icon ▦. Clicking this icon opens up the layer's table. Notice at the top of the table the number of points (or records) listed. This number signifies the number of springs within 500 meters of existing trails within this part of the National Forest.

1.   How many springs are within 500 meters of existing trails in this part of the National Forest?

2.   Provide two examples of when you might use this kind of analysis.

EXERCISE 9
Earth–Sun Relations

Objective:	To review Earth–Sun relations, and the reasons for the change of seasons.
Reference:	Hess, Darrel. *McKnight's Physical Geography*, 12th ed., pp. 17–22.

THE CHANGE OF SEASONS

The changes brought by the annual march of the seasons are due to four characteristics of Earth in its relationship to the Sun.

Rotation: Earth completes one rotation on its axis every 24 hours. The most obvious consequence of this rotation is the daily change from day to night.

Revolution: Earth orbits around the Sun, completing one revolution every $365\frac{1}{4}$ days. The orbit of Earth is not a perfect circle. The distance averages about 150 million kilometers (93 million miles), but Earth is 3.3 percent nearer to the Sun on about January 3 at its closest point (known as **perihelion**), than it is on about July 4 at its farthest point (**aphelion**).

Inclination: Earth's axis is tilted relative to the **plane of the ecliptic**, Earth's orbital path. Earth's axis maintains an inclination of 23.5° from the vertical throughout the year.

Polarity (Parallelism): Earth's axis is always pointing in the same direction (toward the North Star, *Polaris*). Notice in Figure 9-1 that because of polarity, in June the North Pole is leaning most directly toward the Sun, whereas in December it is leaning most directly away from the Sun.

Earth–Sun relations on four special days of the year, the **March equinox** (about March 20), the **June solstice** (about June 21), the **September equinox** (about September 22), and the **December solstice** (about December 21) are shown in Figure 9-1.

On the equinoxes (Figure 9-1c), the **vertical rays of the Sun** are striking the equator. In other words, on an equinox, at the equator, the Sun would be directly overhead in the sky at noon. The **tangent rays of the Sun** (those just skimming past Earth) are striking the North and South Poles. Notice also that the **circle of illumination** (the dividing line between night and day) is bisecting all parallels. This means that all latitudes experience 12 hours of daylight and 12 hours of darkness on this day.

On the solstices, the situation is quite different. On the June solstice (Figure 9-1b), the North Pole is leaning most directly toward the Sun, and the vertical rays of the Sun are striking the Tropic of Cancer at 23.5° N. On the December solstice, six months later (Figure 9-1d), the North Pole is leaning most directly away from the Sun, and the vertical rays of the Sun are striking the Tropic of Capricorn at 23.5° S.

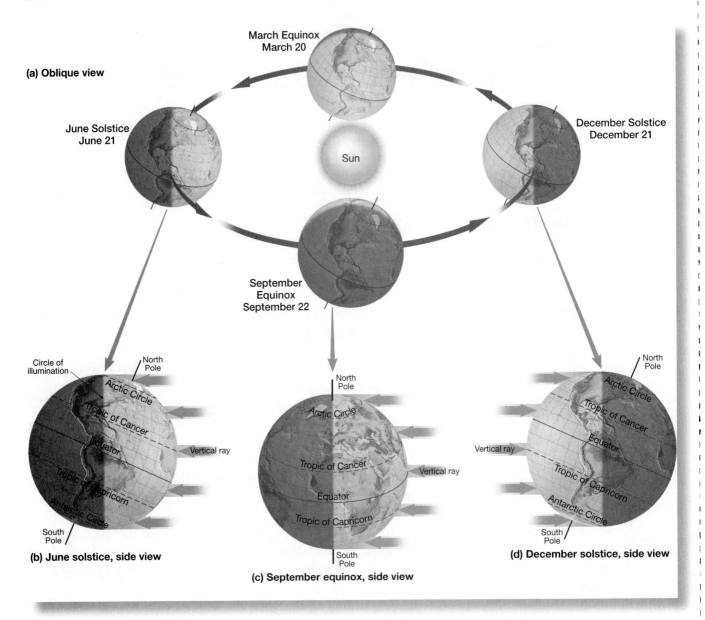

Figure 9-1: (a) The annual march of the seasons showing Earth–Sun relations on the June solstice, September equinox, December solstice, and March equinox. (b) On the June solstice, the vertical rays of the noon Sun strike 23.5° N. (c) On the March equinox and September equinox, the vertical rays of the noon Sun strike the equator. (d) On the December solstice, the vertical rays of the noon Sun strike 23.5° S. (From Hess, *McKnight's Physical Geography*, 12th ed.)

Name _____   Section _____

EXERCISE 9 PROBLEMS—PART I

The following questions are based on Figure 9-2, showing Earth–Sun relations on June 21.

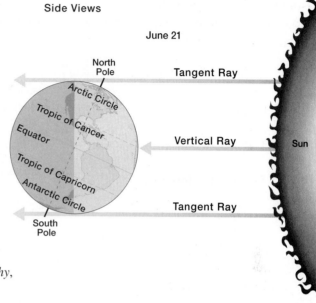

Figure 9-2: Earth–Sun relations on June solstice. (Adapted from McKnight and Hess, *Physical Geography*, 9th ed.)

1. What is the latitude of the vertical (direct) rays of the Sun?   _____

2. (a) What is the latitude of the tangent rays in the Northern Hemisphere?   _____

    (b) What is the latitude of the tangent rays in the Southern Hemisphere?   _____

3. Why is the June solstice associated with the Southern Hemisphere winter?

4. Noting the orientation of the circle of illumination on June 21, explain the following:

    (a) Why does the equator receive equal day and night?

    (b) What happens to the length of day (the number of hours of daylight) as you move north of the equator?

    (c) Within which range of latitudes in the Northern Hemisphere is 24 hours of day-light experienced?

Name _____   Section _____

EXERCISE 9 PROBLEMS—PART II

Using the diagram provided, draw in and label the following as they would appear on the December solstice. You are encouraged to use a straight edge and protractor to increase the accuracy of your diagram.

1.   North and South Poles

2.   Equator

3.   Tropic of Cancer

4.   Tropic of Capricorn

5.   Circle of illumination

6.   Arctic Circle

7.   Antarctic Circle

Using arrows to represent incoming sunlight, show the latitudes of the following:

8.   Vertical rays of the Sun

9.   Tangent rays of the Sun

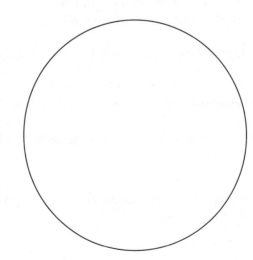

EXERCISE 10

Solar Angle

Objective:	To use the analemma to determine the latitude of the vertical rays of the Sun throughout the year, and to learn to calculate the altitude of the noon Sun at different times of the year.
Reference:	Hess, Darrel. *McKnight's Physical Geography,* 12th ed., pp. 20–21.

DECLINATION OF THE SUN

The **vertical rays of the Sun** at noon (rays coming in perpendicular to the surface; sometimes called the "direct" rays) will strike Earth at the Tropic of Cancer on the June solstice, at the equator on the equinoxes, and at the Tropic of Capricorn on the December solstice. The latitude of the vertical rays of the Sun is known as the **declination of the Sun**.

The changing declination of the Sun throughout the year is shown graphically with the **analemma** (Figure 10-1). On the analemma, the days of the year are shown on the "figure-eight" pattern. The declination of the Sun is read along the vertical axis. The latitudes shown range from 24° N at the top, to 24° S at the bottom, with the equator (0°) in the center. (The horizontal scale showing the "equation of time" will not be dealt with here.)

On the analemma, notice that the declination of the Sun is 0° on about September 22 and March 20, 23.5° N on about June 21, and 23.5° S on about December 21. The declination of the Sun on other days of the year is also easy to determine with the analemma. For example, on October 15, the declination is about 8° S. This means that on October 15, the noon Sun is directly overhead at a latitude of 8° S.

SOLAR ALTITUDE

The **solar altitude** refers to the apparent elevation of the noon Sun in the sky. In other words, the solar altitude is the angle of the noon Sun above the horizon. You can calculate the solar altitude for any latitude on any day of the year with the equation:

$$SA = 90° - AD$$

"SA" is the solar altitude, and "AD" is the "arc distance." The arc distance is the number of degrees of latitude between the latitude in question and the declination of the Sun.

For example, we can calculate the solar altitude at 40° N on the September equinox. On September 22 the declination of the Sun is 0° (as shown on the analemma). The arc distance between the declination of the Sun (0°) and the latitude in question (40°) is 40° (Figure 10-2). The solar altitude is: 90° − 40° = 50°. This means that on September 22, at a latitude of 40° N, the noon Sun will be 50° above the southern horizon.

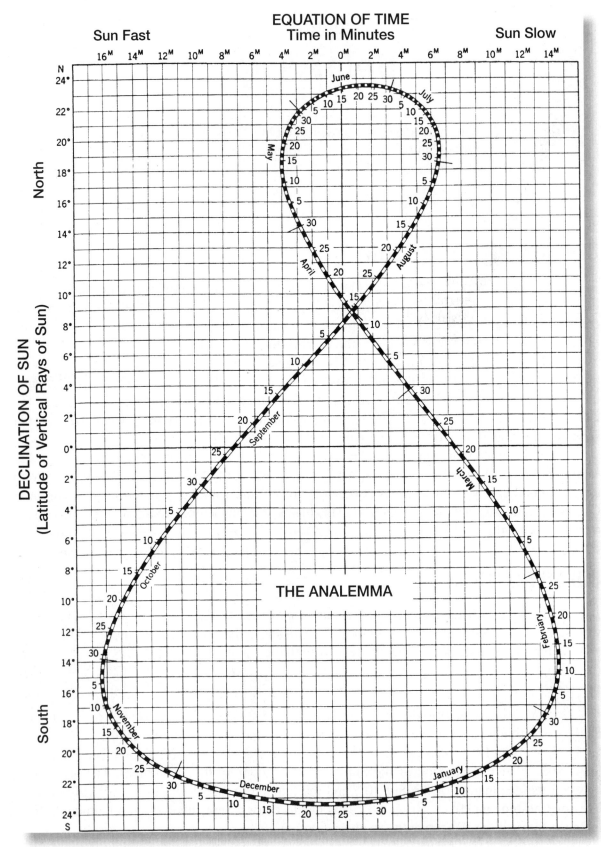

EQUATION OF TIME
Time in Minutes

Sun Fast   Sun Slow

THE ANALEMMA

Figure 10-1: The analemma shows the declination of the Sun and the difference in time between clock noon and sundial noon for each day of the year. (Adapted from U.S. Coast and Geodetic Survey)

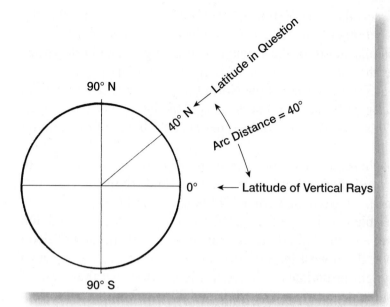

Figure 10-2: Earth–Sun relations on September 22.

Now, let's calculate the solar altitude at 40° N on February 9. The declination of the Sun is 15° S (as shown on the analemma). The arc distance is 55° (40° + 15° = 55°; Figure 10-3). The solar altitude is 90° − 55° = 35°. So, on February 9, at a latitude of 40° N, the noon Sun will be just 35° above the southern horizon.

LENGTH OF DAY

In Exercise 9 (Earth–Sun Relations) we discussed the changes in the length of day throughout the year, focusing on the circumstances during the equinoxes and solstices. With our understanding of the analemma we can now begin to describe the situation on other dates.

Figure 10-4 shows the Earth–Sun relationship on April 23. On this date the declination of the Sun is 12° N (as determined with the analemma). Notice that the **circle of illumination** (the

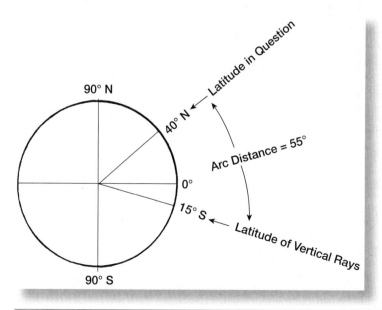

Figure 10-3: Earth–Sun relations on February 9.

dividing line between day and night) bisects the equator, but no other parallels. This means that the equator will have equal day and night, but other latitudes will have *unequal* day and night.

A greater proportion of each parallel north of the equator is in daylight than in darkness, and as we move toward the North Pole, the proportion of each parallel in daylight increases. This means that the length of day increases as we move toward the North Pole. Also notice that we can reach a latitude at which *no* portion of a parallel is in darkness. This is the latitude of the *tangent rays of the Sun* at midnight—north of this latitude, we would experience 24 hours of daylight on this day.

Looking in the Southern Hemisphere, the opposite is true, as we move toward the South Pole, the length of day becomes shorter and shorter, until we reach the latitude of the tangent rays of the Sun at noon—south of this latitude, we would experience 24 hours of darkness.

For any day of the year it is possible to calculate the latitude of the tangent rays of the Sun at midnight in the summer hemisphere, and the latitude of the tangent rays of the Sun at noon in the winter hemisphere, and therefore, the lowest latitude (the latitude closest to the equator) that would experience 24 hours of daylight or darkness. Notice in Figure 10-4 that the circle of illumination extends 90° to the north of the declination of the Sun, and 90° to the south of the declination of the Sun (which is 12° N in our example). This means that the tangent rays of the Sun will strike 12° beyond the North Pole at a latitude of 78° N. Therefore, on April 23, all latitudes north of 78° N will experience 24 hours of daylight. Conversely, all latitudes south of 78° S will experience 24 hours of darkness.

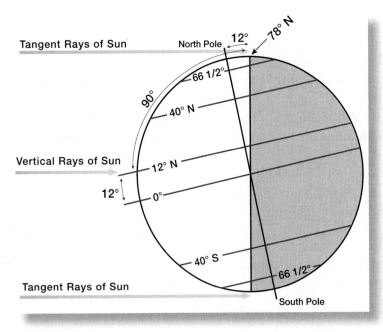

Figure 10-4: Earth–Sun relations on April 23. On this day, the vertical rays of the Sun (the declination of the Sun) are striking 12° north of the equator, so the tangent rays of the Sun will be 12° away from the North Pole.

Name _____   Section _____

EXERCISE 10 PROBLEMS—PART I

1. Using the analemma (Figure 10-1), find the declination of the Sun on the following dates; indicate whether the declination is north latitude or south latitude.

	Date	Declination			Date	Declination
(a)	January 10	_____	(c)		May 9	_____
(b)	March 6	_____	(d)		November 18	_____

2. Using the analemma and the equation $SA = 90° - AD$ calculate the solar altitude of the noon Sun at the following latitudes on the dates given:

	Date	Latitude in Question	Declination of Sun	Arc Distance	Solar Altitude
(a)	January 10	0°			
(b)	March 6	38° N			
(c)	May 9	70° S			

3. In your own words, explain what you have calculated in problem 2c (explain what the answer means, *not how it was calculated*). Be as specific as possible.

4. Determine the latitude of the tangent rays of the Sun in the Northern Hemisphere on the following days of the year, then indicate if latitudes north of the tangent rays in the Northern Hemisphere are receiving 24 hours of daylight or 24 hours of darkness. Remember, the tangent rays and circle of illumination will be the same number of degrees of latitude away from the North Pole as the declination of the Sun is from the equator. If the declination of the Sun is north of the equator, it is the high-Sun season ("summer") in the Northern Hemisphere; if the declination of the Sun is south of the equator, it is the low-Sun season ("winter").

Date	Declination of Sun	Latitude of Tangent Rays of Sun (and the circle of illumination) in Northern Hemisphere	Are latitudes north of the tangent rays receiving 24 hours of daylight or 24 hours of darkness? (Choose either "daylight" or "darkness")
March 29			
July 3			
October 14			

Name _____   Section _____

EXERCISE 10 PROBLEMS—PART II

Using the analemma, answer the following questions:

1. Determine how much the declination of the Sun changes during the following 10-day periods.

Period	Beginning Declination (to nearest $\frac{1}{2}°$)	Ending Declination (to nearest $\frac{1}{2}°$)	Total Latitude Change in Degrees (°)
12/12 to 12/22			
1/15 to 1/25			
3/20 to 3/30			

2. Based on your calculations in problem 1, as well as your general observations of the analemma, answer the following questions:

    (a) During which two times of the year (six months apart) is the declination of the Sun changing most rapidly from one day to the next? (You may name the two months of the year when declination changes most rapidly, or name the two special days of the year around which declination changes most rapidly.)

    (b) During which two times (or months) of the year (six months apart) is the declination of the Sun changing most slowly?

3. Solar altitude and length of day are related to the declination of the Sun. Using the analemma and your answers for problems 1 and 2 for reference, answer the following questions:

    (a) How much does the solar altitude in your city change over the following 10-day periods? (Hint: No new calculations are necessary; see your answers to problem 1.)

    12/12 to 12/22: _____   3/20 to 3/30: _____

    (b) During which two times of the year (six months apart) does solar altitude and length of day change most rapidly from one day to the next?

    (c) During which two times of the year (six months apart) does solar altitude and length of day change most slowly?

4. It is May 5 and you are somewhere in the Northern Hemisphere. If you determine that the noon Sun is 51° above your southern horizon, what is the latitude of your location?

EXERCISE 11

Insolation

Objective:	To study how insolation patterns are influenced by the angle of incidence of the Sun's rays, variations in the length of day, and the obstruction of the atmosphere.
Reference:	Hess, Darrel. *McKnight's Physical Geography*, 12th ed., pp. 90–92.

INSOLATION

In previous exercises, we saw how the changing relationship of Earth to the Sun produces differences in the angle of the incoming solar rays and in the length of day at different latitudes. In this exercise we will explore how these differences influence the average daily **insolation** (*in*coming *sol*ar radi*ation*) patterns at the surface of Earth.

In this exercise, insolation is defined as the rate at which solar energy strikes a surface (this is defining insolation in terms of *intensity*—the amount of energy received during a given period of time, in a given area). **Average daily insolation** refers to the average of this rate over a 24-hour period.

One common way to describe insolation is in watts per square meter (W/m²). One watt is equivalent to one *joule* per second; one joule is equal to about 0.239 calories (one calorie is the amount of energy required to raise the temperature of one gram of water by 1°C).

The insolation at Earth's upper atmosphere is about 1372 W/m²—this is known as the **solar constant** and assumes that the radiation is striking perpendicularly. However, the average daily insolation at Earth's surface is less than this. Three factors—the angle of incidence of the incoming radiation, the length of day, and the obstruction of the atmosphere—determine the actual daily insolation at the surface.

ANGLE OF INCIDENCE

There is a direct relationship between the **angle of incidence** (the angle at which the Sun's rays strike the surface) and the intensity of radiation that reaches the ground. Figure 11-1 shows that a beam of incoming solar radiation will be spread out over an increasingly larger surface area as the angle of incidence decreases from 90° (directly overhead) to lower angles of incidence (when the Sun is close to the horizon). A high angle of incidence, in essence, concentrates the solar energy in a small area, while a low angle of incidence spreads this energy out over a greater area.

Figure 11-2 is a chart showing the angle of incidence (the **solar altitude** of the noon Sun) at the equator, 45° N, and the North Pole on the 22nd day of each month (these solar altitudes were calculated using the method described in Exercise 10). The 22nd day of each month was chosen so that the solstices and equinoxes would fall on a day shown in the table.

Copyright © 2017 Pearson Education, Inc.

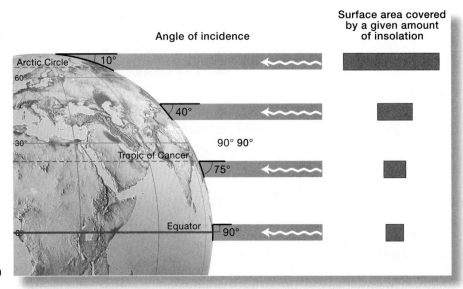

Figure 11-1: Comparative angles of the Sun's rays. (From Hess, *McKnight's Physical Geography*, 12th ed.)

	J	F	M	A	M	J	J	A	S	O	N	D
Equator	70°	80°	90°	78°	70°	66.5°	70°	78°	90°	79°	70°	66.5°
45° N	25°	35°	45°	57°	65°	68.5°	65°	57°	45°	34°	25°	21.5°
90° N	−20°	−10°	0°	12°	20°	23.5°	20°	12°	0°	−11°	−20°	−23.5°

Figure 11-2: Angle of noon Sun above the horizon (solar altitude) on the 22nd day of each month. Negative angles indicate the Sun is below the horizon at noon.

LENGTH OF DAY

The length of day is the second important factor that influences the amount of solar energy received at the surface. Even if the intensity of radiation is low, long hours of daylight significantly increase the total energy received. Figure 11-3 is a chart showing the approximate hours of

	J	F	M	A	M	J	J	A	S	O	N	D
Equator	12	12	12	12	12	12	12	12	12	12	12	12
45° N	9.5	10.5	12	14	15	15.5	15	14	12	10.5	9	8.5
90° N	0	0	12*	24	24	24	24	24	12*	0	0	0

Figure 11-3: Approximate hours of daylight (to the nearest half-hour) on the 22nd day of each month. (* The North Pole receives 12 hours of daylight only on the two equinoxes, *not* for the entire months of March and September.)

	J	F	M	A	M	J	J	A	S	O	N	D
Equator	420	430	440	420	405	395	410	425	440	430	420	410
45° N	150	230	305	405	460	475	455	400	300	200	140	120
90° N	0	0	0*	300	480	520	450	250	0*	0	0	0

Figure 11-4: Average daily insolation at the top of the atmosphere on 22nd day of each month in W/m^2. (* The North Pole receives no insolation in March during the days preceding the equinox, but increasing amounts of insolation following the equinox; in September the North Pole receives decreasing amounts of insolation preceding the equinox, but no insolation following the equinox.)

daylight at the equator, 45° N, and the North Pole on the 22nd day of each month. Notice that the equator receives virtually 12 hours of daylight throughout the year, and that the midlatitudes show moderate variation in the length of day from winter to summer. The North Pole receives either 24 hours of darkness or 24 hours of daylight (with the exception of the two equinoxes—the North Pole receives 12 hours of daylight on March 22 and September 22, but *not* for the entire months of March and September).

Figure 11-4 is a chart showing the average daily insolation at the top of the atmosphere (in W/m^2) at the equator, 45° N, and the North Pole on the 22nd day of each month. These data take into account variations in the angle of incidence and the length of day, but do not take into account the travel of radiation through the atmosphere to the surface. Note that the North Pole *does* receive insolation in March *following* the equinox, and in September *preceding* the equinox.

ATMOSPHERIC OBSTRUCTION

The atmosphere exerts a strong influence over the intensity of radiation received at the surface. Because of processes such as absorption and scattering, when the Sun's rays travel a great distance through the atmosphere, the intensity of radiation reaching the surface will be less than if the rays travel a short distance through the atmosphere.

The length of travel through the atmosphere is determined primarily by the angle of incidence. When the Sun's rays strike perpendicular to Earth's surface (an angle of incidence of 90°), the radiation has the shortest possible travel through the atmosphere, and this results in relatively little decrease in intensity. In contrast, when the Sun's rays strike Earth at a lower angle of incidence, the radiation reaching the surface will be less intense because it must travel through much more atmosphere (Figure 11-5).

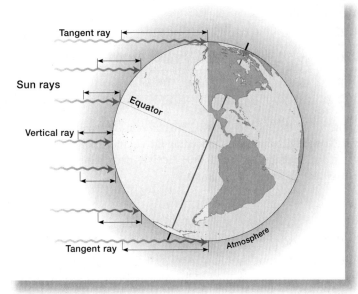

Figure 11-5: Atmospheric obstruction of sunlight. Low-angle rays must pass through more atmosphere than high-angle rays; thus, low-angle rays are subject to more depletion through reflection, scattering, and absorption. The day shown in this diagram is the December solstice. (Adapted from McKnight and Hess, *Physical Geography*, 9th ed.)

This relationship is shown in Figure 11-6. This table shows the approximate percentage of solar radiation reaching the surface for different angles of incidence. These figures are *highly* generalized. The actual effect of the atmosphere varies greatly from place to place, and from time to time (Figure 11-7), and the values in Figure 11-6 can provide only a rough approximation of the attenuating effects of the atmosphere.

Sun's Angle of Incidence	Percentage of Radiation Reaching Surface
90°	75%
70°	74%
50°	69%
30°	56%
20°	43%
10°	20%
5°	5%
0°	0%

Figure 11-6: The approximate percentage of solar radiation reaching the surface through the atmosphere.

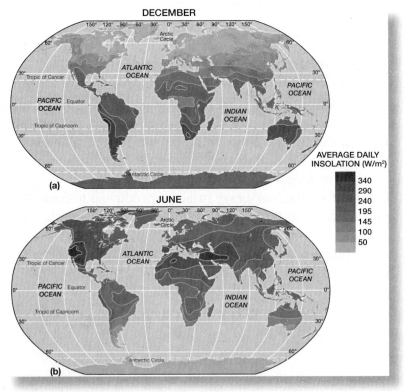

Figure 11-7: Global distribution of average daily insolation in W/m² for (a) December and (b) June. (From Hess, *McKnight's Physical Geography*, 12th ed.)

Name _____   Section _____

EXERCISE 11 PROBLEMS—PART I

1. On the chart, use the data from Figure 11-2 to plot the altitude of the noon Sun on the 22nd day of each month for the equator, 45° N, and 90° N. Connect the values for each latitude with a labeled line (you may also use a red line for the equator, a green line for 45° N, and a blue line for 90° N). Plot negative angles as a solar altitude of 0°.

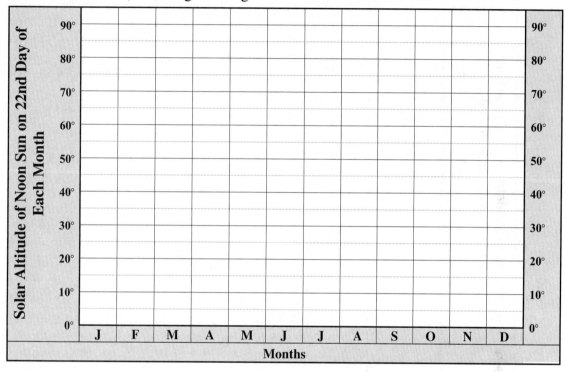

Using the completed chart from problem 1, answer the following questions:

2. (a) On June 22, at which latitude (the equator or 45° N) is the noon Sun highest in the sky?   _____

    (b) Does the noon Sun at the North Pole ever get as high as it is on January 22 at 45° N?   _____

3. If the angle of incidence (based on the solar altitude at noon) were the **only** factor influencing average daily insolation at the surface:

    (a) Which of the three latitudes would receive the highest insolation on 6/22? _____

    (b) Which of the three latitudes would receive the lowest insolation on 6/22? _____

(c)   Which of the three latitudes would receive the highest insolation on 12/22?_____

(d)   Which of the three latitudes would receive the lowest insolation on 12/22? _____

4.   On the chart, use the data from Figure 11-4 to plot the average daily insolation at the top of the atmosphere (in W/m^2) for the 22nd day of each month at the equator, 45° N, and 90° N. Connect the values for each latitude with a labeled line (you may also use a red line for the equator, a green line for 45° N, and a blue line for 90° N).

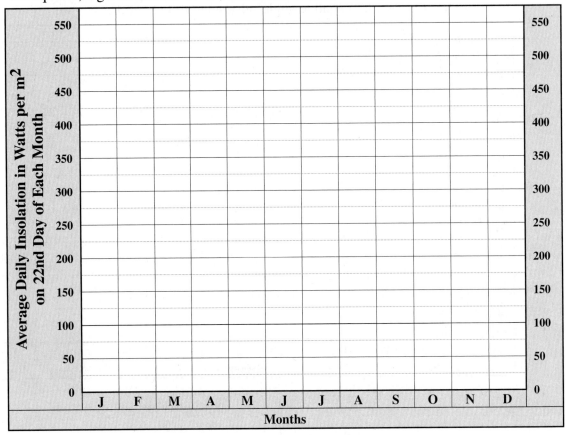

After completing the chart, answer the following questions. Keep in mind that these data show the insolation at the top of the atmosphere—they take into account the angle of incidence of incoming radiation and the length of day, but ignore the effects of the atmosphere.

5.   (a)   During a year, which of the three latitudes experiences the least
                variation in average daily insolation at the top of the atmosphere?   _____

       (b)   Why?

6. (a) During a year, which of the three latitudes experiences the greatest variation in average daily insolation at the top of the atmosphere? _____

    (b) Why?

7. (a) For how many months of the year does the North Pole receive no insolation? (Note: When answering this question you need to consider the *actual* period of time the North Pole goes without sunlight each year—interpret Figure 11-3 and your chart in problem 4 carefully; you should be able to answer this question *without* looking at Figure 11-3 or your chart.)

    _____ months

    (b) Which dates mark the beginning and end of this period of zero insolation at the North Pole?

    Beginning: _____   End: _____

8. (a) In which month does the top of the atmosphere at 45° N receive its highest average daily insolation? _____

    (b) In that same month, does the top of the atmosphere at the equator receive higher or lower average daily insolation than at 45° N?_____

    (c) What explains this?

9. During the year, at the top of the atmosphere the equator experiences two periods (six months apart) of maximum average daily insolation and two periods (six months apart) of minimum average daily insolation.

    (a) When do the two maximums occur? _____

    (b) When do the two minimums occur? _____

    (c) Explain the reason(s) for this pattern.

Name _____   Section _____

EXERCISE 11 PROBLEMS—PART II

Answer the following questions after completing the problems in Part I.

10.   On the chart, use the data from Figure 11-6 to plot the relationship between the angle of incidence of the incoming rays of the Sun and the approximate percentage of radiation reaching the surface through the atmosphere. Connect the values with a line.

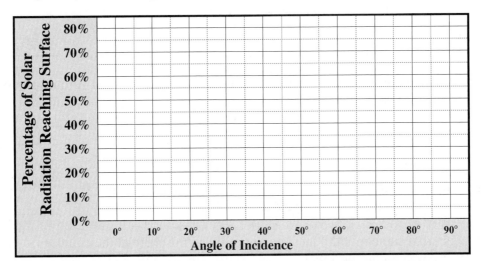

11.   (a)   Use the completed chart from problem 10 and Figure 11-2 (showing the angle of incidence of the noon Sun) to estimate the percentage of radiation passing through the atmosphere to the surface at noon on June 22 at the following latitudes.

	Angle of Incidence on June 22	Percentage of Radiation Received at Surface on June 22
Equator		
45° N		
90° N		

(b)   If the effects of the atmosphere *are* taken into account, which of the three latitudes shown would likely exhibit the greatest overall ***decrease*** in average daily insolation at the surface in June from that shown in problem 4?   _____

(c)   Why?

12.   What kinds of factors might help explain why the actual global patterns of average daily insolation shown in Figure 11-7 are not uniform by latitude? For example, notice equatorial Africa in December and western North America in June.

Temperature Patterns

Objective:	To study global temperature patterns and to explore the reasons for these patterns.
Reference:	Hess, Darrel. *McKnight's Physical Geography*, 12th ed., pp. 92–100.

FACTORS INFLUENCING TEMPERATURE PATTERNS

A number of factors influence the temperature regime of a location. The following factors are among the most important.

Latitude: Latitude is the most basic control of temperature. In general, because of the lower total insolation received at high latitudes compared with low latitudes, temperature decreases as we move away from the equator and toward the poles. In addition, the tropics generally show little temperature change during the year, while the mid- and high latitudes experience variation in temperature from summer to winter. These basic global patterns are apparent in Map T-28a (shown in color at the back of the Lab Manual), showing average sea-level temperatures in January mapped with isotherms, and in Map T-28b, showing average sea-level temperatures in July.

Were latitude the only control of temperature, the isotherms would run exactly east to west, parallel to the lines of latitude. However, this hypothetical pattern is altered by a number of additional factors.

Land-Water Contrasts: Land and water react differently to solar heating, and this exerts a strong influence on the atmosphere. In general, land warms up and cools off faster and to a greater extent than water. This means that the interiors of continents will be hotter in summer and colder in winter than maritime regions at the same latitude. The ocean also significantly moderates the temperatures of the coastal regions of a continent. This is illustrated in Figure 12-1, showing the shift of an isotherm over a continent from winter to summer.

In addition to the lower annual temperature range associated with maritime regions, the ocean also exhibits a lag in reaching its coolest point in winter and its warmest point in summer. This means that coastal regions often reach their temperature extremes several months after interior regions.

Ocean Currents: The general circulation of the ocean is a significant mechanism of global heat transfer. Major surface ocean currents move warm water from the equatorial regions toward the poles, and bring cool water from the poles back toward the equator. In each of the main ocean basins, warm water is moving toward the poles off the east coasts of continents, while cool water is moving toward the equator off the west coasts of continents (Figure 12-2).

Wind Patterns: In many regions of the world, the dominant wind direction strongly influences local temperature patterns. For example, in the midlatitudes the dominant wind direction is from the west, meaning that air masses will tend to move from west to east. As a consequence, the temperature patterns of midlatitude locations along the east coast of a continent can be

---

quite "continental"—the westerlies can bring the seasonal warmth or coldness of the interior of the continent all the way to the east coast.

Altitude: In general, temperature decreases with increased altitude. A high elevation station will have a very similar annual temperature pattern to a nearby lowland station, although the high elevation station will be consistently cooler throughout the year. On average, temperature in the **troposphere** decreases by approximately 6.5°C per 1000 meters of elevation increase (3.6°F/1000 feet)—this is known as the **average lapse rate**.

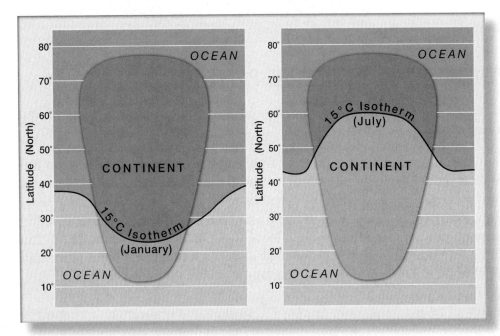

Figure 12-1: Idealized seasonal migration of the 15°C (59°F) isotherm over a hypothetical Northern Hemisphere continent. The isotherm shifts farther south in winter and farther north in summer over land than over the ocean. (From Hess, *McKnight's Physical Geography*, 12th ed.)

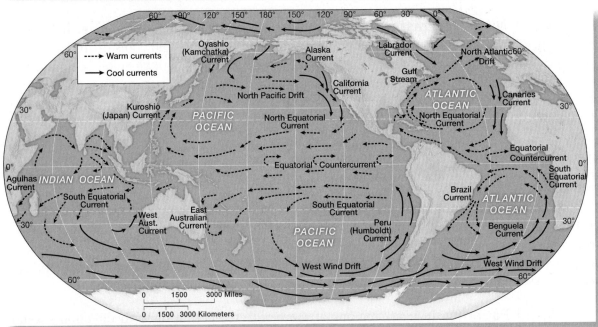

Figure 12-2: Major surface ocean currents. Notice that in the midlatitudes, warm currents flow toward the poles along the eastern coasts of continents, whereas cool currents flow toward the equator along the western coasts. (From Hess, *McKnight's Physical Geography*, 12th ed.)

Name_____   Section _____

EXERCISE 12 PROBLEMS—PART I

The following questions are based on color maps in the back of the Lab Manual showing average January sea-level temperatures (Map T-28a) and average July sea-level temperatures (Map T-28b):

1. Is the temperature contrast between the equator and the
    Arctic region greatest in the winter or summer?   _____

2. (a) Were latitude the only control of temperature, the isotherms would run straight across the maps from east to west. Describe one region of the world where this hypothetical isotherm pattern is actually observed.

    (b) Why is the hypothetical pattern seen here?

3. (a) Is the influence of cool ocean currents on coastal
        temperatures more pronounced in summer or winter?   _____

    (b) Why?

4. (a) Comparing the January map with the July map, describe one region of the world that exhibits a large *annual temperature range* (the difference between the January and July average temperatures).

    (b) What explains this large annual temperature range?

    (c) Describe one region of the world that exhibits a small annual temperature range.

    (d) What explains this small annual temperature range?

Name _____   Section _____

EXERCISE 12 PROBLEMS—PART II

Using a straightedge, draw a line across the July temperature map (Map T-28b; shown in color at the back of the Lab Manual) from point "A" to point "B." This reference line can be thought of as the "hypothetical" position of the 16°C (60°F) isotherm were there no land–water contrasts, ocean currents, and so on. Compare the actual 16°C isotherm, with the line you have just drawn. In places where the actual 16°C isotherm is south of the hypothetical line, temperatures are lower than expected; in places where the actual 16°C isotherm is north of the hypothetical line, temperatures are higher than expected.

1.　　Begin in the west and move across the map to the east, briefly explaining why the actual 16°C (60°F) isotherm deviates from the hypothetical.

2.　　Why isn't the shift of this 16°C isotherm uniform across all of Eurasia?

Name_____   Section _____

EXERCISE 12 PROBLEMS—PART III

Six charts showing the average monthly temperature (in °C and °F) for seven U.S. cities are provided (the first letter of each month of the year is shown along the bottom of the charts). For each of the cities, the latitude and longitude, as well as the elevation, are provided.

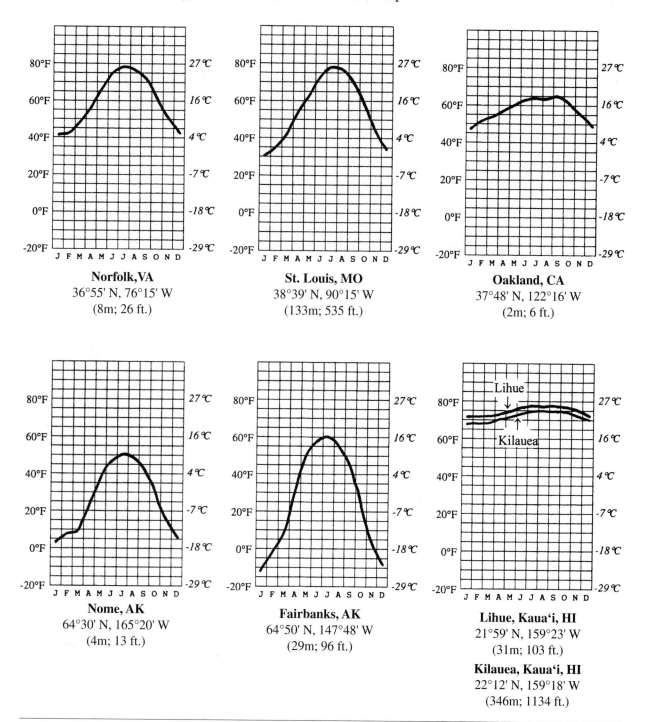

Norfolk, VA
36°55' N, 76°15' W
(8m; 26 ft.)

St. Louis, MO
38°39' N, 90°15' W
(133m; 535 ft.)

Oakland, CA
37°48' N, 122°16' W
(2m; 6 ft.)

Nome, AK
64°30' N, 165°20' W
(4m; 13 ft.)

Fairbanks, AK
64°50' N, 147°48' W
(29m; 96 ft.)

Lihue, Kaua'i, HI
21°59' N, 159°23' W
(31m; 103 ft.)

Kilauea, Kaua'i, HI
22°12' N, 159°18' W
(346m; 1134 ft.)

Answer the following questions by comparing the temperature charts on the previous page. In your answers, consider the *one* **temperature control factor that is *most* responsible** for the patterns shown (choose from latitude, land–water contrasts, wind patterns, or altitude). You may use the same answer for more than one question. You should locate each of the cities on a world map before trying to answer the questions. *If altitude is the main factor cited, calculate the expected temperature difference between the two cities based on the average lapse rate.*

1.   Which factor primarily explains the different temperature patterns of St. Louis and Oakland?

2.   Why is the warmest month of summer different in St. Louis and Oakland?

3.   Why does St. Louis have colder winters than Norfolk?

4.   Although both are coastal cities, compared to Oakland, Norfolk has a very "continental" temperature pattern. Why?

5.   Which factor primarily explains the difference in temperature patterns between Fairbanks and St. Louis?

6.   Which factor primarily explains the difference in temperature patterns between Fairbanks and Nome?

7.   Why does Lihue have a smaller annual temperature range than Oakland?

8.   What explains the difference in temperature patterns between Lihue and Kilauea?

EXERCISE **13**

Air Pressure

Objective:	To introduce the standard weather map station model and to study pressure patterns shown on maps with isobars.
Reference:	Hess, Darrel. *McKnight's Physical Geography*, 12th ed., pp. 110–112.

AIR PRESSURE

The **pressure of air** is the force exerted by the atmosphere on a surface. Gravity pulls the gases of the atmosphere toward Earth. Atmospheric pressure is the force—exerted in all directions—by the weight of these gas molecules on a unit area of Earth's surface.

FACTORS INFLUENCING AIR PRESSURE

Many factors influence air pressure. The pressure, density, and temperature of the air are all closely interrelated. If one factor changes, the other two also tend to change. We can, however, make a few generalizations about the kinds of conditions that tend to produce either high or low pressure near the surface.

The following are generalizations and *not* absolute laws. In practice, however, most surface pressure cells can be explained by the dominance of one of these four conditions.

1. Ascending (rising) air tends to produce **low pressure** near the surface. Lows caused by strongly rising air are sometimes called **dynamic lows**.

2. Warm surface conditions can produce low pressure near the surface. Lows caused by warm surface conditions are sometimes called **thermal lows**.

3. Descending (subsiding) air tends to produce **high pressure** near the surface. Highs produced by strongly descending air are sometimes called **dynamic highs**.

4. Cold surface conditions can produce high pressure near the surface. Highs produced by cold surface conditions are sometimes called **thermal highs**.

MEASURING AIR PRESSURE

There are several measurement systems used to describe air pressure. Although most television and newspaper weather reports use **inches of mercury** (the height of a column of mercury in a liquid barometer), the most common unit of pressure measurement used in meteorology in the United States is the **millibar**. The millibar (mb) is a measure of force per unit area. The definition of 1 millibar is the force of 1000 dynes per square centimeter (1 dyne is the force required to accelerate 1 gram

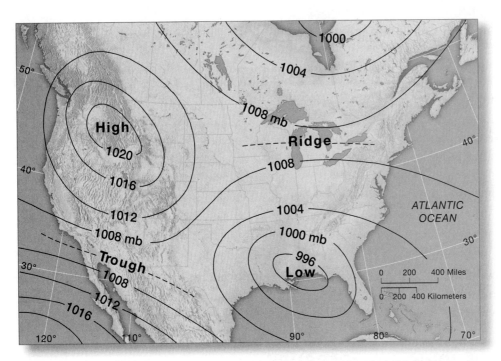

Figure 13-1: Isobar map showing areas of high and low pressure. (From McKnight and Hess, *Physical Geography*, 12th ed.)

of mass 1 centimeter per second per second). In some countries air pressure is described with the *pascal* (Pa; 1 Pa = 1 newton/m^2 [1 newton is the force required to accelerate a 1 kg mass 1 meter per second per second]) or the *kilopascal* (kPa; 1 kPa = 10 mb).

For comparison, the average sea-level pressure is 29.92 inches of mercury, which is equivalent to 1013.25 mb. We are generally interested in relative differences in pressure. For example, at the surface, 1032 mb would usually represent relatively high pressure, whereas 984 mb would represent relatively low pressure (equivalent to 30.47 inches and 29.06 inches of mercury, respectively).

ISOBARS

Differences in pressure can be mapped with isolines called **isobars**. Isobars are lines that connect points of equal atmospheric pressure. Figure 13-1 shows a region where areas of high and low pressure have been mapped with isobars. Note that a **ridge** is an elongated area of relatively high pressure, while a **trough** is an elongated area of relatively low pressure.

THE STATION MODEL

We now introduce the **station model** and the standardized system of data presentation that is used on weather maps around the world. In its bare form, a station model is a circle on a map that shows the location of a weather station (in our scale of analysis, usually a city). Various weather data are then plotted in a specific form, and in a specific location, around (and within) this circle.

In this Lab Manual, we will be using an abbreviated form of the station model, introducing new elements in subsequent exercises. However, we will follow standard station model protocol in the placement of each data element.

Temperature: The current surface temperature of a station is written to the upper left of the circle. In the example here, the temperature is 65° Fahrenheit. Note that neither the degree symbol (°) nor an "F" (for Fahrenheit) is included.

65 ◯

Pressure: Current surface pressure is written to the upper right of the circle. In the example here, the surface pressure is 970.2 mb.

970.2 ◯

On standard weather maps, however, the notation for pressure is abbreviated. To save space, the first 9 or 10 is left off, and the decimal point is removed. So, in a standard station model, a pressure of 970.2 mb would be written as follows:

702 ◯

While 1035.6 mb would be written as follows:

356 ◯

To rewrite these pressures in their full form, simply add the decimal point, and then decide if adding a 9 or 10 makes more sense. Usually this is easy, since surface pressures typically range from about 960.0 mb to 1050.0 mb. For example, if given the pressure shown on the following station model:

237 ◯

first add the decimal point: "_23.7"; then add a 10: "1023.7 mb." Adding a 9 usually would not make sense, because 923.7 mb is lower than expected for a typical surface pressure (a quick way to decide whether to add a 9 or a 10 is to choose the number that makes the pressure value closest to 1000 mb).

Occasionally, the pressure may be higher or lower than is typically encountered—the extremely low pressure associated with the eye of a hurricane is one example of this. In such cases, you will have to decide by the context of nearby stations if adding a 9 or a 10 makes more sense.

Name _____   Section _____

EXERCISE 13 PROBLEMS—PART I

1. Encode the following data around the station model (using the standard abbreviated form for pressure).

	Temperature	Pressure (mb)	Station Model
(a)	42°F	1017.0	◯
(b)	101°F	1003.4	◯
(c)	23°F	1024.9	◯
(d)	54°F	998.2	◯

2. Decode the following station models (be sure to indicate the correct units of measurement for each value).

	Station Model	Temperature	Pressure
(a)	36 ◯ 131		
(b)	76 ◯ 768		
(c)	55 ◯ 380		
(d)	81 ◯ 997		

Name _____   Section _____

EXERCISE 13 PROBLEMS—PART II

This map is a simplified weather map of the United States showing surface pressure conditions in millibars (here written out in full form). Draw in all of the appropriate isobars at 4 mb intervals. Your highest isobar value will be 1028 mb (the sequence of isobars will be 1028 mb, 1024 mb, 1020 mb, 1016 mb, etc.). Label each isobar. There is one high pressure center and one low pressure center on the map—label each appropriately ("H" or "L"). Begin by drawing your isobar lines lightly in pencil until you are certain of their location. For a review of drawing isolines in general, refer to Exercise 6. Hint: First draw in the 1020 mb isobar that will run from northeast to southwest across the middle of the country; next enclose the high pressure center; finally, enclose the low pressure area.

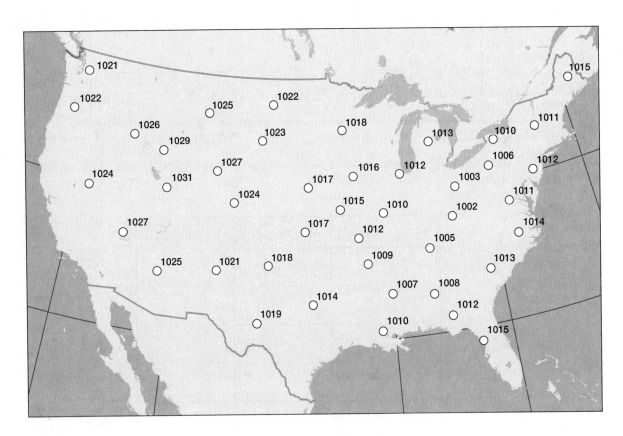

EXERCISE 14
Wind

Objective:	To map and study winds associated with high and low pressure areas.
Resources:	Internet access or mobile device QR (Quick Response) code reader app (optional).
Reference:	Hess, Darrel. *McKnight's Physical Geography*, 12th ed., pp. 112–125.

PRESSURE GRADIENTS

Air moves horizontally when there is a difference in pressure from one place to another. The change in pressure over a given distance is called the **pressure gradient**, and the force exerted by this difference in pressure is known as the **pressure gradient force**.

As we saw in Exercise 13, pressure patterns are mapped with **isobars** (lines of equal pressure). Closely spaced isobars indicate a great change in pressure over a short distance. A "steep" pressure gradient such as this produces a strong pressure gradient force, and so the resulting wind will be relatively strong (fast). On the other hand, isobars spaced far apart indicate a small change in pressure over a given distance. A "gentle" pressure gradient such as this produces a weak pressure gradient force, and so the resulting wind will be relatively weak (slow).

WIND DIRECTION

The initial propelling force behind wind is the pressure gradient force. Air begins to move "down" the pressure gradient, from high to low pressure, at right angles to the isobars (Figure 14-1a). However, the actual direction of the wind is influenced by two other factors: the Coriolis effect and friction.

At high elevations, above about 1000 meters (about 3300 feet), the moving air encounters very little **friction**, but the **Coriolis effect** will deflect the path of the wind. The Coriolis effect (also referred to as the **Coriolis force**) deflects the path of free moving objects to the right in the Northern Hemisphere, and to the left in the Southern Hemisphere. Figure 14-1b shows that while the wind begins to blow down the pressure gradient, the Coriolis effect deflects the path of the wind about 90° to the right (in the Northern Hemisphere). In the upper atmosphere, the balance of the pressure gradient force and the Coriolis effect results in winds that blow approximately parallel to the isobars. These winds are called **geostrophic winds**.[1] In the Northern Hemisphere upper elevation winds tend to blow parallel to the isobars, clockwise around highs, and counterclockwise around lows.

In the lowest parts of the atmosphere, friction becomes important (Figure 14-1c). Friction from the surface slows the wind, and so the Coriolis effect deflection is reduced (rapidly moving objects are deflected more by the Coriolis effect than slowly moving objects).

[1]Strictly speaking, geostrophic wind is found only in areas where the isobars are parallel and straight; the term *gradient wind* is a more general term used to describe wind flowing parallel to the isobars. In this Lab Manual we use the term geostrophic to mean all wind blowing parallel to the isobars.

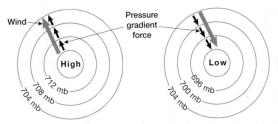

(a) If the pressure gradient force were the only factor, wind would blow "down" the pressure gradient away from high pressure and toward low pressure, crossing the isobars at an angle of 90°.

Figure 14-1: Wind direction is influenced by three factors: pressure gradient force, Coriolis effect, and friction. (a) Hypothetical pattern if pressure gradient force was the only factor. (b) Geostrophic winds in the upper atmosphere (above about 1000 meters [3300 feet]). (c) In the friction layer of the lower atmosphere, friction slows the wind (and so reduces the Coriolis effect deflection), so wind diverges out of a high and converges into a low. (From Hess, *McKnight's Physical Geography*, 12th ed.)

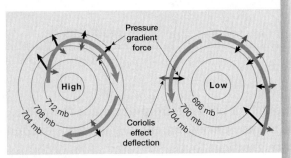

(b) In the upper atmosphere the balance between the pressure gradient force and the Coriolis effect results in geostrophic wind blowing parallel to the isobars.

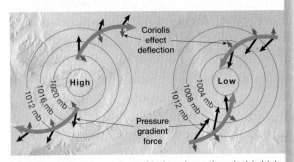

(c) In the lower atmosphere, friction slows the wind (which results in less Coriolis effect deflection) and so wind diverges clockwise out of a high and converges counterclockwise into a low in the Northern Hemisphere.

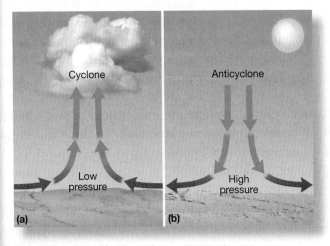

Figure 14-2: (a) In a cyclone (low pressure cell), air converges and rises. (b) In an anticyclone (high pressure cell), air descends and diverges. (From Hess, *McKnight's Physical Geography*, 12th ed.)

Cyclones and Anticyclones: Near the surface, the balance of the pressure gradient force, Coriolis effect, and friction results in winds that diverge clockwise out of **highs** (**anticyclones**) and converge counterclockwise into **lows** (**cyclones**) in the Northern Hemisphere. In the Southern Hemisphere, winds are deflected in the opposite direction.

There is a prominent vertical component of air movement within cyclones and anticyclones (Figure 14-2). In cyclones, air converges at the surface and then rises to the upper atmosphere. In anticyclones, air descends from the upper atmosphere, then diverges at the surface.

GENERAL CIRCULATION PATTERNS

At the global scale, the four most prominent components of the general circulation of the atmosphere are associated with the **Hadley cells** (Figure 14-3). This pair of large convection cells

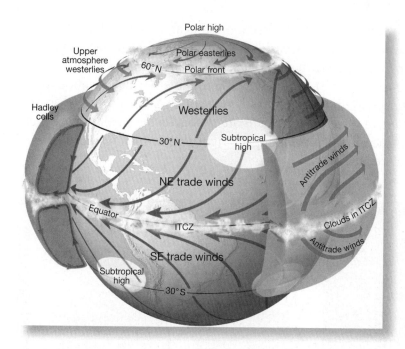

Figure 14-3: Idealized global circulation patterns in the atmosphere. Air rises in the Hadley cells in the ITCZ and descends into the subtropical highs. The subtropical highs are the source of the surface trade winds and westerlies. (From Hess, *McKnight's Physical Geography*, 12th ed.)

is driven by the warm surface conditions in equatorial latitudes. Warm air rises near the equator in the **intertropical convergence zone (ITCZ)**, an area of generally low pressure and cloudy conditions. This air descends at about 30° North and South in the **subtropical highs** (STH), areas of high pressure and clear, dry conditions.

Two surface wind systems diverge from the STHs, the **westerlies** (blowing from the west) in the midlatitudes and the **trade winds** (blowing from the east) within the tropics. The upper atmosphere westerlies, including the high velocity **jet streams**, generally flow from west to east, as do the upper atmosphere trade winds (the **antitrade winds**) near the top of the Hadley cells.

INDICATING WIND ON WEATHER MAPS

Wind direction is indicated on weather map station models with a shaft that points *into* the wind. Wind direction is described as the direction *from* which the wind is blowing, so a "westerly" wind is blowing from the west, and an "easterly" wind is blowing from the east. Occasionally, wind direction is described in terms of an **azimuth**—the number of degrees clockwise from north (so north = 000° [or 360°]; east = 090°; south = 180°; west = 270°; and so on).

The wind speed is shown with "feathers" on the wind direction shaft. The wind speed, in **knots**, is determined by adding up the number of feathers (a "knot" is 1 nautical mile per hour, which equals 1.15 statute mph or 1.85 km/hr). Each full feather represents an increase in wind speed of 10 knots. Half-feathers are used to represent 5 knots, and a solid triangular flag is used to represent 50 knots (see Figure III-7 in Appendix III). For example, the following station model indicates wind from the northwest at 25 knots.

When mapping upper atmosphere winds, wind direction shafts and feathers are often used without the rest of the station model (this is done occasionally when mapping surface wind patterns as well). Figure 14-4 shows a pair of satellite maps of the northeastern Pacific Ocean showing forecast surface and 300 mb (upper atmosphere) winds. The 300 mb level is the elevation in the atmosphere where the pressure has decreased to 300 mb. This typically is at an altitude of around 9 kilometers (30,000 feet). The centers of three pressure cells are shown on the small surface pressure map. The west coast of North America is visible along the right side of each image. The dark areas are ocean, and the white areas are clouds.

Figure 14-4: Forecast upper atmosphere wind at 300 mb level (top image) and surface wind (bottom image) in North Pacific on February 9, 2013. Large pressure centers are shown on the small surface pressure map. Enlarged wind symbols from each labeled region are shown in circles on the left. (Images courtesy of Naval Research Laboratory, Marine Meteorology Division)

Name _____   Section _____

EXERCISE 14 PROBLEMS—PART I

The two maps shown are pressure maps for the United States (pressure is shown in millibars). The top map (a) shows the isobars at an altitude of 2500 meters (8000 feet). The bottom map (b) shows the surface pressure.

On each map, use a colored pencil to draw 1 centimeter ($\frac{1}{2}$ inch) long arrows to show the wind pattern you would expect to observe. Align the arrows in their proper relationship to the isobars. Use 8 to 12 arrows for each map. Remember, in the upper atmosphere all wind is geostrophic, whereas near the surface, none is geostrophic.

(a) Upper Atmosphere
Pressure

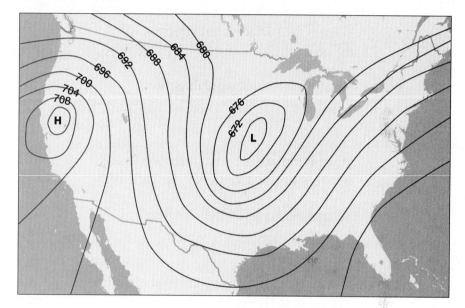

(b) Surface
Pressure

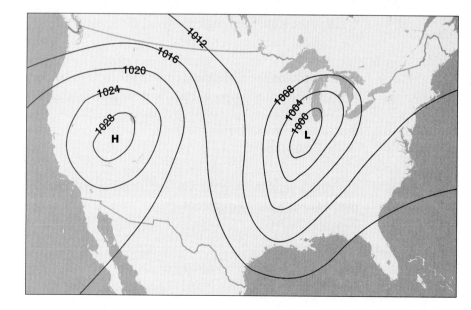

Name _____   Section _____

EXERCISE 14 PROBLEMS—PART II

Two satellite images are reproduced in Figure 14-4. You may view these maps by going to the Hess *Physical Geography Laboratory Manual*, 12th edition, website at **www.MasteringGeography.com,** and then select Exercise 14, or by scanning the QR (Quick Response) code for this exercise. They show the forecast surface and upper atmosphere winds (300 mb level) over the northeastern Pacific Ocean. The west coast of North America is visible along the right side of each image. The dark areas are ocean, and the white areas are clouds. The centers of three pressure cells are shown on the small surface pressure map. Enlarged wind symbols from each labeled region are shown on the left.

1.   (a)   What is the surface wind direction off the coast of California (35° N, 125° W)?   From the _____

     (b)   What explains this wind direction?

2.   (a)   What is the surface wind direction in the northern Pacific Ocean (45° N, 155° W)?   From the _____

     (b)   What explains this wind direction?

3.   (a)   What is the surface wind direction in the tropical Pacific near Hawai'i (20° N, 160° W)?   From the _____

     (b)   What is the upper atmosphere wind direction in the tropical Pacific over Hawai'i?   From the _____

     (c)   What explains the difference in wind direction between the surface and the upper atmosphere in the tropical Pacific near Hawai'i?

     (d)   Suggest a reason why the surface wind speed and upper atmosphere wind speed near Hawai'i are so different.

EXERCISE 15
Humidity

Objective:	To study the relationship between the water vapor content of the air, temperature, and relative humidity.
Materials:	Sling Psychrometer (optional).
Reference:	Hess, Darrel. *McKnight's Physical Geography*, 12th ed., pp. 145–149 and A8–A9.

HUMIDITY

In order to understand cloud formation and precipitation, we must begin by studying water vapor in the atmosphere. There are several ways to describe **humidity**—the amount of **water vapor** in the air—but two are important for us here.

Mixing Ratio: The **mixing ratio** describes the actual amount of water vapor in the air. The mixing ratio is expressed as the mass ("weight") of water vapor in a given mass of dry air, described in grams of water vapor per kilogram of air (g/kg).

In addition to the mixing ratio, there are other ways to describe the actual amount of water vapor in the air. For example, the **specific humidity** is similar to the mixing ratio, except that it describes the number of grams of water vapor per kilogram of air, including water vapor. The **absolute humidity** describes the mass of water vapor in a given volume of air, expressed in grams of water vapor per cubic meter of air (g/m^3).

In many cases, the mixing ratio is more useful to meteorologists than the absolute humidity because the mixing ratio does not change as the volume of air changes (as happens when air rises). As a very rough comparison, at sea level 1 cubic meter of air at room temperature has a mass of about 1.4 kilograms.

Relative Humidity: **Relative humidity** does not describe the actual amount of water vapor in the air. Rather, it is a ratio that compares the actual amount of water vapor in the air (the mixing ratio) to the maximum amount of water vapor that can be in the air at a given temperature—also called the **capacity**.

$$\text{Relative humidity} = \frac{\text{Actual water vapor content}}{\text{Water vapor capacity}}$$

The relative humidity (RH) expresses this degree of **saturation** as a percentage. For example, 50% RH means that the air contains half of the water vapor necessary for saturation; 75% RH means that the air has three-quarters of the water vapor necessary for saturation; 100% RH means that the air is saturated. When air is saturated, **condensation**, and therefore cloud formation, can take place.

Temperature		Saturation Mixing Ratio ("Capacity") g/kg
°F	°C	
15°F	−9.4°C	1.9
20°F	−6.7°C	2.2
25°F	−3.9°C	2.8
30°F	−1.1°C	3.5
35°F	1.7°C	4.3
40°F	4.4°C	5.2
45°F	7.2°C	6.2
50°F	10.0°C	7.6
55°F	12.8°C	9.3
60°F	15.6°C	11.1
65°F	18.3°C	13.2
70°F	21.1°C	15.6
75°F	23.9°C	18.8
80°F	26.7°C	22.3
85°F	29.4°C	26.2
90°F	32.2°C	30.7
95°F	35.0°C	36.5
100°F	37.8°C	43.0

Figure 15-1: Approximate saturation mixing ratios in g/kg at various temperatures (°F and °C). (Note: at temperatures below freezing over ice, the saturation mixing ratios will be slightly lower than indicated here.)

The water vapor capacity of air at a given temperature is also called the **saturation mixing ratio**, because it is the mixing ratio of a saturated parcel of air. The water vapor capacity of air depends almost entirely on temperature. As temperature increases, the water vapor capacity of the air also increases. Figure 15-1 shows the capacity (the saturation mixing ratio) of air at different temperatures.

In popular terms, it is said that warm air can "hold" more water vapor than cold air, but this is somewhat misleading. The air doesn't actually hold water vapor as if it were a sponge. Water vapor is simply one of the gaseous components of the atmosphere—the water vapor capacity of the air is determined by the temperature, which determines the rate of vaporization of water.

CALCULATING RELATIVE HUMIDITY

In this exercise, we will use the mixing ratio to describe the actual water vapor content of the air, and the saturation mixing ratio to describe the water vapor capacity of the air. Relative humidity is calculated with a simple formula:

$$RH = \frac{\text{Mixing ratio}}{\text{Saturation mixing ratio}} \times 100$$

For example, if the mixing ratio is 13.5 g/kg and the saturation mixing ratio is 22.5 g/kg, the relative humidity is

$$\frac{13.5 \text{ g/kg}}{22.5 \text{ g/kg}} \times 100 = 60\% \text{ RH}$$

The key to understanding relative humidity is recognizing the relationship between temperature and the water vapor capacity of air. When temperature changes, relative humidity changes. For example, as the temperature of the air decreases, water vapor capacity decreases. This means that as the temperature decreases, the relative humidity increases. If a parcel of air is cooled enough, its mixing ratio will match its capacity and the air will be saturated (100% relative humidity) and condensation can take place. If cooling continues after a parcel of air has become saturated, the relative humidity will tend to remain at 100%—as the capacity continues to decrease, more and more water vapor will condense out of the air, keeping the mixing ratio of the parcel the same as its capacity and so maintaining 100% relative humidity.[1]

THE DEW POINT TEMPERATURE

The temperature at which a parcel of air reaches 100% relative humidity is called the **dew point temperature** (or simply the **dew point**). This is the temperature at which the water vapor capacity of the air (the saturation mixing ratio) is the same as the actual water vapor content of the air (the mixing ratio). Notice that the dew point is determined by the mixing ratio. For example, the dew point of a parcel of air with a mixing ratio of 11.1 g/kg is always 15.6°C (60°F). Conversely, a parcel of air with a dew point of 15.6°C has a mixing ratio of 11.1 g/kg. This relationship lets us use the table of saturation mixing ratios (Figure 15-1) in several ways:

(1)   *If you know the temperature:* You can determine the water vapor capacity of the air. Read the "Saturation Mixing Ratio" directly from the table.

(2)   *If you know the mixing ratio:* You can determine the dew point temperature of the air. Find the value of the mixing ratio in the "Saturation Mixing Ratio" column—the dew point is the "Temperature" (remember, at the dew point temperature the air is at 100% RH, so the mixing ratio and the saturation mixing ratio are the same).

(3)   *If you know the dew point temperature:* You can determine the mixing ratio (the actual water vapor content) of the air. Find the value of the dew point in the "Temperature" column—the mixing ratio is the same as the "Saturation Mixing Ratio" (again, remember that at the dew point temperature, the mixing ratio, and saturation mixing ratio are the same). In practice, the dew point is often used as a measure of the actual water vapor content of the air.

---

[1]It is possible for air to have a relative humidity greater than 100% without condensation taking place. Such air is called "supersaturated." In this Lab Manual we make the simplifying assumption that condensation begins when air reaches 100% relative humidity.

THE SLING PSYCHROMETER

One common method of determining relative humidity is with an instrument called a **sling psychrometer**. The sling psychrometer consists of two thermometers mounted next to each other. One of the thermometers is an ordinary one that is used to measure the air temperature and is called the **dry-bulb thermometer**. The bulb of the other thermometer is wrapped in cloth that is saturated with room-temperature distilled water before use and is called the **wet-bulb thermometer**.

The instrument is called a "sling" psychrometer because it has a handle that is used to whirl the apparatus around for several minutes. The whirling of the sling psychrometer promotes the evaporation of water from the wet-bulb thermometer. Because evaporation is a cooling process, the temperature of the wet-bulb will decrease. If the air is dry, there will be rapid evaporation and greater cooling, and the temperature of the wet-bulb thermometer will decline more than if the air is relatively moist. In other words, if the temperature of the wet-bulb thermometer is much lower than that of the dry-bulb thermometer, the relative humidity is low. If the temperature of the wet-bulb thermometer is only slightly lower than that of the dry-bulb thermometer, the relative humidity is high. If there is no difference in temperature, the air is saturated, because no net evaporation took place.

After using the sling psychrometer, the relative humidity is determined with a table, shown in Figure 15-2 (°C) or Figure 15-4 (°F). The dry-bulb temperature is the "Air Temperature" (read along the left side of the chart). The "Depression of Wet-Bulb Thermometer" is the difference (in degrees) between the dry-bulb and wet-bulb temperature. Match the "Air Temperature" with the appropriate wet-bulb depression to find the relative humidity, expressed as a percentage.

For example, after spinning the psychrometer, if the dry-bulb temperature is 20°C, and the wet-bulb temperature is 14°C, the wet-bulb depression is 6°C. From the chart in Figure 15-2, under an air temperature of 20°C and a depression of 6°C, you read that the relative humidity is 51%.

A sling psychrometer can also be used to determine the temperature of the dew point. The table in Figure 15-3 (°C) or Figure 15-5 (°F) provides the dew points for various "Dry-Bulb" and "Depression of the Wet-Bulb" readings from a psychrometer. For example, with an air temperature of 20°C and a wet-bulb depression of 6°C, the temperature of the dew point is 10°C; as we saw, from this dew point temperature we can determine that the mixing ratio of the air is 7.6 g/kg.

Tips on Using Sling Psychrometers: After moistening the wet bulb (and being careful not to spill any water on the dry bulb), whirl around the sling psychrometer for about one minute. Stop briefly to check the wet-bulb temperature and begin whirling again. After another minute, stop again to check the wet-bulb temperature. If the temperature is the same as when you first checked, the wet-bulb temperature has stabilized and you can use the reading; if the temperature has continued to decrease, whirl again until the wet-bulb temperature stabilizes. Be sure that the wet bulb has not completely dried out in the process—if so, moisten again and take a new set of readings.

Air Temp. °C	Depression of Wet-Bulb Thermometer (°C)																					
	1	2	3	4	5	6	7	8	9	10	11	12	13	14	15	16	17	18	19	20	21	22
−4	77	54	32	11																		
−2	79	58	37	20	1																	
0	81	63	45	28	11					Relative Humidity (%)												
2	83	67	51	36	20	6																
4	85	70	56	42	27	14																
6	86	72	59	46	35	22	10	0														
8	87	74	62	51	39	28	17	6														
10	88	76	65	54	43	33	24	13	4													
12	88	78	67	57	48	38	28	19	10	2												
14	89	79	69	60	50	41	33	25	16	8	1											
16	90	80	71	62	54	45	37	29	21	14	7	1										
18	91	81	72	64	56	48	40	33	26	19	12	6	0									
20	91	82	74	66	58	51	44	36	30	23	17	11	5									
22	92	83	75	68	60	53	46	40	33	27	21	15	10	4	0							
24	92	84	76	69	62	55	49	42	36	30	25	20	14	9	4	0						
26	92	85	77	70	64	57	51	45	39	34	28	23	18	13	9	5						
28	93	86	78	71	65	59	53	45	42	36	31	26	21	17	12	8	4					
30	93	86	79	72	66	61	55	49	44	39	34	29	25	20	16	12	8	4				
32	93	86	80	73	68	62	56	51	46	41	36	32	27	22	19	14	11	8	4			
34	93	86	81	74	69	63	58	52	48	43	38	34	30	26	22	18	14	11	8	5		
36	94	87	81	75	69	64	59	54	50	44	40	36	32	28	24	21	17	13	10	7	4	
38	94	87	82	76	70	66	60	55	51	46	42	38	34	30	26	23	20	16	13	10	7	5

Figure 15-2: Relative Humidity Psychrometer Tables (°C).

Air Temp. °C	Depression of Wet-Bulb Thermometer (°C)																					
	1	2	3	4	5	6	7	8	9	10	11	12	13	14	15	16	17	18	19	20	21	22
−4	−7	−17	−22	−29																		
−2	−5	−8	−13	−20																		
0	−3	−6	−9	−15	−24					Dew Point (°C)												
2	−1	−3	−6	−11	−17																	
4	1	−1	−4	−7	−11	−19																
6	4	1	−1	−4	−7	−13	−21															
8	6	3	1	−2	−5	−9	−14															
10	8	6	4	1	−2	−5	−9	−14	−28													
12	10	8	6	4	1	−2	−5	−9	−16													
14	12	11	9	6	4	1	−2	−5	−10	−17												
16	14	13	11	9	7	4	1	−1	−6	−10	−17											
18	16	15	13	11	9	7	4	2	−2	−5	−10	−19										
20	19	17	15	14	12	10	7	4	2	−2	−5	−10	−19									
22	21	19	17	16	14	12	10	8	5	3	−1	−5	−10	−19								
24	23	21	20	18	16	14	12	10	8	6	2	−1	−5	−10	−18							
26	25	23	22	20	18	17	15	13	11	9	6	3	0	−4	−9	−18						
28	27	25	24	22	21	19	17	16	14	11	9	7	4	1	−3	−9	−16					
30	29	27	26	24	23	21	19	18	16	14	12	10	8	5	1	−2	−8	−15				
32	31	29	28	27	25	24	22	21	19	17	15	13	11	8	5	2	−2	−7	−14			
34	33	31	30	29	27	26	24	23	21	20	18	16	14	12	9	6	3	−1	−5	−12	−29	
36	35	33	32	31	29	28	27	25	24	22	20	19	17	15	13	10	7	4	0	−4	−10	
38	37	35	34	33	32	30	29	28	26	25	23	21	19	17	15	13	11	8	5	1	−3	−9

Figure 15-3: Dew Point Psychrometer Tables (°C).

Air Temp. °F	Depression of Wet-Bulb Thermometer (°F)																													
	1	2	3	4	5	6	7	8	9	10	11	12	13	14	15	16	17	18	19	20	21	22	23	24	25	26	27	28	29	30
0	67	33	1																											
5	73	46	20																											
10	78	56	34	13	15																									
15	82	64	46	29	11																									
20	85	70	55	40	26	12																								
25	87	74	62	49	37	25	13	1																						
30	89	78	67	56	46	36	26	16	6																					
35	91	81	72	63	54	45	36	27	19	10	2																			
40	92	83	75	68	60	52	45	37	29	22	15	7																		
45	93	86	78	71	64	57	51	44	38	31	25	18	12	6																
50	93	87	74	67	61	55	49	43	38	32	27	21	16	10	5															
55	94	88	82	76	70	65	59	54	49	43	38	33	28	23	19	11	9	5												
60	94	89	83	78	73	68	63	58	53	48	43	39	34	30	26	21	17	13	9	5	1									
65	95	90	85	80	75	70	66	61	56	52	48	44	39	35	31	27	24	20	16	12	9	5	2							
70	95	90	86	81	77	72	68	64	59	55	51	48	44	40	36	33	29	25	22	19	15	12	9	6	3					
75	96	91	86	82	78	74	70	66	62	58	54	51	47	44	40	37	34	30	27	24	21	18	15	12	9	7	4	1		
80	96	91	87	83	79	75	72	68	64	61	57	54	50	47	44	41	38	35	32	29	26	23	20	18	15	12	10	7	5	3
85	96	92	88	84	81	77	73	70	66	63	59	57	53	50	47	44	41	38	36	33	30	27	25	22	20	17	15	13	10	8
90	96	92	89	85	81	78	74	71	68	65	61	58	55	52	49	47	44	41	39	36	34	31	29	26	24	22	19	17	15	13
95	96	93	89	86	82	79	76	73	69	66	63	61	58	55	52	50	47	44	42	39	37	34	32	30	28	25	23	21	19	17
100	96	93	89	86	83	80	77	73	70	68	65	62	59	56	54	51	49	46	44	41	39	37	35	33	30	28	26	24	22	21
105	97	93	90	87	84	81	78	75	72	69	66	64	61	58	56	53	51	49	46	44	42	40	38	36	34	32	30	28	26	24

Relative Humidity (%)

Figure 15-4: Relative Humidity Psychrometer Tables (°F).

Air Temp. °F	Depression of Wet-Bulb Thermometer (°F)																													
	1	2	3	4	5	6	7	8	9	10	11	12	13	14	15	16	17	18	19	20	21	22	23	24	25	26	27	28	29	30
0	−7	−20																												
5	−1	−9	−24																											
10	5	−2	−10	−27																										
15	11	6	0	−9	−26																									
20	16	12	8	2	−7	−21																								
25	22	19	15	10	5	−3	−15	−51																						
30	27	25	21	18	14	8	2	−7	−25																					
35	33	30	28	25	21	17	13	7	0	−11	−41																			
40	38	35	33	30	28	25	21	18	13	7	−1	−14																		
45	43	41	38	36	34	31	28	25	22	18	13	7	−1	−14																
50	48	46	44	42	40	37	34	32	29	26	22	18	13	8	0	−13														
55	53	51	50	48	45	43	41	38	36	33	30	27	24	20	15	9	1	−12	−59											
60	58	57	55	53	51	49	47	45	43	40	38	35	32	29	25	21	17	11	4	−8	−36									
65	63	62	60	59	57	55	53	51	49	47	45	42	40	37	34	31	27	24	19	14	7	−3	−22							
70	69	67	65	64	62	61	59	57	55	53	51	49	47	44	42	39	36	33	30	26	22	17	11	2	−11					
75	74	72	71	69	68	66	64	63	61	59	57	55	54	51	49	47	44	42	39	36	32	29	25	21	15	8	−2	−23		
80	79	77	76	74	73	72	70	68	67	65	63	62	60	58	56	54	52	50	47	44	42	39	36	32	28	24	20	13	6	−7
85	84	82	81	80	78	77	75	74	72	71	69	68	66	64	62	61	59	57	54	52	50	48	45	42	39	36	32	28	24	19
90	89	87	86	85	83	82	81	79	78	76	75	73	72	70	69	67	65	63	61	59	57	55	53	51	48	45	43	39	36	32
95	94	93	91	90	89	87	86	85	83	82	80	79	78	76	74	73	71	70	68	66	64	62	60	58	56	54	52	49	46	43
100	99	98	96	95	94	93	91	90	89	87	86	85	83	82	80	79	77	76	74	72	71	69	67	65	63	61	59	57	55	52
105	104	103	101	100	99	98	96	95	94	93	91	90	89	87	86	84	83	82	80	78	77	75	74	72	70	68	67	65	63	61

Dew Point (°F)

Figure 15-5: Dew Point Psychrometer Tables (°F).

Name _____   Section _____

EXERCISE 15 PROBLEMS—PART I (S.I. Units)

1. Complete the following chart (round off relative humidity to the nearest percent).

	Mixing Ratio (Actual Water Vapor Content) (g/kg)	Air Temperature (°C)	Saturation Mixing Ratio ("Capacity") (g/kg)	Relative Humidity (%)
(a)	2.8	−1.1°C		
(b)	2.8	32.2°C		
(c)	11.1		13.2	
(d)	22.3		36.5	

2. The air inside a room is at a temperature of 18.3°C and has a mixing ratio of 5.2 g/kg.

   (a) What is the relative humidity?   _____ %

   (b) What is the dew point?   _____ °C

   (c) If the mixing ratio remains the same, but the temperature
       of the room increases to 26.7°C, what is the new
       relative humidity?   _____ %

3. The air inside a room is at a temperature of 35°C and has a mixing ratio of 7.6 g/kg.

   (a) What is the relative humidity?   _____ %

   (b) What is the dew point?   _____ °C

   (c) If the room temperature decreases by 5°C per hour, how many
       hours will it take for the air to reach saturation?   _____ hours

   (d) After reaching saturation, if the temperature of the room
       continues to decrease for one more hour, approximately
       how many grams of water vapor (per kg of air) will have
       had to condense out of the air to maintain a relative
       humidity of 100%?   _____ g/kg

Name _____   Section _____

EXERCISE 15 PROBLEMS—PART II (*S.I. Units*)

Using the psychrometer tables (Figures 15-2 and 15-3) and the table of saturation mixing ratios (Figure 15-1), answer the following questions about determining relative humidity with a sling psychrometer.

1.   If the dry-bulb temperature is 32°C and the wet-bulb temperature is 26°C,

    (a)   What is the relative humidity?   _____ %

    (b)   What is the dew point?   _____ °C

    (c)   What is the mixing ratio? (Estimate from Figure 15-1)   _____ g/kg

2.   If the dry-bulb temperature is 14°C and the wet-bulb temperature is 12°C,

    (a)   What is the relative humidity?   _____ %

    (b)   What is the dew point?   _____ °C

    (c)   What is the mixing ratio? (Estimate from Figure 15-1)   _____ g/kg

OPTIONAL

3.   If a sling psychrometer is available in class, determine the following both indoors (in the classroom) and outdoors. Be sure to include the correct units (%, g/kg, etc.) in each answer.

	Indoors	**Outdoors**
Dry-Bulb Temperature		
Wet-Bulb Temperature		
Depression of Wet-Bulb Thermometer		
Relative Humidity		
Dew Point		
Mixing Ratio (Estimate from Figure 15-1)		

Name _____   Section _____

EXERCISE 15 PROBLEMS—PART III *(English Units)*

1. Complete the following chart (round off relative humidity to the nearest percent).

	Mixing Ratio (Actual Water Vapor Content) (g/kg)	Air Temperature (°F)	Saturation Mixing Ratio ("Capacity") (g/kg)	Relative Humidity (%)
(a)	2.8	30°F		
(b)	2.8	90°F		
(c)	11.1		13.2	
(d)	22.3		36.5	

2. The air inside a room is at a temperature of 65°F and has a mixing ratio of 5.2 g/kg.

    (a) What is the relative humidity?   _____ %

    (b) What is the dew point?   _____ °F

    (c) If the mixing ratio remains the same, but the temperature of
         the room increases to 80°F, what is the new relative humidity?   _____ %

3. The air inside a room is at a temperature of 70°F and has a mixing ratio of 7.6 g/kg.

    (a) What is the relative humidity?   _____ %

    (b) What is the dew point?   _____ °F

    (c) If the room temperature decreases by 10°F per hour, how many
         hours will it take for the air to reach saturation?   _____ hours

    (d) After reaching saturation, if the temperature of the room
         continues to decrease for one more hour, how many grams
         of water vapor (per kg of air) will have had to condense
         out of the air to maintain a relative humidity of 100%?   _____ g/kg

Name _____   Section _____

EXERCISE 15 PROBLEMS—PART IV *(English Units)*

Using the psychrometer tables (Figures 15-4 and 15-5) and the table of saturation mixing ratios (Figure 15-1), answer the following questions about determining relative humidity with a sling psychrometer.

1.   If the dry-bulb temperature is 90°F and the wet-bulb temperature is 79°F,

    (a)   What is the relative humidity?   _____ %

    (b)   What is the dew point?   _____ °F

    (c)   What is the mixing ratio?   _____ g/kg

2.   If the dry-bulb temperature is 55°F and the wet-bulb temperature is 52°F,

    (a)   What is the relative humidity?   _____ %

    (b)   What is the dew point?   _____ °F

    (c)   What is the mixing ratio?   _____ g/kg

OPTIONAL

3.   If a sling psychrometer is available in class, determine the following both indoors (in the classroom) and outdoors. Be sure to include the correct units (%, g/kg, etc.) in each answer.

	Indoors	**Outdoors**
Dry-Bulb Temperature		
Wet-Bulb Temperature		
Depression of Wet-Bulb Thermometer		
Relative Humidity		
Dew Point		
Mixing Ratio (Estimate from Figure 15-1)		

Adiabatic Processes

Objective:	To study adiabatic processes in the atmosphere, and to calculate tempera-ture and humidity changes in parcels of moving air.
Reference:	Hess, Darrel. *McKnight's Physical Geography*, 12th ed., pp. 149–151.

ADIABATIC PROCESSES

In Exercise 15 we looked at the relationship between temperature and **relative humidity**, noting that as the temperature of a parcel of air decreases, the relative humidity increases. When a parcel of air has cooled to the **dew point temperature**, it becomes saturated and **condensation** can take place. The most common way that a parcel of air is cooled enough to form clouds and precipitation is through **adiabatic cooling**.

As a parcel of air rises, it comes under lower pressure and expands. As the air expands, it cools adiabatically ("adiabatic" means without the gain or loss of heat). Rising air always cools adiabatically. Conversely, as air descends, it comes under higher pressure and compresses. As the air compresses, it warms adiabatically. Descending air always warms adiabatically.

If a parcel of rising air is unsaturated (the relative humidity is less than 100%) it will cool at the **dry adiabatic rate** (DAR; also called the "dry adiabatic lapse rate") of about 10°C per 1000 meters (5.5°F per 1000 feet). As the air rises and cools, its relative humidity increases. At some point, the parcel of air will have cooled enough to reach its dew point. The elevation at which a parcel of air cools to its dew point temperature is called the **lifting condensation level** (LCL), and at this point, condensation and cloud formation can begin.

If a parcel of air keeps rising while condensation is taking place, the air will continue to cool, but at a lesser rate. Saturated air cools at the **saturated adiabatic rate** (SAR; also called the "wet" or "saturated adiabatic lapse rate") of about 6°C per 1000 meters (3.3°F per 1000 feet). The SAR varies, however, and the rate of cooling may be as little as 4°C per 1000 meters (2.0°F per 1000 feet).

Rising saturated air cools at a lesser rate than rising unsaturated air because of the release of **latent heat** during condensation. **Evaporation** is, in effect, a cooling process because latent heat is stored when water changes from liquid to gas. When the water vapor condenses back to liquid water, this heat is released. As saturated air rises, it expands and cools adiabatically, but the latent heat released during condensation counteracts some of this cooling.

Figure 16-1 shows the temperature changes in a parcel of air as it rises up and over a 4000-meter-high mountain. In this hypothetical example, the dew point of the parcel is 5°C and the lifting condensation level is 2000 meters.

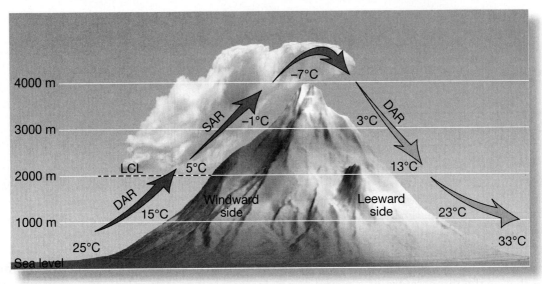

Figure 16-1: Temperature changes in a hypothetical parcel of air passing over a 4000-meter-high mountain (assuming no evaporation as the air descends down the lee side of the mountain). The lifting condensation level (LCL) of the parcel is 2000 meters, the dry adiabatic rate (DAR) is 10°C/1000 m, and the saturated adiabatic rate (SAR) is 6°C/1000 m. Notice that because of the release of latent heat during condensation on the windward side of the mountain, by the time the air has descended back down to sea level on the leeward side, it is warmer than before it started up the windward side. (From Hess, *McKnight's Physical Geography*, 12th ed.)

Notice that the air descending down the lee side of the mountain warms at the DAR. Descending air generally warms at the DAR, because as air warms, its capacity increases and so it cannot be saturated.[1]

Adiabatic temperature changes may lead to changes in both the relative humidity and the **water vapor** content—the **mixing ratio**—of a parcel of air. For example, as unsaturated air rises or descends (and the temperature decreases or increases), its capacity changes. Because of this, the relative humidity of the parcel will change.

On the other hand, as rising saturated air cools adiabatically, the relative humidity of the parcel generally remains at about 100% as condensation takes place—to avoid "supersaturation" (relative humidity greater than 100%), water vapor must condense out of the air, thus keeping the mixing ratio and the saturation mixing ratio equal. As the air continues to rise and water vapor is lost through condensation, the water vapor content (the mixing ratio) of the parcel will change.

---

[1]Although descending air usually warms at the dry adiabatic rate, there is a circumstance when this may not be the case. If air descends through a cloud, some water droplets may evaporate and the evaporative cooling will counteract some of the **adiabatic warming**. As a result, such descending air can warm at a rate very close to the saturated adiabatic rate. As soon as evaporation of water droplets ceases, this descending air will warm at the dry adiabatic rate as usual.

Name _____   Section _____

EXERCISE 16 PROBLEMS—PART I (S.I. Units)

Assume that a parcel of air is forced to rise up and over a 4000-meter-high mountain (as shown). The initial temperature of the parcel at sea level is 30°C, and the lifting condensation level (LCL) of the parcel is 2000 meters. The DAR is 10°C/1000 m and the SAR is 6°C/1000 m. Assume that condensation begins at 100% relative humidity and that no evaporation takes place as the parcel descends.

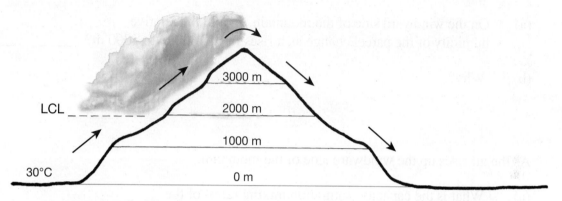

1. Calculate the temperature of the parcel at the following elevations as it rises up the windward side of the mountain.

   (a) 1000 m _____ °C   (b) 2000 m _____ °C   (c) 4000 m _____ °C

2. (a) After the parcel of air has descended down the lee side of the
        mountain to sea level, what is the temperature of the parcel?   _____ °C

   (b) Why is the parcel now warmer than it was at sea level on the windward side
        (what is the source of the heat energy)?

3. (a) On the windward side of the mountain, is the relative
        humidity of the parcel increasing or decreasing as it
        rises from sea level to 2000 meters?   _____

   (b) Why?

4. (a) On the lee side of the mountain, is the relative humidity
        of the parcel increasing or decreasing as it descends from
        4000 meters to sea level?   _____

   (b) Why?

Name _____   Section _____

EXERCISE 16 PROBLEMS—PART II (S.I. Units)

Answer the following questions after completing the problems in Part I. You will also need to refer to the chart of Saturation Mixing Ratios in Figure 15-1; interpolate from the chart as needed. Assume that condensation begins at 100% relative humidity and that no evaporation takes place as the parcel descends.

5.   (a)   On the windward side of the mountain, should the relative
             humidity of the parcel change as it rises from 2000 m to 4000 m? _____

    (b)   Why?

6.   As the air rises up the windward side of the mountain,

    (a)   What is the capacity (saturation mixing ratio) of the
             rising air at 2000 meters? _____ g/kg

    (b)   What is the capacity of the air at 4000 meters? _____ g/kg

7.   What is the capacity of the air after it has descended back down to
    sea level on the lee side of the mountain? _____ g/kg

8.   (a)   Assuming that *no* water vapor is added as the parcel descends
             down the lee side of the mountain to sea level, is the water vapor
             content (the mixing ratio) of the parcel higher or lower than
             before it began to rise over the mountain? _____

    (b)   Why?

    (c)   What is the lifting condensation level of this parcel now, after
             descending to sea level on the lee side of the mountain? _____ meters

Name _____   Section _____

EXERCISE 16 PROBLEMS—PART III *(English Units)*

Assume that a parcel of air is forced to rise up and over a 6000-foot-high mountain (as shown). The initial temperature of the parcel at sea level is 76.5°F, and the lifting condensation level (LCL) of the parcel is 3000 feet. The DAR is 5.5°F/1000 feet and the SAR is 3.3°F/1000 feet. Assume that condensation begins at 100% relative humidity and that no evaporation takes place as the parcel descends. Indicate calculated temperatures to 1 decimal place.

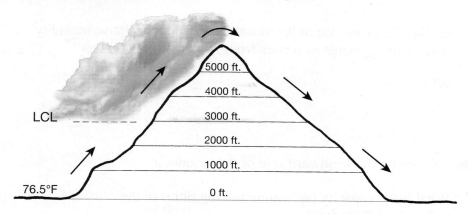

1.   Calculate the temperature of the parcel at the following elevations as it rises up the windward side of the mountain.

    (a) 1000 feet _____ °F   (b) 3000 feet _____ °F   (c) 6000 feet _____ °F

2.   (a)   After the parcel of air has descended down the lee side of the
        mountain to sea level, what is the temperature of the parcel?   _____ °F

    (b)   Why is the parcel now warmer than it was at sea level on the windward side
        (what is the source of the heat energy)?

3.   (a)   On the windward side of the mountain, is the relative humidity
        of the parcel increasing or decreasing as it rises from sea level
        to 3000 feet?   _____

    (b)   Why?

4.   (a)   On the lee side of the mountain, is the relative humidity
        of the parcel increasing or decreasing as it descends from
        6000 feet to sea level?   _____

    (b)   Why?

Name _____   Section _____

EXERCISE 16 PROBLEMS—PART IV (*English Units*)

Answer the following questions after completing the problems in Part III. You will also need to refer to the table of Saturation Mixing Ratios in Figure 15-1; interpolate from the chart as needed. Assume that condensation begins at 100% relative humidity and that no evaporation takes place as the parcel descends.

5.   (a)   On the windward side of the mountain, should the relative humidity of the parcel change as it rises from 3000 feet to 6000 feet?   _____

        (b)   Why?

6.   As the air rises up the windward side of the mountain

        (a)   What is the capacity (saturation mixing ratio) of the rising air at 3000 feet?   _____ g/kg

        (b)   What is the capacity of the air at 6000 feet?   _____ g/kg

7.   What is the capacity of the air after it has descended back down to sea level on the lee side of the mountain?   _____ g/kg

8.   (a)   Assuming that *no* water vapor is added as the parcel descends down the lee side of the mountain to sea level, is the water vapor content (the mixing ratio) of the parcel higher or lower than before it began to rise over the mountain?   _____

        (b)   Why?

        (c)   What is the lifting condensation level of this parcel now, after descending to sea level on the lee side of the mountain?   _____ feet

EXERCISE 17
Stability

Objective:	To illustrate the concept of stability in the atmosphere.
Reference:	Hess, Darrel. *McKnight's Physical Geography*, 12th ed., pp. 155–158.

STABILITY

The stability of air is an important characteristic of the atmosphere. Air is **unstable** if it rises on its own. Air is **stable** if it resists upward vertical motion and will rise only when forced.

The temperature of a parcel of air, relative to the temperature of the surrounding air, determines stability for the most part. A parcel of air will be unstable if it is warmer than the surrounding air. A parcel of air will be stable if it is the same temperature, or cooler, than the surrounding air.

LAPSE RATES

In order to understand stability, we must distinguish between the "environmental" lapse rate and the "adiabatic" lapse rates.

The **environmental lapse rate** (ELR) reflects the temperature of the atmosphere at different altitudes—sometimes called the vertical temperature profile or vertical temperature gradient. The ELR averages about 6.5°C per 1000 meters (3.6°F per 1000 feet) within the troposphere (the lowest layer of the atmosphere)—this rate is called the **average lapse rate**. This means that, on average, as we move up through the troposphere, the temperature will be about 6.5°C cooler for each 1000 meters we climb. However, from day to day and from place to place, the ELR frequently deviates from this average rate. Changes in the ELR are also often observed from day to night. Further, there are times when the temperature will actually increase as we move up through the atmosphere—a situation known as a **temperature inversion**.

The **dry adiabatic rate** (DAR; also called the "dry adiabatic lapse rate") and the **saturated adiabatic rate** (SAR; also called the "saturated adiabatic lapse rate") reflect the temperature change within a specific parcel of moving air (see Exercise 16 for a discussion of adiabatic temperature changes). The DAR averages about 10°C per 1000 meters (5.5°F per 1000 feet), whereas the SAR averages about 6°C per 1000 meters (3.3°F per 1000 feet) but can vary significantly from this.

In the context of stability, we can think of the ELR as showing the temperature change of the surrounding air through which a parcel of air is moving and changing temperature following the dry adiabatic rate or the saturated adiabatic rate.

Stable Air: Figure 17-1 shows the temperature changes in a parcel of rising air. On the left, the temperature changes of the parcel relative to the temperature of the surrounding air are shown in diagram form, and on the right, these same changes are shown in graph form. In this hypothetical example, the ELR of the surrounding air is 5°C per 1000 meters. When the parcel rises, it cools first at the DAR of 10°C per 1000 meters, and after the **lifting condensation level** (LCL)

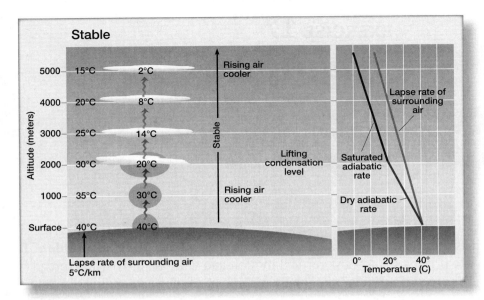

Figure 17-1: Temperature changes in a rising parcel of stable air (ELR = 5°C/1000 m; DAR = 10°C/1000 m; SAR = 6°C/1000 m; LCL = 2000 m). In this example, at all elevations the rising parcel of air is cooler than the surrounding air and so is stable and must be forced to rise. (From Hess, *McKnight's Physical Geography*, 12th ed.)

of 2000 meters is reached, it cools at the SAR of 6°C per 1000 meters. Notice that at all elevations the rising parcel of air is cooler than the surrounding air. In this case, the rising air is stable and must be forced to rise.

Unstable Air: A different situation is shown in Figure 17-2. In this example, the ELR is 12°C per 1000 meters. Notice that at all elevations the rising parcel of air is warmer than the surrounding air. In this case, the rising air is unstable and will rise on its own.

Conditional Instability: A third situation is possible. With an intermediate ELR (such as 8°C per 1000 meters), rising air may be stable when it is first forced to rise from the surface, but above the LCL the release of latent heat during condensation may cause the rising parcel to become warmer than the surrounding air—in this case, the parcel becomes unstable and will continue to rise on its own. This circumstance is known as **conditional instability** because the rising parcel becomes unstable only after condensation releases latent heat.

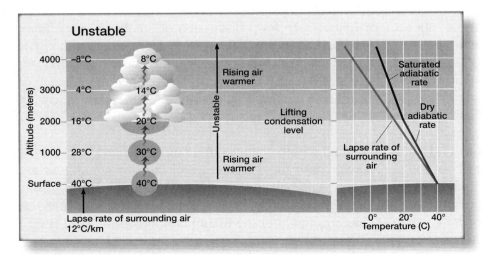

Figure 17-2: Temperature changes in a rising parcel of unstable air (ELR = 12°C/1000 m; DAR = 10°C/1000 m; SAR = 6°C/1000 m; LCL = 2000 m). In this example, at all elevations the rising air is warmer than the surrounding air, so the parcel is unstable and will rise on its own. (From Hess, *McKnight's Physical Geography*, 12th ed.)

Name _____   Section _____

EXERCISE 17 PROBLEMS—PART I (S.I. Units)

1. On the following chart, use the sets of hypothetical data to plot the vertical temperature profile (the environmental lapse rate) of the atmosphere in two locations. Using a straight-edge, connect the temperature points for Location A with a blue line, and for Location B with a green line (if you do not use colored lines, label each line clearly). After completing the temperature profiles, answer the questions on the following page.

Elevation (meters)	Temperature (°C)	
	Location A	Location B
5000	−10°C	−8°C
3500	−4°C	4°C
2500	5°C	14°C
1500	14°C	26°C
1000	9°C	30°C
0	15°C	35°C

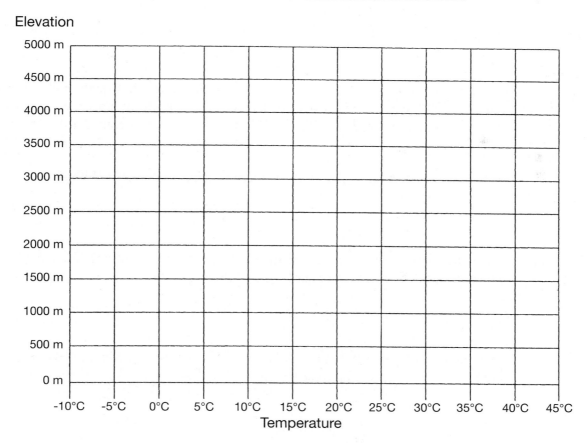

Elevation

Remember, the vertical temperature profiles show the temperature of the surrounding air through which parcels of air can move. For the following questions, assume that the DAR is 10°C/1000 meters and that the SAR is 6°C/1000 meters.

2. A parcel of air with an initial temperature of 15°C begins to rise in Location A. The LCL of the parcel is 1000 meters. With a red (or labeled) line, carefully draw the temperature decrease of this parcel of air as it rises to 3500 meters. Be sure to consider the LCL, and the DAR and SAR.

    (a) Describe the stability pattern of this parcel of air.

    (b) What is the general name for the change observed in the vertical temperature profile between 1000 and 1500 meters?

    (c) Does the parcel of rising air become highly stable or highly unstable between 1000 and 1500 meters? Why?

3. A parcel of air with an initial temperature of 35°C begins to rise in Location B. The LCL of the parcel is 2000 meters. With a red (or labeled) line, carefully draw the temperature decrease of this parcel of air as it rises to 5000 meters. Be sure to consider the LCL, and the DAR and SAR.

    (a) Will this parcel of air begin to rise from the surface on its own? Why?

    (b) Does the stability of this parcel change with increased elevation? If so, at what elevation does this change occur?

    (c) How would the pattern of stability below 5000 meters be different if the lifting condensation level was not reached until 4500 meters?

Name _____   Section _____

EXERCISE 17 PROBLEMS—PART II (English Units)

1. On the following chart, use the sets of hypothetical data to plot the vertical tempera-
    ture profile (the environmental lapse rate) of the atmosphere in two locations. Using a
    straightedge, connect the temperature points for Location A with blue line, and for Loca-
    tion B with a green line (if you do not use colored lines, label each line clearly). After
    completing the temperature profiles, answer the questions on the following page.

Elevation (feet)	Temperature (°F)	
	Location A	Location B
10,000	30°F	33°F
7000	34°F	47°F
5000	41°F	59°F
3000	51°F	67°F
2000	46°F	72°F
0	55°F	80°F

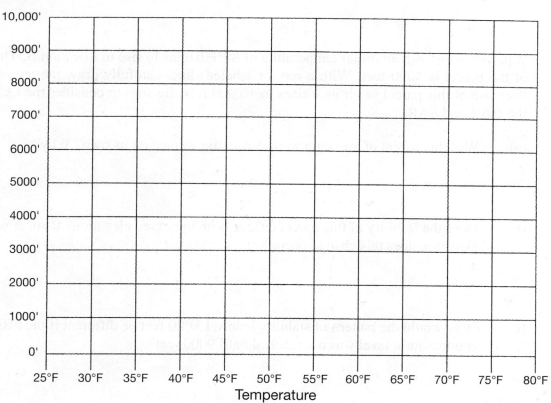

Remember, the vertical temperature profiles show the temperature of the surrounding air through which parcels of air can move. For the following questions, assume that the DAR is 5.5°F/1000 feet and that the SAR is 3.0°F/1000 feet.

2. A parcel of air with an initial temperature of 55°F begins to rise in Location A. The LCL of the parcel is 2000 feet. With a red (or labeled) line, carefully draw the temperature decrease of this parcel of air as it rises to 8000 feet. Be sure to consider the LCL, and the DAR and SAR.

    (a) Describe the stability pattern of this parcel of air.

    (b) What is the general name for the change observed in the vertical temperature profile between 2000 and 3000 feet?

    (c) Does the parcel of rising air become highly stable or highly unstable between 2000 and 3000 feet? Why?

3. A parcel of air with an initial temperature of 80°F begins to rise in Location B. The LCL of the parcel is 4000 feet. With a red (or labeled) line, carefully draw the temperature decrease of this parcel of air as it rises to 10,000 feet. Be sure to consider the LCL, and the DAR and SAR.

    (a) Will this parcel of air begin to rise from the surface on its own? Why?

    (b) Does the stability of this parcel change with increased elevation? If so, at what elevation does this change occur?

    (c) How would the pattern of stability below 10,000 feet be different if the lifting condensation level was not reached until 9000 feet?

Midlatitude Cyclones

Objective:	To study the pressure, wind, and temperature patterns of midlatitude cyclones.
Resources:	Internet access (optional).
Reference:	Hess, Darrel. *McKnight's Physical Geography*, 12th ed., pp. 180–185.

THE MIDLATITUDE CYCLONE

The **midlatitude cyclone** is the most important storm of the midlatitudes. At the heart of a midlatitude cyclone is an area of low pressure, as much as 1600 kilometers (1000 miles) across.

The low pressure cell produces a converging counterclockwise wind flow that pulls together two unlike **air masses** (the wind flow is converging clockwise in the Southern Hemisphere). Relatively cool air from the high latitudes is brought together with relatively warm air from the subtropics. These unlike air masses do not mix readily. Instead, abrupt transition zones known as **fronts** develop between the air masses. At the surface, a mature midlatitude cyclone has a "cool sector" and a "warm sector," separated by a **cold front** (cold air advancing under the warm) and a **warm front** (warm air advancing over the cold).

Figure 18-1 shows a typical well-developed midlatitude cyclone in the Northern Hemisphere, mapped with **isobars**. The lowest pressure is at the heart of the storm, but a **trough** of low pressure extends down the length of the cold front as well. As the whole storm migrates eastward in the flow of the **westerlies** (left to right in this diagram), air converges counterclockwise into the low. The cold front typically advances faster than the storm itself and eventually catches up with the warm front. A cross section through the storm is shown in Figure 18-1b.

Figure 18-2 shows the life cycle of a midlatitude cyclone, beginning with the early development of the storm along the **polar front**, through maturity, and finally the process of **occlusion**, in which the cold front catches up with the warm front, lifting all of the warm air off the ground. After occlusion, the storm generally begins to lose strength and die.

The cross sections shown in Figure 18-2 help illustrate the reasons for the weather typically brought by these storms. Generally, the heaviest precipitation is associated with the cold front. The abrupt uplift of the warm air along the advancing, steeply sloping cold front causes the **adiabatic cooling** needed to produce clouds and precipitation. Because of the more gentle slope of the warm front, this region of the storm is usually associated with more widespread but less intense precipitation than the cold front.

FRONTS ON WEATHER MAPS

There are four common kinds of fronts. Cold fronts develop where the cold air is actively advancing under warm air. Warm fronts occur when the warm air is actively advancing over cold air. **Occluded fronts** develop when the cold front catches up with a warm front. **Stationary fronts** represent boundaries between unlike air masses, but neither air mass is actively advancing. Figure 18-3 shows the commonly used weather map symbols for these four kinds of fronts.

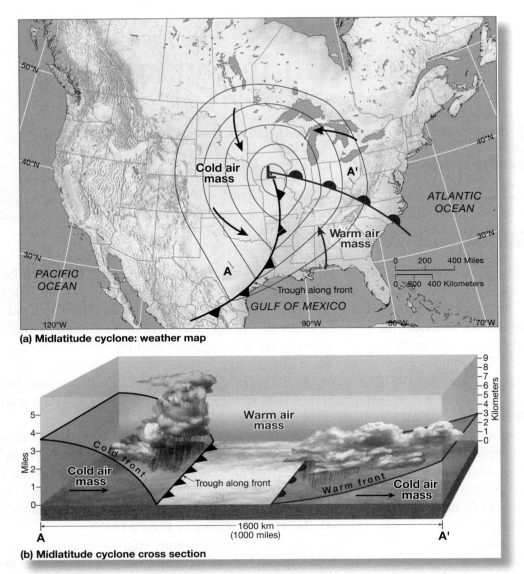

(a) Midlatitude cyclone: weather map

(b) Midlatitude cyclone cross section

Figure 18-1: A map (a) and a cross section (b) of a typical mature midlatitude cyclone. Arrows in (b) indicate the direction of frontal movement. (From Hess, *McKnight's Physical Geography*, 12th ed.)

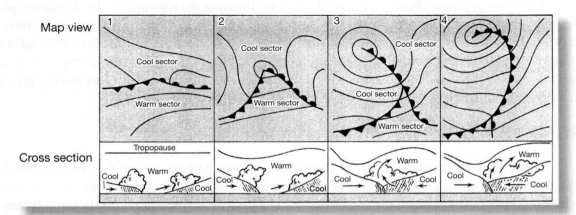

Figure 18-2: Stages in the life of a midlatitude cyclone: (1) early development, (2) maturity, (3) partial occlusion, (4) full occlusion. (Adapted from McKnight, *Physical Geography*, 4th ed.)

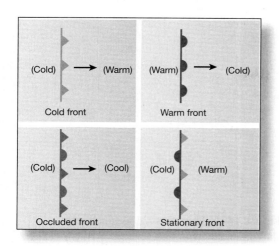

Figure 18-3: Fronts as shown on weather maps. Arrows indicate the direction of movement of the front. The relative temperature of air masses at the surface is shown in parentheses.

Although it might seem that the most obvious way to recognize a front would be an abrupt change in temperature from one meteorological data station to the next, such changes are not always obvious on weather maps. Fronts often represent transition zones that may be 15 kilometers (about 10 miles) or more wide. It is quite possible that the spacing of meteorological stations is such that a sharp difference in temperature is not clearly visible on a weather map.

Figure 18-4 shows a section of a hypothetical weather map in the Northern Hemisphere (top of map is north) showing isobars, a cold front, and 10 meteorological stations. The weather map **station model** was introduced in Exercises 13 and 14. In the simplified form used here, the model shows the temperature, **dew point**, and wind direction. For example,

In this case, the temperature is 40°F, the dew point is 27°F, and the wind is blowing from the northeast at 15 knots (the "feathers" on the wind shaft point *into* the wind). We will take a more complete look at the station model in Exercise 19.

The pattern of dew points may be helpful in locating the position of a front. Dew points are usually lower in relatively cold air than in warm air because cold air is often drier than warm air, and so generally there is a drop in dew points across a front.

Wind direction is another useful indication of the location of a front. Notice in Figure 18-4 that a wind direction shift is observed from one side of the front to the other. In this example, the wind direction in the cold sector suggests that the cold air is advancing, and therefore, pushing the position of the cold front toward the southeast.

Also notice the "kink" in the isobars at the position of the front. A cold front is associated with a trough of low pressure. As a cold front passes, the pressure trend changes from falling to rising.

METEOGRAMS

Meteograms are charts that plot changes in a wide range of weather conditions for a location over a 25-hour period. Meteograms may appear in several different formats, but all contain the same general information. Figure 18-5 is a typical meteogram. The top chart shows temperature ("TMPF"), dew point ("DWPF"), and relative humidity ("RELH"). Below the temperature charts, information such as current weather conditions ("WSYM" or "WX"; see Figure III-3 in Appendix III

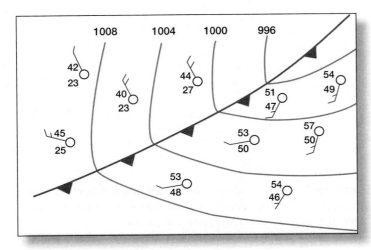

Figure 18-4: Hypothetical weather map showing a cold front.

for an explanation of the symbols used) and wind direction and speed are shown. A middle chart shows the elevation of the cloud base and visibility ("VSBY"), whereas precipitation amounts ("P06I" or "PREC") are shown below. In the bottom chart, atmospheric pressure ("PMSL") is plotted. The date and time of the meteogram is in **Zulu time** or UTC (Universal Time Coordinated; GMT) (see Exercise 3 for a discussion of time zones and Zulu time).

Meteograms clearly show changing trends in weather, such as those associated with the passing of a midlatitude cyclone. For example, in Figure 18-5 notice the change in wind direction, the drop in temperature, the decreasing visibility and lower cloud cover, and the onset of precipitation associated with the passing of a front. The trough of the front passed through Dallas-Fort Worth, Texas, at about 0600Z on December 28, 2012.

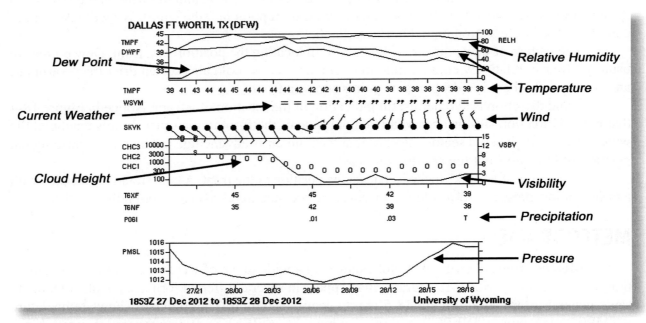

Figure 18-5: Meteogram for Dallas-Fort Worth, Texas, on December 27–28, 2012. (Meteogram courtesy of the University of Wyoming)

Name _____   Section _____

EXERCISE 18 PROBLEMS—PART I

The following questions are based on this hypothetical weather map in the Northern Hemisphere showing isobars and the positions of a cold front and a warm front (top of map is north). Six locations are marked on the map (points A, B, C, D, E, and F). A cross-section diagram along points A, B, C, D, E, and F is shown below the map.

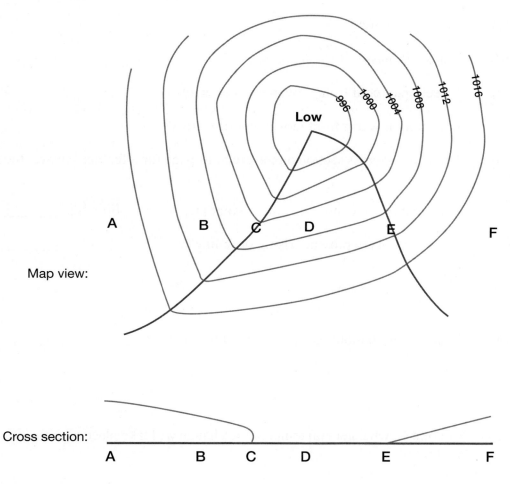

1.   On the map, label the following:

   (a)   Cold front (use standard weather map symbols)

   (b)   Warm front (use standard weather map symbols)

   (c)   Cool sector of storm

   (d)   Warm sector of storm

2.   On the hypothetical weather map shown on the previous page, use arrows to show the wind direction in the western, southern, eastern, and northern parts of the storm.

3.   In which direction is the storm as a whole moving?   From _____ to _____

4.   On the cross-section diagram, label the following:

(a)   Cold front

(b)   Warm front

(c)   Cold air mass(es)

(d)   Warm air mass

(e)   Direction of cold front movement (use arrow)

(f)   Direction of warm front movement (use arrow)

Using your labeled map and cross section on the previous page for reference, answer the following questions:

5.   What is the most likely wind direction at Point D?   From the _____

6.   (a)   At point D, is the pressure rising or falling?   _____

(b)   Why?

7.   (a)   Is precipitation more likely at Point D or Point C?   _____

(b)   Why?

8.   (a)   At Point C, what general temperature change will take place with the passing of the cold front?

(b)   Why?

9.   What is the most likely wind direction at Point B?   From the _____

10.   (a)   At point B, is the pressure rising or falling?   _____

(b)   Why?

Name _____   Section _____

EXERCISE 18 PROBLEMS—PART II

1. Two hypothetical weather maps in the Northern Hemisphere are shown here. Using the appropriate symbols (see Figure 18-3), draw in the position of the front on each map (one map shows a cold front, and the other a warm front). Top of each map is north. (Hint: Each of the fronts can be drawn with a nearly straight line.)

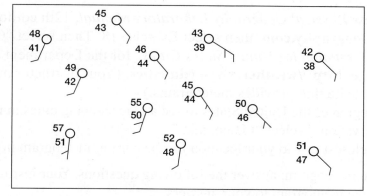

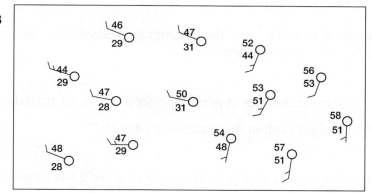

2. Is a cold front or warm front shown in Map A?   _____

3. (a)   In which direction is the front in Map A advancing?   To the _____

    (b)   How can you tell?

4. Is a cold front or warm front shown in Map B?   _____

5. In which direction is the front in Map B advancing?   To the _____

Name _____   Section _____

EXERCISE 18 PROBLEMS—PART III—INTERNET

In this exercise, you will use a meteogram to study the weather changes brought by the passing of a midlatitude cyclone. This exercise will work best about 12 hours after a midlatitude cyclone or front passes through your area. If no storms are currently in your area, your instructor may have you choose another city that has experienced a passing storm within the last day or use the sample meteogram for Dallas–Fort Worth, Texas, shown in Figure 18-5.

- Go to the Hess *Physical Geography Laboratory Manual*, 12th edition, website at **www .MasteringGeography.com**, then select Exercise 18. Then select "Go to *University of Wyoming, Information for United States Cities*" for the Department of Atmospheric Science Web page, **http://weather.uwyo.edu/cities**. (Your instructor may recommend a different Internet site that provides meteograms.)
- Select your region of the United States to see a map showing cities in the area.
- Under "Observations" select "Meteogram."
- Click on the closest city to your location for the current meteogram in that city.

After viewing the meteogram, answer the following questions. Your instructor may ask that you attach a copy of the meteogram to your answers.

1.   Which city did you study?   _____

2.   What was the date and time of the meteogram studied? (Be sure to also indicate the *local* day and time of the meteogram.)

3.   (a)   Describe the changes in pressure over the 25-hour period. _____

     (b)   What might explain these pressure changes?

4.   (a)   Describe the changes in temperature over the 25-hour period. _____

     (b)   What might explain these temperature changes?

5.   (a)   Describe the changes in wind direction over the 25-hour period. _____

     (b)   What might explain these wind direction changes?

6.   Did any precipitation take place during the 25-hour period? If so, how much and when?

7.   (a)   Based on the information in the meteogram, what time did the front(s) and/or storm pass through your city?   _____

     (b)   How can you tell?

EXERCISE 19
Weather Maps

Objective:	To study weather patterns shown on standard weather maps produced by the National Weather Service.
Resources:	Internet access or mobile device QR (Quick Response) code reader app (optional).
Reference:	Hess, Darrel. *McKnight's Physical Geography*, 12th ed., pp. A13–A18.

THE STATION MODEL

In Exercises 13, 14, and 18 we used a simplified version of the weather map **station model**. We will now introduce the complete station model used on standard National Weather Service maps. Figure 19-1 is a sample station model, with labeled information. A complete description of all standard U.S. Weather Service station model codes is found in Appendix III in the back of the Lab Manual.

In earlier exercises, we introduced the format and position of data to indicate current temperature (upper left of station circle), dew point (lower left of station circle), wind direction, wind speed, and atmospheric pressure.

Wind direction is indicated with a shaft that points *into* the wind, and wind speed is determined by adding up the "feathers" on the wind shaft: $\frac{1}{2}$ feather = 5 knots; full feather = 10 knots; triangular flag = 50 knots (1 knot equals 1 nautical mile per hour, 1.15 statute mph, and 1.85 km/hr). For example, the station model in Figure 19-1 shows a 15-knot wind from the north-northeast.

Barometric pressure is shown in an abbreviated form on the station model. The first 9 or 10 is left off, and the decimal point removed, so a pressure of 1021.3 is written simply as "213"; and so "147" indicates a pressure of 1014.7 mb. Immediately below the pressure, the change in pressure in tenths of a millibar (+ or −) over the last 3 hours is given, and immediately to the right of this is a simple line graph showing the pressure tendency over the last 3 hours.

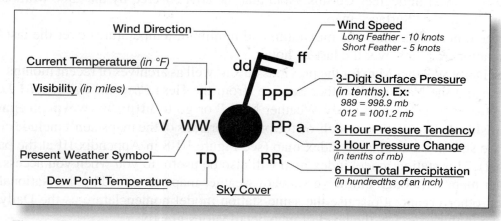

Figure 19-1: Sample weather station model. (Adapted from National Weather Service)

Every station model on a weather map may not contain all of the information shown in the sample. The position of data around a station model, however, does not vary. For example, in the abbreviated station model below, the temperature is 73°F, the dew point is 47°F, the pressure is 995.3 mb and has decreased 3.1 mb over the last 3 hours (falling for 2 hours, then steady for the last hour), and the wind is 25 knots from the southeast.

U.S. WEATHER SERVICE MAPS

Figure 19-2 shows a standard set of online weather maps produced by the National Weather Service, shown here reduced in size. (You may also view the maps in this exercise online by going to the Hess *Physical Geography Laboratory Manual*, 12th edition, website at **www .MasteringGeography.com**, then select Exercise 19, or by scanning the QR (Quick Response) code for this exercise.) These maps were published as part of its *Daily Weather Maps: Weekly Series*. These sets contain weather maps for a seven-day period (Monday through Sunday).

Included for each day is a "Surface Weather Map" showing conditions at 7:00 A.M. Eastern Standard Time. Figure 19-3 is the more-detailed full-size version of the surface map in Figure 19-2, also available online from the National Weather Service. The surface map provides weather data with station models, gives the position of fronts, and uses shading to indicate areas of precipitation. Isobars are used to map surface pressure (in millibars). Dashed isotherms are provided for 32°F and 0°F. Thick dashed lines show the locations of troughs.

The "500-Millibar Height Contours Map" illustrates the conditions in the upper atmosphere. (Figure 19-4 in Exercise 19 Problems, Part III, is the detailed, full-size version of this map.) It shows the altitude of 500 millibars of pressure, given in dekameters above sea level (1 dekameter = 10 meters). High 500-millibar altitudes indicate relatively high pressure below, whereas low 500-millibar altitudes indicate relatively low pressure below (notice also that 500-millibar altitudes generally decrease with increasing latitude due to cooler surface conditions). For reference, a pressure of 500 millibars is found at an average altitude of about 5600 meters (18,400 feet). This map also shows the general trend and speed of upper-elevation winds, locations of upper elevation pressure centers, and isotherms (dashed lines) in degrees Celsius. Note that the area covered by the "500-Millibar" map is greater than that of the "Surface" map.

Three other maps show the maximum and minimum temperatures over the last 24 hours, and precipitation amounts over the last 24 hours.

The Daily Weather Map for the previous day, as well as archives of recent months, are available online from the National Weather Service (from the Hess *Physical Geography Laboratory Manual* website, select "Go to Daily Weather Maps" or go to **http://www.wpc.ncep.noaa.gov/ dwm/dwm.shtml**). The station models on the online versions of the maps don't include city names, but these are shown on a separate index map (see Figure III-8 in Appendix III at the back of the Lab Manual). This station model index map will also be useful to you when you access other online weather maps. Although online versions of weather maps produced by the National Weather Service and other organizations use the same station model nomenclature as the Daily Weather Maps shown in this exercise, the level of map detail varies greatly.

Daily Weather Maps
WEDNESDAY NOVEMBER 9, 2011

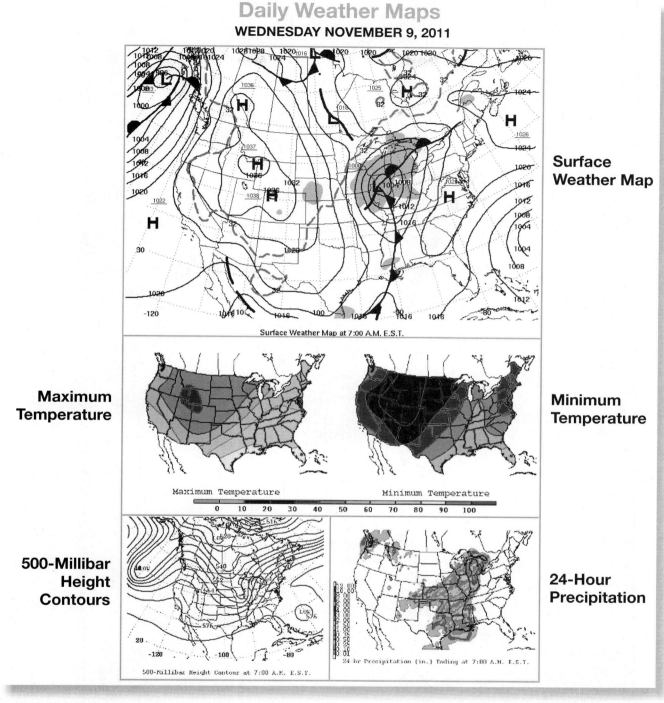

Surface Weather Map

Maximum Temperature

Minimum Temperature

500-Millibar Height Contours

24-Hour Precipitation

Figure 19-2: Online weather maps for November 9, 2011. (From *Daily Weather Maps*, National Weather Service)

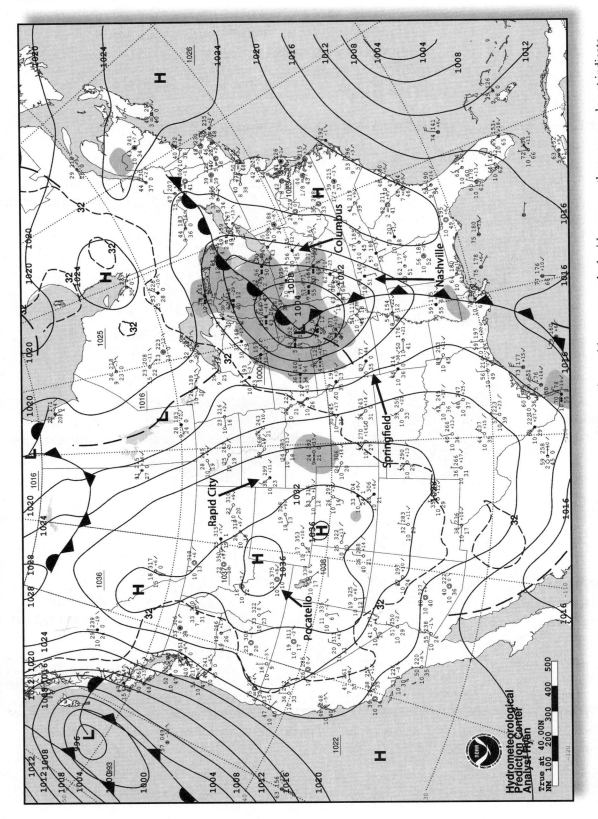

Figure 19-3: Surface Weather Map for November 9, 2011. Cities noted in exercise problems are marked with arrows—these arrows do not indicate the wind direction in those cities. (From *Daily Weather Maps*, National Weather Service)

Name _____   Section _____

EXERCISE 19 PROBLEMS—PART I

The following questions are based on the surface weather map for November 9, 2011 (Figure 19-3). This day was chosen because of the weather conditions present. There is a well-developed midlatitude cyclone centered near Chicago, with a cold front extending to the southwest and a warm front extending to the east. Several areas of high pressure are found along the east coast and in the west.

1. Describe the following weather conditions at 7:00 A.M. EST on November 9, 2011, in Nashville, Tennessee (36° N, 87° W; an enlarged copy of the station model is shown here).

    (a)   Temperature: _____ °F

    (b)   Dew point: _____ °F

    (c)   Wind speed: _____ knots

    (d)   Wind direction: _____

    (e)   Pressure: _____ mb

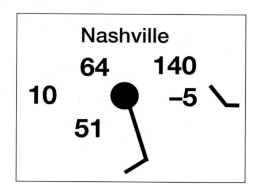

    (f)   Pressure change over last 3 hours: _____ mb

    (g)   What explains this change in pressure?

    (h)   What explains the wind direction?

2. (a)   Which city is more likely to receive precipitation during the next 12 hours, Columbus, Ohio (40° N, 83° W) or Rapid City, South Dakota (44° N, 103° W)?   _____

    (b)   Why?

Name _____   Section _____

EXERCISE 19 PROBLEMS—PART II

The following questions are based on the surface weather map shown in Figure 19-3.

1. Describe the following weather conditions at 7:00 A.M. EST on November 9, 2011, in Springfield, Missouri (37° N, 93° W; an enlarged copy of the station model is shown here).

    (a)   Temperature: _____ °F

    (b)   Wind direction: _____

    (c)   Pressure: _____ mb

    (d)   Pressure change over last 3 hours:
          _____ mb

    (e)   What explains this change in pressure?

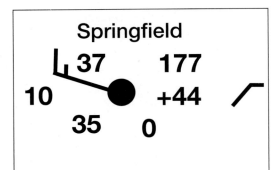

    (f)   What explains the difference in temperature between Springfield and Nashville?

2. (a)   What general kind of weather (wet or dry) should Pocatello, Idaho (43° N, 112° W), expect over the next few hours?   _____

    (b)   Why?

3. (a)   Use the graphic map scale (in nautical miles [NM]) to estimate the distance from the center of the storm south of Chicago (at 40° N, 90° W) to the east coast of the United States (measure along the 40th parallel).   _____ nautical miles

    (b)   Assuming that the storm travels to the east at a speed of 30 knots, how many hours will it take for the center of the low to reach the east coast?   _____ hours

Name _____   Section _____

EXERCISE 19 PROBLEMS—PART III

The following questions are based on Figure 19-4, the 500-Millibar map.

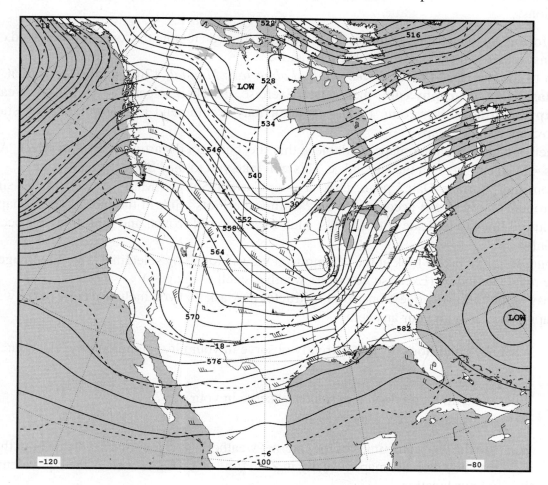

Figure 19-4: 500-Millibar Height Contours map for November 9, 2011. (From *Daily Weather Maps*, National Weather Service)

1. What is the highest recorded velocity of upper elevation
    winds shown on the 500-Millibar map?   _____ knots

2. With a blue pencil, draw a series of arrows across the map from west to east to indicate the
    approximate position and wind direction of the jet stream.

3. Locate the high pressure cell centered over Idaho near Pocatello (shown in Figure 19-3
    at about 45° N, 110° W). Based on the upper elevation wind patterns, in which direction
    would you expect this pressure cell to travel over the next few hours? Look carefully at the
    parallels and meridians, and be as specific as possible.

                                                                        To the _____

Name _____   Section _____

EXERCISE 19 PROBLEMS—PART IV—INTERNET

In this exercise, you will use weather maps from the Internet to explain current local weather conditions.

- Before using the weather maps on the Internet, go outside to note the local sky and wind conditions so that you can answer problems 1 and 2.
- Go to the the Hess *Physical Geography Laboratory Manual*, 12th edition, website at **www .MasteringGeography.com**, then select Exercise 19. Then select "Go to Unisys Weather," **http://weather.unisys.com/**. (Your instructor may recommend a different Internet site that provides weather maps.)
- There are many weather maps available at this site, but a good starting point is to select "Analyses—Surface Data" on the site index. This will take you to the "Surface Data" page.
- From the "Surface Data" page, look at the "Current Surface Map" as well as the simpler "Fronts" map you can access from the "Composite Plots" link in the index. These maps use bold red lines to indicate warm fronts, blue lines for cold fronts, and pink lines for occluded fronts.
- You also might take a look at the "Composite Views" found by selecting "Satellite Images" on the main site index. (Exercise 20 covers weather satellite images.)
- Answer the following questions. Your instructor may ask that you include a copy of the weather maps you used with your answers.

1. What sky conditions do you currently observe in your city (cloudy, rainy, clear, etc.)?

2. What wind direction do you currently observe in your city?

3. What are the dates and times of the Internet weather maps you used? Indicate both the date and time shown on the maps (usually in Zulu time or UTC), as well as the equivalent local time in your location.

    _____ Zulu/UTC = _____ local time.

4. Based on the weather maps, what might explain the current local weather conditions you observed? (Consider the position of pressure cells, fronts, etc.)

5. Are nearby cities experiencing the same general weather? If not, suggest a reason why.

6. How has the weather in your city changed over the last 12 hours, and what might explain these changes?

EXERCISE 20
Weather Satellite Images

Lab Exercise
20

https://goo.gl/uou4Ah

Objective:	To interpret visible light images, infrared images, water vapor images, and movie loops from weather satellites.
Resources:	Internet access or mobile device QR (Quick Response) code reader app (optional).
Reference:	Hess, Darrel. *McKnight's Physical Geography*, 12th ed., p. 162.

GOES WEATHER SATELLITES

The satellite images we commonly see on television weather reports come from satellites operated by the National Oceanic and Atmospheric Administration (NOAA). They are known as **GOES**, or **Geostationary Operational Environmental Satellites**. The GOES satellites orbit at a distance of about 35,800 kilometers (22,300 miles) in fixed locations relative to the surface of Earth below. GOES-East (GOES-13) orbits above the equator in South America (75° W), where it can see the conterminous United States, as well as much of the north and south Atlantic Ocean. GOES-West (GOES-15) orbits above the equator in the Pacific (135° W), where it can see most of the Pacific Ocean from Alaska to New Zealand. (As of this writing, GOES-14 is in orbit on standby as a replacement for either GOES-East or GOES-West; a new GOES satellite is scheduled for launch in late 2016.) GOES satellites (as well as satellites operated by other countries, such as Europe's Meteosat satellites) send back several images each hour, and are important weather forecasting tools. Figures 20-1, 20-2, and 20-3 showing the west coast of North America, were taken by GOES-West. You may also view all of these images online by going to the Hess *Physical Geography Laboratory Manual,* 12th edition, website at **www.MasteringGeography.com**, then select Exercise 20.

The GOES satellites have sensors that can detect radiation in several different bands of wavelengths in the electromagnetic spectrum. The intensity of radiation in each wavelength band is recorded and used to produce the images we see. This exercise will focus on satellite images from three wavelength bands—visible light, thermal infrared, and a portion of the infrared spectrum that allows the creation of "water vapor" images.

VISIBLE LIGHT IMAGES

Visible light satellite images show light that has been reflected off the surface of Earth, or by clouds in the atmosphere. In gray-scale ("black-and-white") images, the intensity of reflected visible light is shown with shades of gray, with white (bright) areas representing surfaces reflecting a great deal of visible light, and black (dark) areas reflecting very little light.

The brightness of a surface depends both upon its **albedo** (reflectance) and the angle of the light striking it. The brightest (high albedo) surfaces in visible light are typically the tops of clouds and snow or ice-covered surfaces. The darkest (low albedo) surfaces are typically land areas (especially unvegetated land surfaces) and the oceans—which are usually the darkest surfaces

Figure 20-1: GOES-West visible light image of northern Pacific Ocean on December 29, 2012, 2100Z. (Image courtesy of Naval Research Laboratory, Marine Meteorology Division)

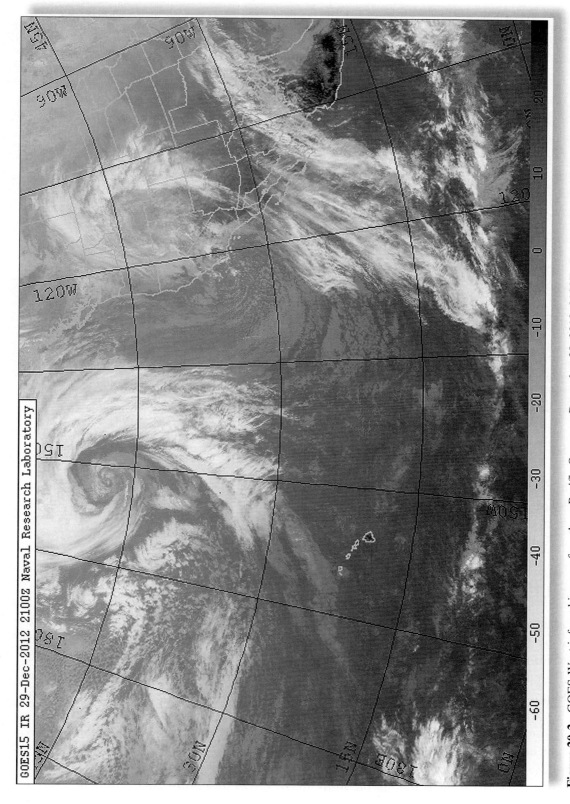

Figure 20-2: GOES-West infrared image of northern Pacific Ocean on December 29, 2012, 2100Z. (Image courtesy of Naval Research Laboratory, Marine Meteorology Division)

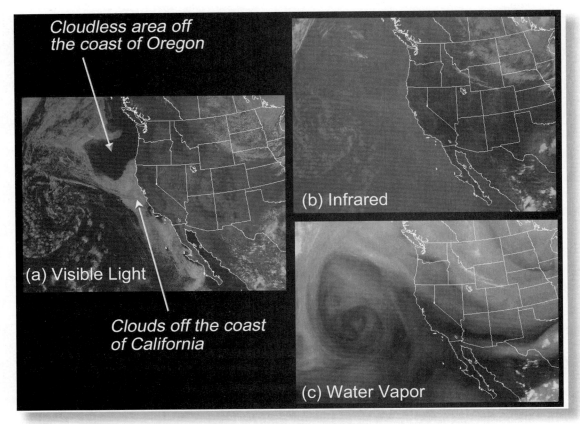

Figure 20-3: GOES-West visible light (a), infrared (b), and water vapor (c) images taken on September 26, 2003. Locations noted in problems shown with arrows. (Images courtesy of NOAA)

seen on visible light satellite images. When a region of Earth is experiencing early morning or late afternoon, the amount of visible light reflected generally will be reduced, producing a much darker image than at midday. Figure 20-1 is a visible light image of the northern Pacific Ocean and the west coast of North America.

INFRARED IMAGES

Infrared satellite images show the **longwave radiation** (or "thermal infrared") that has been emitted by the surface of Earth or by clouds in the atmosphere. Warm objects emit more longwave radiation than cold objects, and so infrared images show us differences in temperature. Although infrared radiation is invisible to the human eye, differences in its intensity are shown on gray-scale satellite images so that cooler surfaces (those emitting relatively little longwave radiation) are shown in white, and warmer surfaces (those emitting a great deal of longwave radiation) are shown in black. Because clouds and the surface emit longwave radiation continuously, infrared images are produced by satellites both day and night.

Figure 20-2 is an infrared image of the northern Pacific Ocean and the west coast of North America taken at the same time as the visible light image shown in Figure 20-1.

In order to distinguish very small differences in temperature, infrared satellite images are frequently computer enhanced with colors to highlight the lowest temperatures. In computer-enhanced infrared images, the lowest temperatures—such as would be found at the tops of the highest clouds—are usually shown in color as shades of green, blue, and purple. These colors help meteorologists recognize the location of the very cold, high tops of the massive **cumulonimbus** clouds associated with thunderstorms and hurricanes. Such detail may be lost on ordinary gray-scale infrared images.

INTERPRETING VISIBLE AND INFRARED IMAGES

Both visible and infrared satellite images are useful in weather forecasting because they provide different kinds of information about the surface of Earth and the atmosphere—visible light images show differences in the reflectance of light, whereas infrared images show differences in temperature.

In infrared images, the tops of high clouds are easy to distinguish from low clouds and fog. The tops of high clouds, such as massive cumulonimbus clouds, are much colder than low clouds and fog, and so will appear brighter (white) on gray-scale infrared images. Low clouds and fog tend to have similar temperatures to that of the surface, and so will appear as nearly the same shade of gray as the surface below.

On the other hand, in visible light both high clouds and low clouds may appear equally bright (because they tend to reflect similar amounts of light). It also may be difficult to tell the difference between snow-covered surfaces and clouds in visible light images. By comparing visible and infrared images of the same area, it is possible to distinguish between high clouds and low clouds, as well as to recognize snow- or ice-covered surfaces.

WATER VAPOR IMAGES

In ordinary infrared satellite images, the wavelengths of infrared radiation emitted by the surface and atmosphere are in the portion of the electromagnetic spectrum called the **atmospheric window**. These are the wavelengths of infrared radiation that can transmit through the atmosphere without being absorbed (wavelengths between approximately 8 and 12 micrometers). Although such infrared satellite images tell us about the temperature of the surface and cloud tops, they may not tell us much about the characteristics of the air itself.

However, some gases in the atmosphere, such as water vapor and carbon dioxide, absorb and emit wavelengths of infrared radiation that are outside the atmospheric window. By detecting wavelengths of infrared radiation outside the atmospheric window that are emitted by water vapor (specifically, wavelengths of 6.7 micrometers and 7.3 micrometers), we have a way to determine the relative amount of water vapor in the atmosphere—water vapor that may be invisible in visible light and ordinary infrared satellite images. Regions with high emission of infrared wavelengths emitted by water vapor contain relatively large amounts of water vapor, whereas regions with low emission of these wavelengths contain relatively small amounts of water vapor. Such satellite images are called **water vapor images**.

Water vapor images let us recognize regions of dry air and moist air. In gray-scale water vapor images, darker shades of gray show areas of relatively dry air, whereas lighter shades of gray show areas of relatively moist air (in color water vapor images, relatively dry air is often shown as dark blue and relatively moist air is shown in shades of orange and red). Figures 20-3a, 20-3b, and 20-3c are satellite images of western North America taken on the same day showing visible light, infrared, and water vapor, respectively.

Because water vapor images can only detect infrared emission from water vapor in the middle and upper **troposphere** (above about 4500 meters or 15,000 feet), in addition to telling us about the moisture content of the atmosphere, these images also give us information about wind movement in the upper atmosphere, especially when time-sequence "movie loops" are used.

SATELLITE MOVIE LOOPS

Time-sequence movie loops (or "animations" as they are called on some websites) are very useful tools in meteorology. By viewing a sequence of satellite images taken over several hours (or several days in some cases), we can track the development, movement, and changes in air masses, storms, and wind patterns (Figure 20-4). Because clouds are associated with cyclonic storms, the wind patterns and movement of midlatitude cyclones and hurricanes are usually quite obvious on satellite movie loops. However, by viewing movie loops, the wind patterns and movement in cloudless areas (such as in an anticyclone) may also be determined by viewing the movement of clouds around the margins of these areas, or by viewing the movement of water vapor.

TIME MARKS ON SATELLITE IMAGES

The time and date of a satellite image are usually indicated along the top or bottom of the frame. Because satellite images cover such large areas (and therefore, many time zones), it is impractical to label them using local time. Instead, the images are labeled with the date and time at Greenwich, England. This is known as **Universal Time Coordinated (UTC)** or **Zulu time**. The times are usually given using a 24-hour clock, with a "Z" to indicate Zulu time. For example, "1800Z" would be 6:00 P.M. Greenwich time.

To convert Zulu time to local standard times in North America, subtract 5 hours for Eastern Standard Time, subtract 6 hours for Central Standard Time, subtract 7 hours for Mountain Standard Time, and subtract 8 hours for Pacific Standard Time. To convert Zulu time to daylight-saving times, subtract one less hour than indicated above.

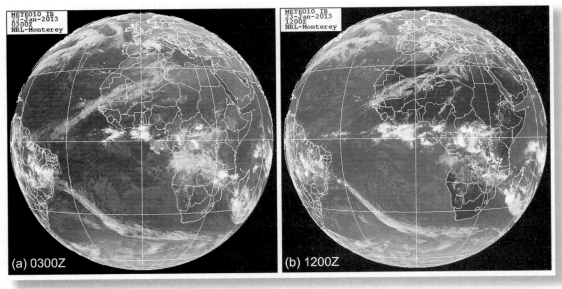

Figure 20-4: METEO-10 infrared satellite images taken 9 hours apart on January 23, 2013. (Images courtesy of Naval Research Laboratory, Marine Meteorology Division)

Name _____   Section _____

EXERCISE 20 PROBLEMS—PART I

The following questions are based on Figure 20-1, a visible light image of the northern Pacific Ocean and the west coast of North America, and Figure 20-2, an infrared image of the same region taken on the same day and time. A well-developed midlatitude cyclone is in the northern Pacific Ocean.

1. (a)   Are the clouds along the cold front of the midlatitude cyclone
          (at about 45° N, 140° W) high clouds or low clouds?   _____

    (b)   How can you tell?

2.   In the infrared image (Figure 20-2), why does south-central Mexico (20° N, 100° W) appear black?

The following questions are based on Figure 20-3 showing visible light, infrared, and water vapor images of western North America and the northern Pacific Ocean basin taken on the same day and time.

3. (a)   In the visible light image, notice the clouds along the coast of
          California and Baja, Mexico. Are these high clouds or low
          clouds? (Hint: You will need to view the infrared image
          to determine the relative altitude of these clouds.)   _____

    (b)   How can you tell?

4. (a)   In the visible light image, find the cloudless area off the
          coast of Oregon and Washington. Does this region contain
          relatively large amounts of water vapor, or relatively little
          water vapor? (Hint: Compare the visible light and water
          vapor images.)   _____

    (b)   What could explain the lack of clouds here?

Name _____   Section _____

EXERCISE 20 PROBLEMS—PART II

The following questions are based on Figure 20-4, infrared images of Africa taken 9 hours apart.

1. Why do northern and southern Africa appear much darker in Figure 20-4b than in 20-4a?

2. (a) Has the ocean around the continent changed temperature
        during the time between the first and second image?   _____

    (b) Why?

3. (a) Which feature of the general circulation of the
        atmosphere explains the band of clouds near the equator?   _____

    (b) Why are these clouds much more obvious in Figure 20-4b than in Figure 20-4a?

EXERCISE 20 PROBLEMS—PART III—INTERNET

Scan the QR (Quick Response) code for this exercise, or go to the Hess *Physical Geography Laboratory Manual*, 12th edition, website at **www.MasteringGeography.com**, to view Figure 20-5 and Figure 20-6, two infrared satellite image movie loops.

Problems 1 and 2 are based on Figure 20-5: "West Coast Infrared Satellite Movie Loop."

1. (a) Is high or low pressure found northeast of Hawai'i (30° N, 140° W)? _____

    (b) How can you tell? _____

2. (a) Is high or low pressure found west of California (35° N, 125° W)? _____

    (b) How can you tell? _____

Problems 3 and 4 are based on Figure 20-6: "Africa 4-Day Infrared Satellite Movie Loop."

3. What explains the general direction of movement of clouds in the region of the Mediterranean Sea and western Eurasia?

4. What explains the growth of clouds just south of the equator as the movie loop proceeds?

Name _____ Section _____

EXERCISE 20 PROBLEMS—PART IV—INTERNET

In this exercise, you will interpret satellite images found on the Internet.

- Before viewing the satellite images go outside and note the general weather conditions. If your school has a weather station, you will be able to provide more information for problem 1 than if you are simply relying on your own observations.
- Go to the Hess *Physical Geography Laboratory Manual*, 12th edition, website at **www .MasteringGeography.com**. Select Exercise 20, and go to the Naval Research Laboratory at **www.nrlmry.navy.mil/sat_products.html**. (Your instructor may recommend a different source of satellite images, and may ask that you print out and attach the images you view.)
- Begin with "West Coast & EPAC" if you live on the West Coast and "US CONUS" if you live in the Midwest or East Coast. View the infrared, visible light and water vapor images; you may make movie loops by clicking on the "animate" button in the menu above most images.

1.   Describe the current weather conditions in your location.

(a)   What is the date and local time of your observations? _____

(b)   What are the general sky conditions you observe outside (cloudy, rainy, clear, etc.)?

(c)   What is the current wind direction? _____

(d)   Has there been any precipitation over the last 24 hours?
       If so, how much? (Optional)   _____

(e)   What is the current temperature? (Optional)   _____

(f)   What is the barometric trend (rising, falling, etc.)? (Optional)   _____

View the latest weather satellite images showing your city and region of the country, and then answer the following questions.

2.   (a)   What is the date and time stamp on the latest satellite images you are viewing?

       Date: _____   Time: _____ Z

(b)   What is the day and *local time in your city* of these images?

       Date: _____   Local Time: _____

3. What kind of sky conditions (high clouds, low clouds, scattered clouds, clear, etc.) do the satellite images suggest are in your immediate location? (You may need to look at both the visible light and infrared images to determine this.)

4. Do your own observations of the sky (from problem 1b) and the satellite images match closely? If not, suggest a reason why.

5. Based on the satellite images and movie loops (which can help you identify wind direction and major pressure cells), what explains your observations of the current sky conditions, wind direction, pressure trend, and/or recent precipitation in problem 1?

EXERCISE 20 PROBLEMS—PART V—INTERNET

Answer the following questions after completing the problems in Part IV.

6. Describe the location of any midlatitude cyclones or tropical cyclones you see over (or near) North America in the satellite images (describe the locations with latitude and longitude, or by their position over states/provinces). Watching for cloud movement by making a time-lapse movie loop using the "animate" feature on the NRL website will be helpful with this.

7. (a) Describe the location of any cloud-free regions you see near or over North America in the satellite images (describe with latitude and longitude or by their position over states/provinces).

    (b) Do you think that any of these cloud-free regions are in areas of high pressure? If so, describe which one(s) and explain why you think this is the case.

OPTIONAL

8. View a current online weather map of the United States (such as those you accessed for Exercise 19, Part IV). Were your interpretations of the locations of pressure cells (from problems 5, 6, and 7) correct? If not, suggest a reason why.

EXERCISE 21
Doppler Weather Radar

Lab Exercise
21

https://goo.gl/4uizs4

Objective:	To use Doppler weather radar images to study storm movement and precipitation patterns.
Resources:	Internet access or mobile device QR (Quick Response) code reader app (optional).
Reference:	Hess, Darrel. *McKnight's Physical Geography*, 12th ed., p. 200.

DOPPLER RADAR

Radar was developed and used primarily for military purposes during World War II. Since the late 1940s, however, radar has been used by meteorologists to detect precipitation and severe weather such as thunderstorms and tornadoes.

Radar is an acronym for *ra*dio *d*etection *a*nd *r*anging. The principle of radar is simple. A rotating antenna emits short bursts of radio waves that transmit through the atmosphere until they reflect off an object and return to the antenna where they are detected. The time lag between the emission of a radio wave pulse and the return of its "echo" is used to calculate the distance to the reflecting object—the greater the time lag, the greater the distance. In meteorological applications, radar "targets" in the atmosphere are mostly raindrops and hailstones—generally the larger the droplets, the stronger the return echo.

Modern meteorological radar units emit pulses of radio waves lasting only about 0.0000016 second, followed by a 0.00019-second listening period for the echo. This process is repeated as many as 1300 times each second (because of the extremely short pulse of radio waves emitted relative to the listening period, a radar antenna is actually transmitting for only a total of about 7 seconds each hour). Meteorological radar typically emits pulses of "microwaves" with wavelengths of about 10 centimeters.

Because of the curvature of Earth and the reduction of signal strength over distance, most weather radar is limited to a distance of about 260 kilometers (about 160 miles) from the antenna; light rain can be detected perhaps only 150 kilometers (90 miles) away from the antenna. Generally speaking, the closer a storm is to the radar station, the higher the resolution of the radar image. In most cases, meteorologists like to view radar images of storms from two or more radar locations in order to overcome deficiencies of coverage and resolution.

Not only can radar detect distances to reflecting objects, but by taking advantage of the **Doppler effect**, meteorological radar can help determine the relative speed and direction of water droplets within a storm. You've experienced the Doppler effect if you have ever listened to an approaching emergency vehicle siren: When the siren is approaching, the sound waves are compressed—this leads to a higher frequency and, therefore, a higher pitch of the sound waves. As the siren passes and moves away, the sound waves are stretched—this leads to a lower frequency and lower pitch of the sound waves. Doppler radar detects changes in the "pitch" of radio wave echoes. When an object is moving away from the radar antenna, the radio waves of the returning echo are stretched, leading to a lower frequency; when an object is moving toward the radar antenna, the radio waves of the echo are compressed, leading to a higher frequency.

TYPES OF RADAR IMAGES

There are three main types of radar images used in meteorology: reflectivity, velocity, and precipitation totals. Current weather radar images for the United States may be viewed at Internet sites such as **http://radar.weather.gov/**. (All of the images shown in this exercise of the Lab Manual, along with supplementary images, may be viewed online on the Hess *Physical Geography Laboratory Manual*, 12th edition, website at **www.MasteringGeography.com,** then select Exercise 21, or by scanning the QR [Quick Response] code for this exercise.)

Reflectivity: Reflectivity images display the intensity of the return echoes, usually measured in decibels of "Z"—the reflected energy (dBZ). In most reflectivity images, low reflectivity is indicated by the colors blue and green, with increasing reflectivity shown as yellow, red, and fuchsia. **Base reflectivity** images show the intensity of echoes when the radar is scanning the atmosphere very close to the horizon—usually about 0.5° above the horizon. Base reflectivity images are helpful in showing precipitation reaching the surface, but do not show activity high in storms or in storms that are distant from the radar station.

By making a series of sweeps at increasingly higher angles (up to about 19.5° above the horizon), a radar unit can detect precipitation high above the ground. A **composite reflectivity** radar image displays the strongest echo detected in each direction from the radar unit—showing precipitation and storm structure well above the horizon (Figure 21-1). Because it takes about 5 or 6 minutes for a radar unit to complete a full set of sweeps to complete a composite image, the time of the base reflectivity image and the composite reflectivity image may be different. As with weather satellite images, radar images are given a date and time stamp using **UTC** or **Zulu time**; for example, 1800Z indicates an image taken at 6:00 P.M. UTC or Greenwich Mean Time (for a review of time zones, see Exercise 3).

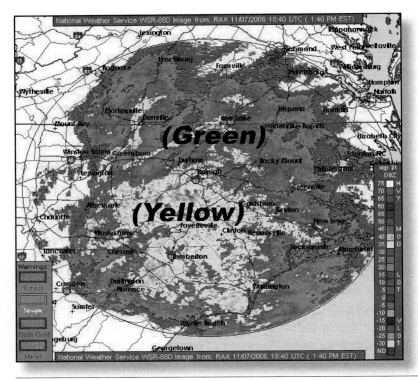

Figure 21-1: Composite reflectivity radar image for Raleigh, North Carolina (in center of image), on November 7, 2006, at 18:40 UTC (1840Z). Areas labeled "green" display low radar reflectivity and areas labeled "yellow" display higher reflectivity. (Image courtesy of National Weather Service)

132

Both base reflectivity and composite reflectivity images are useful to meteorologists. For example, strong echoes in the composite reflectivity image that are lacking in the base reflectivity image may indicate areas within a storm where falling precipitation evaporates before reaching the surface, or where strong updrafts of hail or rain may be present.

Velocity: Radar velocity images indicate the strength of wind moving directly toward or directly away from the radar unit. The color red indicates wind moving away from the radar antenna, the color green indicates movement toward the antenna (red represents longer or "stretched" wavelengths, and green represents shorter or "compressed" wavelengths), and the gray areas are transition zones between incoming and outgoing wind. The color purple indicates areas where the radar is unable to determine velocity (what are called "folded" areas). **Base velocity** images show the overall pattern of movement within a storm (Figure 21-2), whereas **storm relative motion** images show the winds as if the storm were stationary (Figure 21-3)—such images can reveal circulation within a storm. Strong local circulation, such as is associated with the formation of tornadoes, is indicated by strong inbound wind (shown as green) adjacent to an area of strong outbound wind (shown as red). Such areas would be carefully watched by meteorologists for the development of tornadoes.

Precipitation: Within a storm, raindrops and hail are good reflectors of radar waves, so the relative strength of a radar echo can be an indication of the amount of precipitation that is falling within a storm. Generally, an echo value of 20 dBZ indicates that light rain is falling, whereas values of 60 dBZ or greater indicate that large hail (hail with a diameter of more than about 2 centimeters [3/4 inch]) can occur—conditions associated with severe weather. Not only can the echo intensity indicate the size of precipitation that is falling, but it can also indicate the rate of precipitation. For example, a 20 dBZ value suggests that only a trace of rain may fall over the period of an hour, whereas a 47 dBZ value indicates a rain rate of about 3.2 centimeters per hour (1.25 inches/hour) and 60 dBZ indicates a rain rate of over 20 centimeters per hour (8.0 inches/hour).

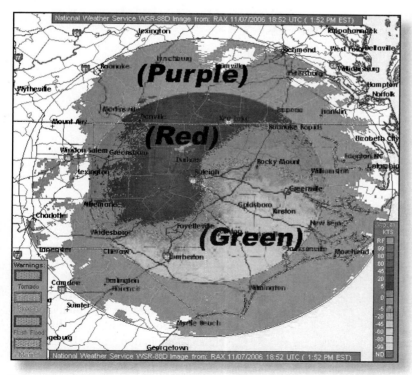

Figure 21-2: Base velocity radar image for Raleigh, North Carolina, on November 7, 2006, at 18:52 UTC (1852Z). Areas labeled "green" indicate movement toward the radar antenna and areas labeled "red" indicate movement away from the antenna; in "purple" areas the direction of motion cannot be determined. (Image courtesy of National Weather Service)

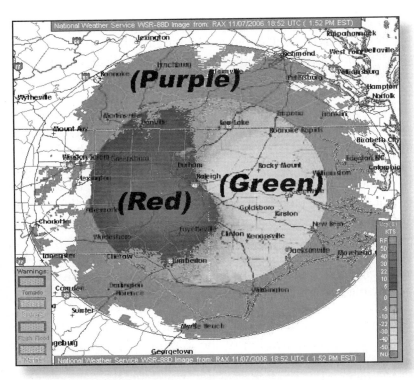

Figure 21-3: Storm relative motion radar image for Raleigh, North Carolina, on November 7, 2006, at 18:52 UTC (1852Z). Areas labeled "green" indicate movement toward the radar antenna and areas labeled "red" indicate movement away from the antenna; in "purple" areas the direction of motion cannot be determined. (Image courtesy of National Weather Service)

One-hour precipitation radar images provide estimates of precipitation that has fallen over the last hour (Figure 21-4). Such images are used to assess the potential for flash floods in a region. Trace amounts of precipitation are shown in light blue, with increasing amounts of one-hour precipitation indicated with dark blue, green, orange, and red. **Storm total precipitation** radar images estimate the total amount of precipitation that has fallen since the last one-hour pause in rainfall.

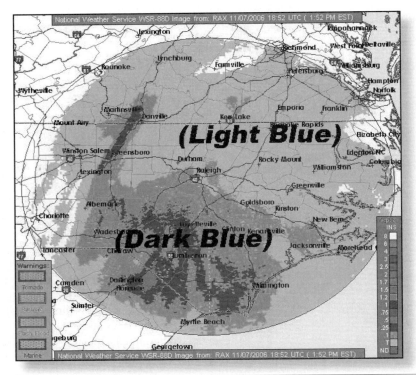

Figure 21-4: One-hour precipitation radar image for Raleigh, North Carolina, on November 7, 2006, at 18:52 UTC (1852Z). Areas labeled "light blue" display low one-hour precipitation totals and areas labeled "dark blue" display higher precipitation totals. (Image courtesy of National Weather Service)

Name _____   Section _____

EXERCISE 21 PROBLEMS—PART I

The following questions are based on Figures 21-1, 21-2, 21-3, and 21-4, radar images of Raleigh, North Carolina, on November 7, 2006. The Doppler radar site in Raleigh is in the center of the image; north is to the top of each radar image. Full-color versions of these figures may be viewed on the Hess *Physical Geography Laboratory Manual*, 12th edition, website at **www .MasteringGeography.com**, then select Exercise 21, or by scanning the QR (Quick Response) code for this exercise.

1.   Where is the area of greatest precipitation around Raleigh?_____

2.   Is the rain generally moving toward Raleigh or away? _____

3.   (a)   Is the direction of wind within the storm the
            same as the overall direction of movement of the storm itself? _____

      (b)   How do you know? _____

EXERCISE 21 PROBLEMS—PART II—INTERNET

The following questions are based on radar images and satellite movie loops available on the Hess *Physical Geography Laboratory Manual*, 12th edition, website at **www.MasteringGeography.com**, then select Exercise 21, or by scanning the QR (Quick Response) code for this exercise. View Figures 21-5 and 21-6, radar movie loops showing base reflectivity and composite reflectivity for Raleigh, North Carolina, on November 7, 2006 between 18:06 and 18:46 UTC (1806Z to 1846Z).

1.   Which radar movie loop—base reflectivity or composite
      reflectivity—shows precipitation over Raleigh?   _____

2.   Why don't both radar image loops show precipitation taking place in the same location?

View Figure 21-7, an infrared satellite movie loop recorded at about the same time as the radar images of Raleigh.

3.   Does the cloud movement over Raleigh in the satellite movie loop match
      the direction of movement in the base velocity radar image (Figure 21-2)? _____

4.   What do you see in the satellite movie loop that helps explain the difference in motion
      shown in Figure 21-2 (base velocity) and Figure 21-3 (storm relative motion)?

Name _____   Section _____

EXERCISE 21 PROBLEMS—PART III—INTERNET

In this exercise, you will compare current weather radar and satellite images. Your instructor may have you complete this problem set in conjunction with Exercise 20 Problems Parts IV and V (Weather Satellite Images).

- Go to the National Weather Service Enhanced Radar Image Loop for the United States at **http://radar.weather.gov/ridge/Conus/index_loop.php**. Here you see a mosaic of Doppler radar base reflectivity across the United States during the last hour.

- Zoom in on your region of the country by clicking on the thumbnail below the U.S. map. If there are no strong radar returns in your region, your instructor may have you select a different part of the country.

1. National Weather Service radar mosaic sector viewed: _____

2. Time and date of latest image: _____ UTC = _____ Local Time

3. Describe the extent and intensity of radar returns in your region.

4. What is the general direction of movement of precipitation shown by the radar image loop?

Go to the Naval Research Laboratory at **www.nrlmry.navy.mil/sat_products.html**. Select the latest infrared (IR) satellite image for your part of the country: "West Coast & EPAC (for the western U.S.) or "CONUS" (for the central and eastern U.S.).

5. Time and date of satellite image: _____ UTC = _____ Local Time

6. What kinds of clouds are shown in the area of most intense radar returns (high, low, etc.)?

7. Are intense radar returns found throughout all areas of clouds? If not, suggest a reason why this might be the case.

Use the "Animate" function at the top of the Naval Research Laboratory Web page to make an infrared satellite image movie loop.

8. Does the direction of cloud movement in the satellite image movie loop match that shown in the radar image loop? _____

9. What explains the direction of movement of the precipitation in this region?

EXERCISE 22
Hurricanes

Objective:	To use weather maps and satellite images to study the weather patterns of hurricanes.
Resources:	Internet access or mobile device QR (Quick Response) code reader app (optional).
Reference:	Hess, Darrel. *McKnight's Physical Geography*, 12th ed., pp. 187–194.

HURRICANES

   Tropical cyclones are intense, low pressure disturbances that develop in the tropics and occasionally move up into the midlatitudes. When wind speed reaches 64 knots (64 nautical miles per hour; 119 kilometers per hour or 74 statute miles per hour) tropical cyclones are officially classified as **hurricanes** in North America and the Caribbean and typhoons in eastern Eurasia (in this exercise we use "hurricane" as a generic term for all such storms).

   At the center of a hurricane is an intense low pressure cell with a steep pressure gradient, which produces a converging counterclockwise wind flow pattern in the Northern Hemisphere (the wind flow is converging clockwise in the Southern Hemisphere). The converging wind pattern pulls in warm, moist air—the "fuel" for a hurricane. After air spirals into the storm at the surface, it rises in intense updrafts within towering **cumulonimbus clouds**, often reaching altitudes of 15 kilometers (10 miles) before flowing out into the upper atmosphere—typically in a clockwise direction in the Northern Hemisphere (Figure 22-1).

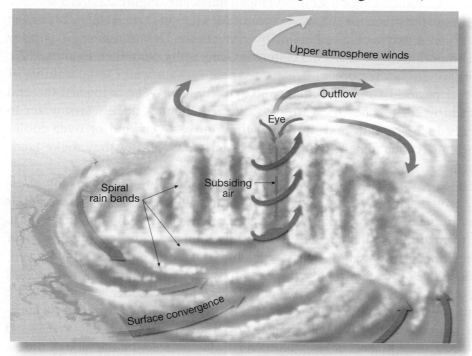

Figure 22-1: Idealized cross section through a well-developed hurricane. Air spirals into the storm horizontally and rises rapidly to produce towering cumulonimbus clouds that produce torrential rainfall. In the center of the storm is the eye, where air movement is downward. (From Hess, *McKnight's Physical Geography*, 12th ed.)

Figure 22-2: MODIS image from NASA's Terra satellite of Hurricane Patricia on October 23, 2015, as it approached the west coast of Mexico. (Image courtesy of NASA)

As the air rises, **adiabatic cooling** causes water vapor to condense out of the air. Condensation provides the moisture for the massive cloud development and heavy rain associated with hurricanes. Condensation also releases **latent heat**. This heat powers the storm by increasing the instability of the air. As latent heat is released into the storm, air rises faster, intensifying the low pressure cell—which in turn pulls in more warm, moist air.

The **eye** in the center of a hurricane is a bit of a paradox. The eye is an area of relative calm and often clear skies, while just beyond the eye wall is the most intense part of the storm. The eye is associated with a downdraft of air (the opposite direction of air flow in a typical low-pressure cell). Hurricane diameters range from 160 to 1000 kilometers (100–600 miles)—much smaller than midlatitude cyclones—with eye diameters ranging from about 16 to 40 kilometers (10–25 miles). A well-defined eye is usually an indication of a well-developed hurricane. Figure 22-2 is a satellite image of Hurricane Patricia as it approached the west coast of Mexico on October 23, 2015. (The images and maps for this exercise may be viewed online by going to the Hess *Physical Geography Laboratory Manual*, 12th edition, website at **www.MasteringGeography.com**, then Exercise 22.)

Hurricanes develop over warm, tropical oceans—typically with water temperatures of at least 26.5°C (80°F). The period of greatest hurricane activity each year is toward the end of summer, when the ocean water reaches its warmest point. Hurricanes need warm water to survive—they will quickly lose strength and die when they move over land or over cooler water (although powerful hurricanes may remain very destructive even after they make landfall and begin to weaken).

Hurricanes generally develop 10–15° poleward of the equator out of preexisting low pressure disturbances moving from east to west in the band of the trade winds. Not all of these minor low pressure disturbances (some of which are known as **easterly waves**) develop into hurricanes. Conditions for hurricane formation also include a deep layer of warm ocean water and the lack of vertical **wind shear** in the atmosphere—wind shear can tear apart a storm before it strengthens.

Although hurricanes develop in the tropics, they can move poleward into the midlatitudes along the eastern sides of continents where warm ocean currents are found; hurricanes are very rare in the midlatitudes along the western sides of continents because of the cool ocean currents. As hurricanes move poleward into the midlatitudes, they will often begin to move from west to east in the band of the westerlies; in some cases, the remnants of hurricanes will turn into milder midlatitude cyclones.

HURRICANE KATRINA

Hurricane Katrina devastated the city of New Orleans and parts of the Gulf Coast of the United States in August 2005. The surface weather maps in Figure 22-3 show Hurricane Katrina

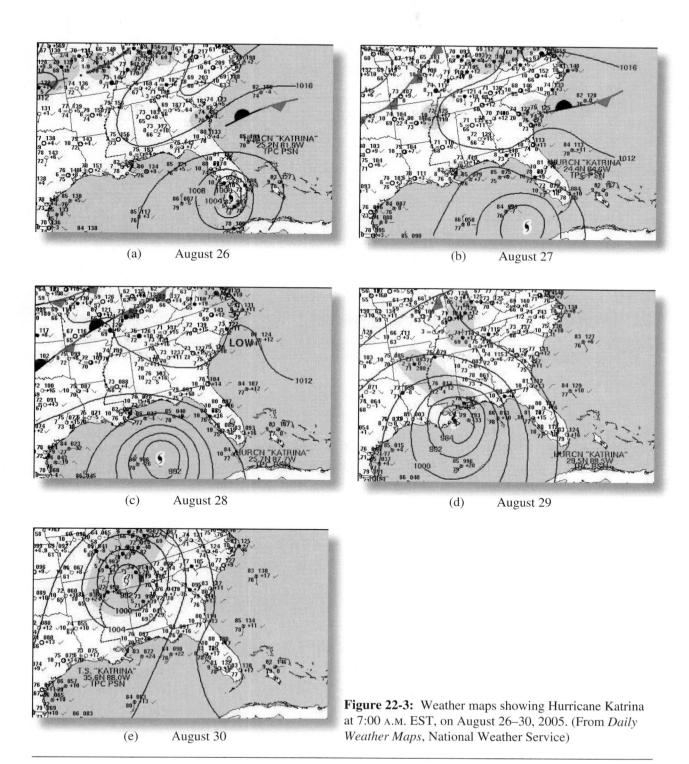

(a)   August 26

(b)   August 27

(c)   August 28

(d)   August 29

(e)   August 30

Figure 22-3: Weather maps showing Hurricane Katrina at 7:00 A.M. EST, on August 26–30, 2005. (From *Daily Weather Maps*, National Weather Service)

Hurricane Katrina August 24 to August 30, 2005			
Date	**Time**	**Central Pressure (millibars)**	**Sustained Wind Speed (knots)**
August 24	0300Z	1007	30
	1500Z	1006	35
August 25	0300Z	1001	45
	1500Z	997	50
August 26	0300Z	984	65
	1500Z	981	70
August 27	0300Z	965	90
	1500Z	940	100
August 28	0300Z	939	100
	1500Z	907	150
August 29	0300Z	904	140
	1500Z	927	110
August 30	0300Z	973	50
	1500Z	985	30

Figure 22-4: Central pressure and sustained wind speed of Hurricane Katrina between August 24 and August 30, 2005. (Data compiled from National Weather Service/National Hurricane Center Forecast Advisories)

at 7:00 A.M. Eastern Standard Time on August 26, 27, 28, 29, and 30. Figure 22-4 is a table listing the central pressure and sustained wind speed of Hurricane Katrina between August 24 and August 30, 2005.

Name _____   Section _____

EXERCISE 22 PROBLEMS—PART I

The following questions are based on Figure 22-3, weather maps showing Hurricane Katrina between August 26 and August 30, 2005 (the full maps may be viewed in color online by going to the Hess *Physical Geography Laboratory Manual*, 12th edition, website at **www.MasteringGeography .com**, then select Exercise 22), and Figure 22-4, a table showing the central pressure and sustained wind speed of the storm between August 24 and August 30.

1. What is the general relationship between the central pressure of Hurricane Katrina and its sustained wind speed?

2. What explains the increase in the intensity of Hurricane Katrina (as reflected in the central pressure) between August 26 and 28?

3. What explains the decrease in the intensity of Hurricane Katrina (as reflected in the central pressure) between August 29 and 30?

4. The following question is based on the enlarged portion of the weather map for August 29 shown here. Three weather stations, marked Land 1 (Lake Charles, LA), Gulf 1, and Gulf 2 (weather buoys in the Gulf of Mexico), are labeled. What explains the difference in wind direction between the three weather stations?

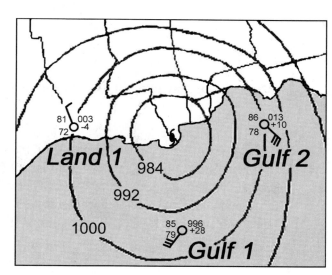

(Adapted from *Daily Weather Maps*, National Weather Service)

141

Name _____   Section _____

EXERCISE 22 PROBLEMS—PART II—INTERNET

The following questions are based on satellite movie loops available on the Hess *Physical Geography Laboratory Manual*, 12th edition, website at **www.MasteringGeography.com**, then select Exercise 22, or by scanning the QR (Quick Response) code for this exercise. Click on Figure 22-5, "Hurricane Katrina Movie Loop," to see an infrared satellite movie loop of Hurricane Katrina from August 24 to August 30, 2005, as it moves across Florida into the Gulf of Mexico and finally comes onshore near New Orleans. You may also want to view Figure 22-6, a close-up movie loop of Hurricane Katrina on August 29. After viewing the satellite movie loops several times, answer the following questions:

1.   (a)   Assuming that a well-defined eye and a tightly circular shape are indications of a strong hurricane, what generally happens to the strength of Hurricane Katrina between August 24 and August 28?

     (b)   What can explain this change in strength?

2.   (a)   What is the general direction of movement of Hurricane Katrina on August 24 (at the beginning of the movie loop)?

                To the _____

     (b)   What is the general direction of movement of Hurricane Katrina on August 30 (at the end of the movie loop)?

                To the _____

     (c)   What may explain this change in direction?

3.   Look at Hurricane Katrina on August 29 (the Figure 22-6 movie loop will be especially helpful here). Although the overall pattern of cloud movement is converging counterclockwise, notice that some of the clouds (such as those you see just to the west of Florida) are diverging clockwise. What explains these diverging clouds?

Name _____   Section _____

EXERCISE 22 PROBLEMS—PART III—INTERNET

In this exercise, you will log the changes in a hurricane over several days by gathering data on the Internet. This exercise works best during the peak of hurricane season (from August through November in the Northern Hemisphere and from February through May in the Southern Hemisphere), when you are most likely to encounter well-developed tropical cyclones.

You will choose an active tropical cyclone (preferably of hurricane strength) and follow its progress over a period of three to seven days. During the period of your study, you will log changes in position, as well as changes in the weather associated with the storm. Because of the many variables in this exercise, your instructor may offer further guidance in gathering information and in reporting the progress of your storm.

- Go to the Hess *Physical Geography Laboratory Manual*, 12th edition, website at **www .MasteringGeography.com**, then select Exercise 22. Click on "Go to *Tropical Cyclones website, Naval Research Laboratory*." This will take you to the Naval Research Laboratory, Marine Meteorology Division, in Monterey, California: **http://www.nrlmry.navy.mil/TC.html**. This site lists all currently active tropical cyclones in the world. Generally the most powerful— or most noteworthy—storm will be shown when the page first opens, but you may choose any storm from the lists on the left side of the page. Before choosing a storm for your study, spend a few minutes reviewing the kind of information that is available. Note that in addition to visible light and infrared satellite images, this website provides tracking maps, information about official watches and warnings, satellite movie loops, and satellite image overlays that show wind speed and precipitation rates.

- Another good starting point is the *National Hurricane Center, Tropical Prediction Center*, website: **http://www.nhc.noaa.gov/**.

- To gather additional information about your storm, such as current eye pressure or speed of movement, you may want to go to general weather Internet sites, such as *Unisys Weather*: **http://weather.unisys.com** (also see Exercise 19).

- After choosing a storm, use the charts on the following page to log your information (attach additional sheets if necessary). Depending on the storm you choose, the information available may vary. Your instructor may specify additional information to log. Your instructor may also ask that you download images and/or tracking maps of your storm.

Storm Name: _____

Days of Study: From _____ to _____

Day 1 Date: _____ Time(s) of Observation: _____

Latitude: _____ Longitude: _____ Wind speed: _____ knots

Heading: _____ at _____ knots (speed) Eye Pressure: _____ mb

Additional information:

_____

Day 2 Date: _____ Time(s) of Observation: _____

Latitude: _____ Longitude: _____ Wind speed: _____ knots

Heading: _____ at _____ knots (speed) Eye Pressure: _____ mb

Additional information or significant changes since previous day:

_____

Day 3 Date: _____ Time(s) of Observation: _____

Latitude: _____ Longitude: _____ Wind speed: _____ knots

Heading: _____ at _____ knots (speed) Eye Pressure: _____ mb

Additional information or significant changes since previous day:

_____

Day 4 Date: _____ Time(s) of Observation: _____

Latitude: _____ Longitude: _____ Wind speed: _____ knots

Heading: _____ at _____ knots (speed) Eye Pressure: _____ mb

Additional information or significant changes since previous day:

_____

Day 5 Date: _____ Time(s) of Observation: _____

Latitude: _____ Longitude: _____ Wind speed: _____ knots

Heading: _____ at _____ knots (speed) Eye Pressure: _____ mb

Additional information or significant changes since previous day:

_____

Climate Classification

Objective:	To use average monthly temperature and precipitation data to classify climates with the Köppen climate classification system.
Resource:	Internet access (optional).
Reference:	Hess, Darrel. *McKnight's Physical Geography*, 12th ed., pp. 205–232.

KÖPPEN CLIMATE CLASSIFICATION SYSTEM

The modified **Köppen system** is the most widely used climate classification system. With the Köppen system, all climates of the world can be grouped into just 15 types, based simply on **average monthly temperature** and **average monthly precipitation**. (Color Maps T-28a and T-28b in the back of Lab Manual show global temperature patterns; a map showing global precipitation is on the inside front cover of the Lab Manual.)

In the Köppen system, each climate type is given a descriptive name, as well as a code based on two or three letters. The first letter refers to the major climate group, the second letter generally refers to the precipitation pattern, and the third letter generally refers to the temperature pattern.

There are actually several different versions of the modified Köppen system in use—the definitions of some climate types vary slightly from version to version. Also, while there are specific boundaries for each climate type, in reality the borders between climates should be thought of as transition zones, rather than sharp boundaries. Map T-29, a color map showing the generalized global distribution of climate, is found in the back of the Lab Manual.

CLIMOGRAPHS

One of the key tools used in climate study is the **climograph** or **climatic diagram** (Figure 23-1). In a single chart, the climatic regime of a location can be summarized. The months of the year are indicated along the bottom. The average monthly temperature is shown with a solid line (the temperature scale is along the left side of chart), and the average monthly precipitation is indicated with bars (the precipitation scale is along the right side of chart).

In the sample diagram, notice that in St. Louis the average temperature in January is about −1°C (30°F), whereas in July the average temperature is 27°C (80°F). Precipitation is evenly distributed throughout the year, each month receiving approximately 8 to 13 centimeters (3–5 inches).

USING THE MODIFIED KÖPPEN SYSTEM CHARTS

Each of the climate types in the Köppen system has a specific definition. The "Modified Köppen System Charts" on the following pages of the Lab Manual provide concise definitions for 14 climate types (plus the special category of "Highland" climate). ***In order to use these charts you must follow the procedure for classifying climates listed below.***

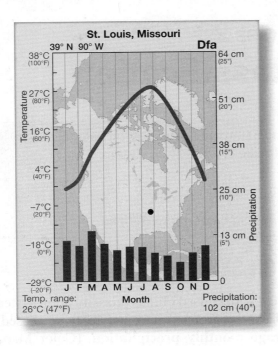

Figure 23-1: Climograph for St. Louis, Missouri. (From Hess, *McKnight's Physical Geography*, 12th ed.)

At the top of each chart, the basic definition for the major climate group is given. Next, the different climate types found within the major group are listed (in some cases a climate type is represented by several different letter combinations). Finally, the specific definitions of the second and third letters for each climate type are provided. Note that the "C" and "D" climates have been grouped together and that not all second and third letters can be combined with both C and D.

PROCEDURE FOR CLASSIFYING CLIMATES

Construct a climograph for the location by plotting the average monthly temperature and average monthly precipitation (this step is actually optional, but it usually makes classification easier). Then, calculate the average annual temperature and average annual precipitation for the location. (This has been done for you in the problems for this exercise.)

Next, determine the major climate group. If you go through the following steps *in sequence*, you will find the correct major climate group. With experience, you will learn shortcuts to narrow your choices more quickly.

1.   If the average temperature of *every* month is below 10°C (50°F), go to the "E" climate chart. (Note: Some "H" climates may also exhibit this temperature pattern.)

2.   If the total annual precipitation is *more* than 89 centimeters (35 inches), continue to step #4.

3.   If the total annual precipitation is *less* than 89 centimeters (35 inches), determine if it is a dry climate by using the "Dry Climate Boundary Charts" under "B—Dry Climates." If it is a dry climate, continue with the "B" climate chart; if not a dry climate, continue to step #4. (A detailed description of dry climates and using these charts is given below.)

4.   If the average temperature of *every* month is above 18°C (64.4°F), go to the "A" climate chart.

5.   If at least one winter month is colder than −3°C (26.6°F), go to the "D" climate chart.

6.   If the coldest winter month is between −3°C (26.6°F) and 18°C (64.4°F), go to the "C" climate chart.

After establishing the major climate group, determine which climate type is correct by checking the definitions of the second, and if necessary, third letters. When assessing seasonal patterns of temperature or precipitation, be sure to consider if the station is in the Northern or Southern Hemisphere. If you are unsure of the hemisphere of the station in question, look at the temperature pattern. If the coolest months are in December, January, and February, it is in the Northern Hemisphere. If the coolest months are in June, July, and August, it is in the Southern Hemisphere.

CLASSIFYING DRY CLIMATES IN THE KÖPPEN SYSTEM

The basic definition of a dry ("B") climate in the Köppen system is one in which the **potential evaporation** exceeds **precipitation**. There is a complex relationship between temperature, precipitation, and the dryness of a region. For example, because of the lower potential for evaporation, a very cold region with an average annual precipitation of 25 centimeters (10 inches) would not be classified as a dry climate, whereas a hot region receiving 25 centimeters (10 inches) of precipitation would be. These relationships are shown on the "Dry Climate Boundary Charts."

In order to use the "Dry Climate Boundary Charts," you need to know the average annual precipitation, the average annual temperature, and if there is a "seasonal concentration" of precipitation. A seasonal concentration means that more than 70% of the precipitation comes in either the six summer months or six winter months. For purposes of classification, April to September are considered the six summer months in the Northern Hemisphere. (These would be the six winter months in the Southern Hemisphere.)

To determine if a seasonal concentration is present, add up the precipitation amounts for the months April to September and divide this by the total annual precipitation. For example, if a Northern Hemisphere location with annual precipitation of 38 centimeters (15 inches) receives 30 centimeters (12 inches) of rain between April and September, make the following calculation: 30 cm/38 cm = 0.8 or 80% (12 inches/15 inches = 0.8 or 80%). Because more than 70% of the precipitation comes between April and September, this location has a "summer concentration" of rainfall.

Modified Köppen System Charts

To use these charts, go through the following steps *in sequence:*

1. If average temperature of every month is below 10°C (50°F), go to "E" climate chart.

2. If total annual precipitation is more than 89 centimeters (35 inches), continue to #4.

3. If total annual precipitation is less than 89 centimeters (35 inches), determine if it is a dry climate by using "Dry Climate Boundary Charts" under "B—Dry Climates." If a dry climate, continue with "B" climate chart; if not a dry climate, continue to #4.

4. If average temperature of every month is above 18°C (64.4°F), go to "A" climate chart.

5. If at least one winter month is colder than −3°C (26.6°F), go to "D" climate chart.

6. If coldest winter month is between −3°C (26.6°F) and 18°C (64.4°F), go to "C" climate chart.

A—TROPICAL HUMID: Temperature of every month above 18°C (64.4°F).

Group A Climate Types
Af — Tropical Wet
Am — Tropical Monsoon
Aw — Tropical Savanna

Second Letters	Definition
f — Wet All Year	Every month has at least 6 cm (2.4″) of rainfall.
m — Monsoon Pattern	Short dry season; pronounced rainy season.*
w — Winter Dry	Winter dry season of 3 to 6 months.*

*To calculate boundaries between "Am" and "Aw," determine the average rainfall and the average rainfall of the driest month; then use the chart below. For example, a location with an average annual rainfall of 200 cm (about 80″) and 5 cm (2″) of rainfall in the driest month is an "Am" climate.

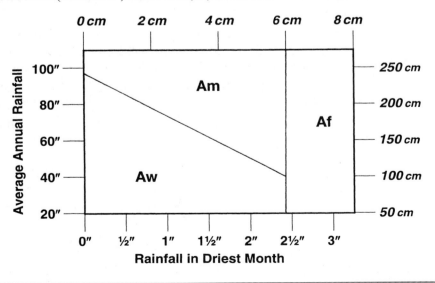

B—DRY CLIMATES: Evaporation exceeds precipitation.

Group B Climate Types

BWh — Subtropical Desert   (Avg. annual temperature above 18°C [64.4°F]).

BSh — Subtropical Steppe   "   "   "   "   "

BWk — Midlatitude Desert   (Avg. annual temperature below 18°C [64.4°F]).

BSk — Midlatitude Steppe   "   "   "   "   "

DRY CLIMATE BOUNDARY CHARTS

1. Determine the average annual temperature and average annual precipitation.

2. Determine if the precipitation is distributed evenly during the year:

    (a) If more than 70% of the precipitation comes in the six summer months (April to September in Northern Hemisphere), use "Summer Concentration" chart.

    (b) If more than 70% of the precipitation comes in the six winter months (October to March in Northern Hemisphere), use "Winter Concentration" chart.

    (c) If precipitation is evenly distributed throughout year (neither a or b), use "Even Distribution" chart.

3. Line up the average annual temperature with the average annual precipitation to find the climate.

Example: If a location has an annual precipitation of 38 cm (15 inches) with a winter concentration, use the "Winter Concentration" chart. If the average annual temperature is 21°C (70°F), the climate is BSh; if the average annual temperature is 10°C (50°F), it is not a dry climate.

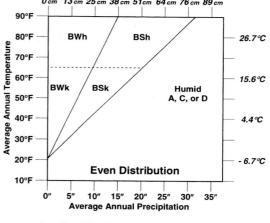

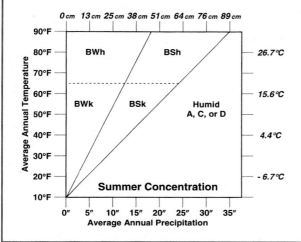

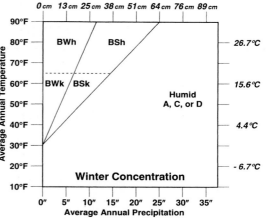

C—MILD MIDLATITUDE:

Temperature of warmest month above 10°C (50°F); coldest month between −3°C (26.6°F) and 18°C (64.4°F).

D—SEVERE MIDLATITUDE:

Warmest month above 10°C (50°F); coldest month below −3°C (26.6°F).

Group C & D Climate Types

Cs — Mediterranean (includes Csa and Csb)
Cfa — Humid Subtropical (also includes Cwa)
Cfb — Marine West Coast (also includes Cfc)
Dfa — Humid Continental (also includes Dwa, Dfb, and Dwb)
Dfc — Subarctic (also includes Dwc, Dfd, and Dwd)

Second Letters	Definition
s—Summer Dry	Wettest winter month has at least 3× precipitation of driest summer month.
w—Winter Dry	Wettest summer month has at least 10× precipitation of driest winter month.
f—Wet All Year	Neither "s" nor "w" above.

Third Letters	Definition
a—Hot Summer	Warmest month above 22°C (71.6°F).
b—Warm Summer	Warmest month below 22°C (71.6°F); at least 4 months above 10°C (50°F).
c—Cool Summer	Warmest month below 22°C (71.6°F); 1 to 3 months above 10°C (50°F); coldest month above −38°C (−36.4°F).
d—Severe Winter	Coldest month below −38°C (−36.4°F).

E—POLAR CLIMATES: Temperature of every month below 10°C (50°F).

Group E Climate Types
ET—Tundra
EF—Ice Cap

Second Letters	Definition
T—Tundra	At least one month above 0°C (32°F).
F—Ice Cap	All months below 0°C (32°F).

H—HIGHLAND CLIMATES: Significant variation or modification of a climate type due to high elevation.

Highland climates are not defined in the same way as other climates in the Köppen system. Rather, these are regions in high mountain areas where the climate has been significantly modified from the adjacent lowlands by high elevation.

FINAL SUGGESTIONS ON CLIMATE CLASSIFICATION

You may also want to compare the climograph of the station in question with those in your textbook. This is a quick way to determine if your classification is reasonable. If you know the location of the station in question, look at the generalized map of climate, color Map T-29 in the back of the Lab Manual. This map may help narrow down the climate type to several possibilities. However, because there are local variations in climate, this map alone is *not* enough to accurately determine all climates. You will need to use the charts defining each climate type to verify your answer.

KÖPPEN CLASSIFICATION AND CLIMATE CONTROLS

Köppen climate classification is based solely on temperature and precipitation patterns. Although the Köppen system does not consider the origin of a climate, the location and dominant controls of each climate type are quite predictable.

This regularity is illustrated with the hypothetical continent shown in Figure 23-2. This idealized distribution pattern predicts quite closely the actual arrangement of climate types on the continents and reflects the dominant controls that produce each of these climates.

The dominant controls of climate include:

1. **Latitude:** Temperature generally decreases moving away from the equator and toward the poles; wind and pressure patterns as well as precipitation also vary with latitude (see color Maps T-28a and T-28b in the back of the Lab Manual).

2. **Land-water contrasts:** Inland areas generally exhibit larger seasonal and daily temperature variations than maritime areas (Maps T-28a and T-28b).

3. **Ocean currents:** In the midlatitudes, cool ocean currents flow equatorward along the west coasts of continents, whereas warm currents flow poleward along the east coasts (see Figure 12-2 in Exercise 12).

4. **General circulation of the atmosphere:** Rising air near the equator brings rain to the ITCZ; easterly trade winds blow in the tropics; descending air brings dry conditions to the subtropical highs; westerly winds blow in the midlatitudes (see Figure 14-3 in Exercise 14). The seasonal shifts of the ITCZ and the subtropical highs are especially important influences on the seasonality of rainfall and climate types along the west coasts of continents (Figure 23-3).

5. **Altitude:** Temperature generally decreases with increased elevation.

6. **Mountain barriers:** The windward slopes of mountains are generally wetter than the leeward slopes.

7. **Storms:** Midlatitude cyclones bring precipitation as they move west to east in the westerlies; hurricanes bring summer precipitation to many tropical regions and to many midlatitude east coasts.

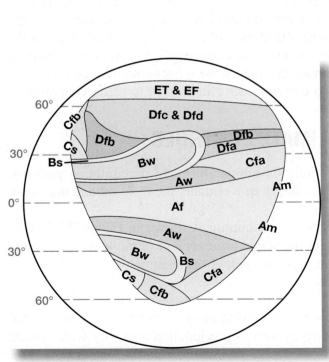

Figure 23-2: The presumed arrangement of Köppen climatic types on a hypothetical continent. (Adapted from McKnight, *Physical Geography*, 4th ed.)

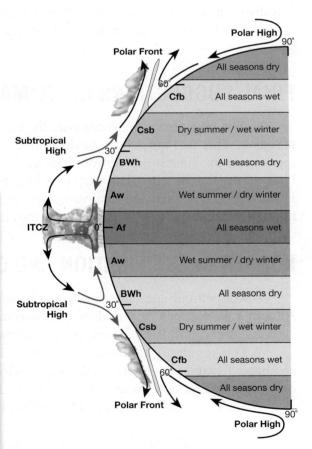

Figure 23-3: Idealized seasonal precipitation patterns and climates along the west coasts of continents. Much of this pattern is due to the seasonal shifts of the ITCZ and the subtropical highs. (From Hess, *McKnight's Physical Geography*, 12th ed.)

Name _____   Section _____

EXERCISE 23 PROBLEMS—PART I

For each of the following six locations, complete the climograph using the average monthly temperature ("Temp") given in degrees Celsius and Fahrenheit, and the average monthly precipitation ("Precp") given in centimeters and inches. The average annual temperature and precipitation are provided for you. After completing the climographs, answer the questions at the end of Part I. You may plot data on the climographs using either S.I. or English (note that the English unit and S.I. unit scales on the climographs are not exactly equivalent). It may be helpful to locate each of these stations on a map (the latitude and longitude of each location is provided). No "H" climates are given.

1. Cuiabá, Brazil (18°S, 57°W)   Average Annual: 26°C (78°F); 138.8 cm (54.6″)

	JAN	FEB	MAR	APR	MAY	JUN	JUL	AUG	SEP	OCT	NOV	DEC
Temp	81°F 27°C	80°F 27°C	81°F 27°C	80°F 27°C	75°F 24°C	72°F 22°C	73°F 23°C	75°F 24°C	79°F 26°C	82°F 28°C	81°F 27°C	81°F 27°C
Precp	9.6″ 24.4 cm	8.9″ 22.6 cm	8.1″ 20.6 cm	4.1″ 10.4 cm	2.0″ 5.1 cm	0.3″ 0.8 cm	0.2″ 0.5 cm	1.1″ 2.8 cm	2.0″ 5.1 cm	4.4″ 11.2 cm	6.0″ 15.2 cm	7.9″ 20.1 cm

2. Kashi (Kashgar), China (39°N, 76°E)   Average Annual: 12°C (54°F); 8.7 cm (3.4″)

	JAN	FEB	MAR	APR	MAY	JUN	JUL	AUG	SEP	OCT	NOV	DEC
Temp	22°F −6°C	34°F 1°C	47°F 8°C	61°F 16°C	70°F 21°C	77°F 25°C	80°F 27°C	76°F 24°C	59°F 15°C	56°F 13°C	40°F 4°C	26°F −3°C
Precp	0.3″ 0.8 cm	0.0″ 0.0 cm	0.2″ 0.5 cm	0.2″ 0.5 cm	0.8″ 2.0 cm	0.4″ 1.0 cm	0.3″ 0.8 cm	0.7″ 1.8 cm	0.3″ 0.8 cm	0.0″ 0.0 cm	0.0″ 0.0 cm	0.2″ 0.5 cm

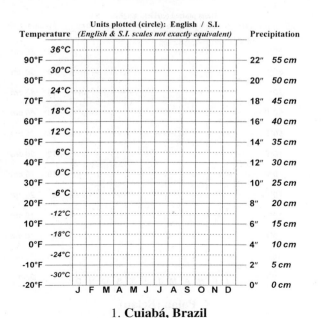

1. Cuiabá, Brazil

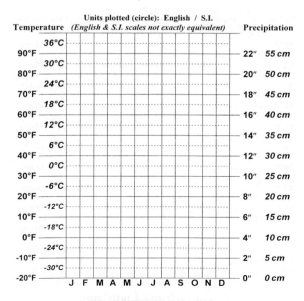

2. Kashi (Kashgar), China

153

3. New Orleans, Louisiana (30°N, 90°W) — Average Annual: 21°C (70°F); 161.8 cm (63.6″)

	JAN	FEB	MAR	APR	MAY	JUN	JUL	AUG	SEP	OCT	NOV	DEC
Temp	56°F 13°C	58°F 14°C	63°F 17°C	70°F 21°C	76°F 24°C	82°F 28°C	83°F 28°C	83°F 28°C	80°F 27°C	73°F 23°C	62°F 17°C	57°F 14°C
Precp	4.8″ 12.2 cm	4.2″ 10.7 cm	6.6″ 16.8 cm	5.4″ 13.7 cm	5.4″ 13.7 cm	5.6″ 14.2 cm	7.1″ 18.0 cm	6.4″ 16.3 cm	5.8″ 14.7 cm	3.7″ 9.4 cm	4.0″ 10.2 cm	4.6″ 11.9 cm

4. Palau (Belau), Caroline Islands (5°N, 137°E) — Average Annual: 27°C (81°F); 396.2 cm (155.9″)

	JAN	FEB	MAR	APR	MAY	JUN	JUL	AUG	SEP	OCT	NOV	DEC
Temp	81°F 27°C	80°F 27°C	81°F 27°C	82°F 28°C	82°F 28°C	82°F 28°C	81°F 27°C	81°F 27°C	81°F 27°C	81°F 27°C	81°F 27°C	81°F 27°C
Precp	15.3″ 38.9 cm	9.4″ 23.9 cm	6.8″ 17.3 cm	7.6″ 19.3 cm	15.5″ 39.4 cm	12.4″ 31.5 cm	19.9″ 50.5 cm	14.0″ 35.6 cm	15.7″ 39.9 cm	14.8″ 37.6 cm	11.8″ 30.0 cm	12.7″ 32.3 cm

5. Irkutsk, Siberia (52°N, 104°E) — Average Annual: 0°C (31°F); 37.0 cm (14.6″)

	JAN	FEB	MAR	APR	MAY	JUN	JUL	AUG	SEP	OCT	NOV	DEC
Temp	−5°F −21°C	1°F −17°C	17°F −8°C	37°F 3°C	48°F 9°C	59°F 15°C	65°F 18°C	60°F 16°C	48°F 9°C	33°F 1°C	13°F −11°C	1°F −17°C
Precp	0.6″ 1.5 cm	0.5″ 1.3 cm	0.4″ 1.0 cm	0.6″ 1.5 cm	1.2″ 3.0 cm	2.3″ 5.8 cm	2.9″ 7.4 cm	2.4″ 6.1 cm	1.6″ 4.1 cm	0.7″ 1.8 cm	0.6″ 1.5 cm	0.8″ 2.0 cm

6. Dublin, Ireland (53°N, 6°W) — Average Annual: 9°C (48°F); 70.4 cm (27.7″)

	JAN	FEB	MAR	APR	MAY	JUN	JUL	AUG	SEP	OCT	NOV	DEC
Temp	40°F 4°C	41°F 5°C	42°F 6°C	45°F 7°C	49°F 9°C	55°F 13°C	58°F 14°C	57°F 14°C	54°F 12°C	48°F 9°C	44°F 7°C	41°F 5°C
Precp	2.2″ 5.6 cm	1.9″ 4.8 cm	1.9″ 4.8 cm	1.9″ 4.8 cm	2.1″ 5.3 cm	2.0″ 5.1 cm	2.6″ 6.6 cm	3.1″ 7.9 cm	2.0″ 5.1 cm	2.6″ 6.6 cm	2.9″ 7.4 cm	2.5″ 6.4 cm

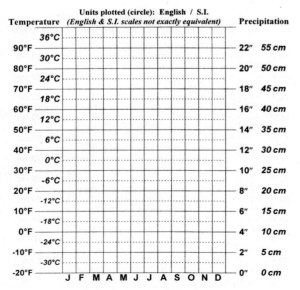

3. **New Orleans, Louisiana**

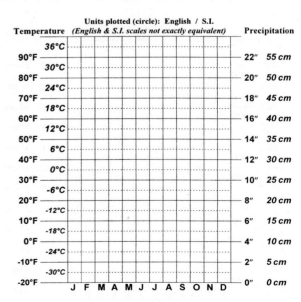

4. **Palau (Belau)**

Name _____   (Ex. 23—Part I)

Units plotted (circle): English / S.I.
Temperature *(English & S.I. scales not exactly equivalent)* Precipitation

Temperature	°C		Precipitation	
90°F	36°C		22″	55 cm
80°F	30°C		20″	50 cm
70°F	24°C		18″	45 cm
60°F	18°C		16″	40 cm
50°F	12°C		14″	35 cm
40°F	6°C		12″	30 cm
30°F	0°C		10″	25 cm
20°F	-6°C		8″	20 cm
10°F	-12°C		6″	15 cm
0°F	-18°C		4″	10 cm
-10°F	-24°C		2″	5 cm
-20°F	-30°C		0″	0 cm

J F M A M J J A S O N D

5. **Irkutsk, Siberia**

Units plotted (circle): English / S.I.
Temperature *(English & S.I. scales not exactly equivalent)* Precipitation

Temperature	°C		Precipitation	
90°F	36°C		22″	55 cm
80°F	30°C		20″	50 cm
70°F	24°C		18″	45 cm
60°F	18°C		16″	40 cm
50°F	12°C		14″	35 cm
40°F	6°C		12″	30 cm
30°F	0°C		10″	25 cm
20°F	-6°C		8″	20 cm
10°F	-12°C		6″	15 cm
0°F	-18°C		4″	10 cm
-10°F	-24°C		2″	5 cm
-20°F	-30°C		0″	0 cm

J F M A M J J A S O N D

6. **Dublin, Ireland**

After completing the climographs, answer the following questions about each location.

1. **Cuiabá, Brazil:**

   (a) Köppen climate type: Letter code: _____

        Descriptive name: _____

   (b) Dominant climate controls for this location:

2. **Kashi (Kashgar), China:**

   (a) Köppen climate type: Letter code: _____

        Descriptive name: _____

   (b) Dominant climate controls for this location:

3. **New Orleans, Louisiana:**

   (a)   Köppen climate type: Letter code: _____

          Descriptive name: _____

   (b)   Dominant climate controls for this location:

4. **Palau (Belau)**

   (a)   Köppen climate type: Letter code: _____

          Descriptive name: _____

   (b)   Dominant climate controls for this location:

5. **Irkutsk, Siberia:**

   (a)   Köppen climate type: Letter code: _____

          Descriptive name: _____

   (b)   Dominant climate controls for this location:

6. **Dublin, Ireland:**

   (a)   Köppen climate type: Letter code: _____

          Descriptive name: _____

   (b)   Dominant climate controls for this location:

Name _____   Section _____

EXERCISE 23 PROBLEMS—PART II

For each of the following six locations, complete the climograph using the average monthly temperature ("Temp") given in degrees Celsius and Fahrenheit, and the average monthly precipitation ("Precp") given in centimeters and inches. The average annual temperature and precipitation are provided for you. After completing the climographs, answer the questions at the end of Part II. You may plot data on the climographs using either S.I. or English units (note that the English unit and S.I. unit scales on the climographs are not exactly equivalent). No "H" climates are given.

1.   Average Annual: 10°C (50°F); 31.4 cm (12.3″)

	JAN	FEB	MAR	APR	MAY	JUN	JUL	AUG	SEP	OCT	NOV	DEC
Temp	21°F −6°C	30°F −1°C	42°F 6°C	54°F 12°C	63°F 17°C	70°F 21°C	73°F 23°C	71°F 22°C	61°F 16°C	51°F 11°C	35°F 2°C	23°F −5°C
Precp	0.1″ 0.3 cm	0.1″ 0.3 cm	0.3″ 0.8 cm	0.5″ 1.3 cm	0.7″ 1.8 cm	1.5″ 3.8 cm	2.6″ 6.6 cm	3.6″ 9.1 cm	2.2″ 5.6 cm	0.6″ 1.5 cm	0.1″ 0.3 cm	0.0″ 0.0 cm

2.   Average Annual: −12°C (10°F); 13.4 cm (5.2″)

	JAN	FEB	MAR	APR	MAY	JUN	JUL	AUG	SEP	OCT	NOV	DEC
Temp	−20°F −29°C	−13°F −25°C	−13°F −25°C	−2°F −19°C	22°F −6°C	35°F 2°C	41°F 5°C	39°F 4°C	32°F 0°C	16°F −9°C	0°F −18°C	−15°F −26°C
Precp	0.1″ 0.3 cm	0.4″ 1.0 cm	0.2″ 0.5 cm	0.3″ 0.8 cm	0.3″ 0.8 cm	0.8″ 2.0 cm	0.3″ 0.8 cm	0.9″ 2.3 cm	0.5″ 1.3 cm	0.7″ 1.8 cm	0.3″ 0.8 cm	0.4″ 1.0 cm

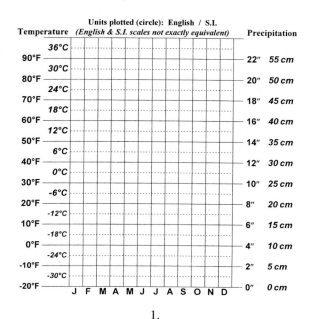

1.

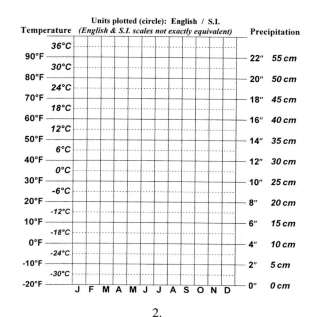

2.

3. Average Annual: 16°C (61°F); 52.6 cm (20.7")

	JAN	FEB	MAR	APR	MAY	JUN	JUL	AUG	SEP	OCT	NOV	DEC
Temp	69°F 21°C	68°F 20°C	66°F 19°C	61°F 16°C	57°F 14°C	55°F 13°C	53°F 12°C	54°F 12°C	57°F 14°C	59°F 15°C	64°F 18°C	67°F 19°C
Precp	0.4" 1.0 cm	0.6" 1.5 cm	0.5" 1.3 cm	2.1" 5.3 cm	3.5" 8.9 cm	3.3" 8.4 cm	3.3" 8.4 cm	2.9" 7.4 cm	1.8" 4.6 cm	1.2" 3.0 cm	0.7" 1.8 cm	0.4" 1.0 cm

4. Average Annual: 27°C (81°F); 291.4 cm (114.7")

	JAN	FEB	MAR	APR	MAY	JUN	JUL	AUG	SEP	OCT	NOV	DEC
Temp	81°F 27°C	82°F 28°C	84°F 29°C	85°F 29°C	84°F 29°C	80°F 27°C	79°F 26°C	79°F 26°C	80°F 27°C	80°F 27°C	81°F 27°C	81°F 27°C
Precp	0.8" 2.0 cm	0.8" 2.0 cm	1.7" 4.3 cm	3.7" 9.4 cm	11.4" 29.0 cm	27.8" 70.6 cm	25.3" 64.3 cm	12.5" 31.8 cm	9.22" 23.4 cm	12.9" 32.8 cm	6.7" 17.0 cm	1.9" 4.8 cm

5. Average Annual: 10°C (50°F); 83.6 cm (32.9")

	JAN	FEB	MAR	APR	MAY	JUN	JUL	AUG	SEP	OCT	NOV	DEC
Temp	25°F −4°C	27°F −3°C	36°F 2°C	48°F 9°C	58°F 14°C	68°F 20°C	74°F 23°C	72°F 22°C	66°F 19°C	54°F 12°C	40°F 4°C	30°F −1°C
Precp	1.9" 4.8 cm	1.9" 4.8 cm	2.7" 6.9 cm	2.9" 7.4 cm	3.5" 8.9 cm	3.7" 9.4 cm	3.3" 8.4 cm	3.1" 7.9 cm	3.0" 7.6 cm	2.6" 6.6 cm	2.3" 5.8 cm	2.0" 5.1 cm

6. Average Annual: 21°C (70°F); 25.2 cm (9.9")

	JAN	FEB	MAR	APR	MAY	JUN	JUL	AUG	SEP	OCT	NOV	DEC
Temp	84°F 29°C	82°F 28°C	77°F 25°C	68°F 20°C	60°F 16°C	54°F 12°C	53°F 12°C	58°F 14°C	65°F 18°C	73°F 23°C	79°F 26°C	82°F 28°C
Precp	1.7" 4.3 cm	1.3" 3.3 cm	1.1" 2.8 cm	0.4" 1.0 cm	0.6" 1.5 cm	0.5" 1.3 cm	0.3" 0.8 cm	0.3" 0.8 cm	0.3" 0.8 cm	0.7" 1.8 cm	1.2" 3.0 cm	1.5" 3.8 cm

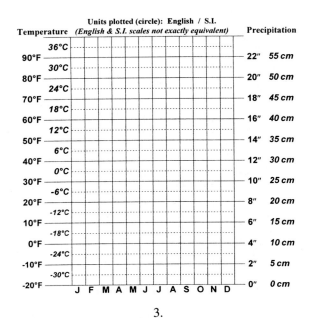

3.

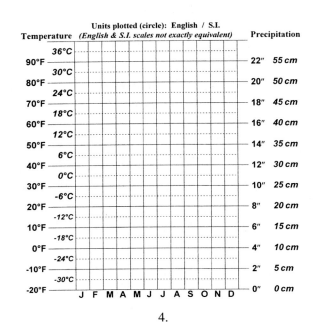

4.

Name _____   (Ex. 23—Part II)

Temperature Units plotted (circle): English / S.I. **Precipitation**
(English & S.I. scales not exactly equivalent)

5.

Temperature Units plotted (circle): English / S.I. **Precipitation**
(English & S.I. scales not exactly equivalent)

6.

After completing the climographs, assign a Köppen letter code and descriptive climate name to each location.

1.   Köppen climate type: Letter code: _____

     Descriptive name: _____

2.   Köppen climate type: Letter code: _____

     Descriptive name: _____

3.   Köppen climate type: Letter code: _____

     Descriptive name: _____

4.   Köppen climate type: Letter code: _____

     Descriptive name: _____

5.   Köppen climate type: Letter code: _____

     Descriptive name: _____

6.   Köppen climate type: Letter code: _____

     Descriptive name: _____

Name _____   Section _____

EXERCISE 23 PROBLEMS—PART III

Before answering the following questions, assign Köppen climate types to each of the six locations in Part II. Match each of the locations in Part II (1 through 6) with its most likely city from the list below (the latitude and longitude of each is provided for you):

Alice Springs, Australia (24°S, 134°E)   Cochin, India (10°N, 76°E)
Cape Town, South Africa (34°S, 18°E)   Lanzhou, China (36°N, 104°E)
Chicago, Illinois (42°N, 88°W)   Barrow, Alaska (71°N, 157°W)

1.   (a)   Most likely city: _____

     (b)   Why is this the most likely city?

2.   (a)   Most likely city: _____

     (b)   Why is this the most likely city?

3.   (a)   Most likely city: _____

     (b)   Why is this the most likely city?

4.   (a)   Most likely city: _____

     (b)   Why is this the most likely city?

5.   (a)   Most likely city: _____

     (b)   Why is this the most likely city?

6.   (a)   Most likely city: _____

     (b)   Why is this the most likely city?

Name _____   Section _____

EXERCISE 23 PROBLEMS—PART IV

The following questions are based on the diagram of the Köppen climate distribution on a hypothetical continent (Figure 23-2). It may also be helpful to compare the hypothetical continent
with the map of actual climate distribution color Map T-29 in the back of the Lab Manual and
Figure 23-3 showing the influence of the seasonal shifts of the ITCZ and subtropical highs along
the west coasts of continents. In answering the questions, consider both the characteristics of a
climate and the dominant controls producing that climate.

1.   Why are Aw (tropical savanna) climates found in bands north and south of the Af (tropical
      wet) climates?

2.   Why do the Af climates extend farther toward the poles along the east coast than along the
      west coast?

3.   What explains the distribution of BW (desert) climates centered at about 25° to 30° North
      and South along the west coast?

4.   On the hypothetical continent, why does the BW climate extend farther inland in the
      Northern Hemisphere than in the Southern Hemisphere?

5.   What explains the distribution of BS (steppe) climates?

6. What explains the narrow coastal band of Cs (mediterranean) climates at about 35° North and South along the west coast?

7. Why do the Cfb (marine west coast) climates, just poleward of the dry summer Cs climates, receive rain all year?

8. Why do Cfa (humid subtropical) climates along the east coast receive rain all year, but at the same latitude along the west coast, the Cs climates have dry summers?

9. Why is the Dfb (humid continental) climate in a band just north of the band of Dfa climate?

10. Why is the high latitude interior of the continent dominated by Dfc and Dfd (subarctic) climates?

11. Why are no D or E climates shown in the Southern Hemisphere?

Name _____   Section _____

EXERCISE 23 PROBLEMS—PART V—INTERNET

In this exercise, you will use the Internet to find the climate record of the city where you live, and then classify its climate with the Köppen system.

- Go to the Hess *Physical Geography Laboratory Manual*, 12th edition, website at **www.Mastering Geography.com**, then select Exercise 23. Select "Go to *Regional Climate Centers*" under NOAA: **http://www.ncdc.noaa.gov/oa/climate/regionalclimatecenters.html**. (Your instructor may recommend a different Internet site that provides climate data.)

- From the map of the United States, select the Regional Climate Center for your state, and then look for the historical climate summary for your city.

- To properly classify the climate of your city, you will need the average monthly temperature and average monthly precipitation. (If "Average" or "Mean" Monthly Temperature is not given, it may be calculated by taking the average of a month's "Mean Maximum Temperature" and its "Mean Minimum Temperature.") The averages should be based on weather data over at least a 25-year period. If you choose not to print out the data, write down the information in the chart on the following page. Also note the years on which the averages are based.

- Indicate if your data are in S.I. or English units.

- Note (or calculate) the average annual temperature and precipitation.

- Using the graph on the following page, complete a climograph for your city.

- Use the "Modified Köppen System Charts" to classify the climate of your city.

As an alternative assignment, your instructor may ask that you classify the climate of a city other than where you live.

City: _____

Years of climate record: From _____ to _____

Units used: S.I. or English (circle)

	JAN	FEB	MAR	APR	MAY	JUN	JUL	AUG	SEP	OCT	NOV	DEC
Temp												
Precp												

Average Annual Temperature: _____

Average Annual Precipitation: _____

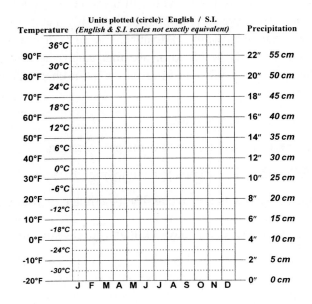

Köppen climate type: Letter code: _____

Descriptive name: _____

Weather Variability
and Climate Change

Objective:	To study some of the causes and measures of weather variability and climate change.
Reference:	Hess, Darrel. *McKnight's Physical Geography*, 12th ed., pp. 232–247.

WEATHER VARIABILITY AND CLIMATE CHANGE

Weather and climate are related but different concepts. **Weather** refers to the conditions in the atmosphere over a relatively short period of time, whereas **climate** refers to the composite of weather conditions over an extended period of time—30 years being the minimum period typically used to define the "normal" (average) temperature and precipitation conditions for a location. However, the climate of a location does not simply refer to average weather conditions—weather extremes and year-to-year fluctuations are also part of climate, and in any given year the precipitation and temperatures experienced may deviate significantly from long-term averages.

Adding to the complexity are cyclical phenomena in the atmosphere and ocean, such as *El Niño* and the *Pacific Decadal Oscillation*, that influence year-to-year or even decade-to-decade weather. Because of such variation, it is often difficult to distinguish the "signal" of long-term climate change from the "noise" of weather variability or cyclical phenomena.

In this exercise, we look at variations in the atmosphere over two very different time scales: changes that have occurred over the past 800,000 years and those that have occurred over the last century. We begin with a review of some factors that influence long-term patterns of weather and climate.

FACTORS INFLUENCING LONG-TERM ATMOSPHERIC PATTERNS

Many factors can lead to temperature and precipitation variation in the long-term atmospheric record. Some of these factors include:

Variations in Solar Output: The energy output of the Sun—sometimes referred to as *total solar irradiance*—varies slightly (by perhaps 0.1% to 0.2%) during its well-documented 11-year sunspot cycle, and some evidence points to temperature fluctuations of as much as 0.2°C (0.4°F) in some parts of the world as a result.

Variations in Earth–Sun Relations: Over periods of many thousands of years, the relationship between Earth and the Sun varies slightly through a series of predictable cycles known as *Milankovitch cycles*—named for the early twentieth-century Yugoslavian astronomer, Milutin Milankovitch, who first established the cycles' significance to climate change. The shape of Earth's orbit, or *eccentricity*, varies in a series of cycles lasting about 100,000 years; the inclination of Earth's axis, or *obliquity*, varies in a cycle lasting about 41,000 years; and the orientation of Earth's axis relative to the stars varies in a 25,800-year cycle called *precession*. As these cycles "overlap" there are periods of thousands of years when significantly less insolation reaches Earth's surface, and these variations are one key to the general timing of the glacial and interglacial periods that occurred over the last few million years.

Atmospheric Aerosols: A large increase in atmospheric **aerosols** from major volcanic eruptions, such as El Chichón in 1982 and Mount Pinatubo in 1991, can lead to temperature changes. For example, following the eruption of Mount Pinatubo in the Philippines (the largest volcanic eruption in the world since 1912), average global temperature dropped by about 0.5°C (0.9°F) for the following year.

Greenhouse Gas Concentration: **Greenhouse gases**, especially water vapor and carbon dioxide (CO_2), play important roles in regulating the temperature of the atmosphere (Figure 24-1). Generally, higher greenhouse gas concentrations are associated with higher global temperatures. For example, if greenhouse gas concentrations increase, a greater amount of the **longwave radiation** emitted by Earth's surface will be absorbed by the atmosphere, and so temperatures can increase.

The concentration of CO_2 in the atmosphere varies naturally. There are seasonal fluctuations from summer to winter: Plants absorb CO_2 through *photosynthesis* and release CO_2 through *plant respiration*; during the summer growing season, photosynthesis (especially by the large forests of the Northern Hemisphere) dominates and so CO_2 levels drop until winter, when plant growth diminishes and respiration dominates (see Figures 24-8 and 24-9 on page 172).

Over periods of many thousands of years, the overall amount of CO_2 in the atmosphere has also fluctuated (see Figure 24-5 on page 169). For example, when climate is warm, feedback loops (such as increased activity of microbes in the soil) tend to increase the amount of CO_2 in the atmosphere.

The sharp increase in atmospheric carbon dioxide since the Industrial Revolution (see Figure 24-1), however, has a chemical "fingerprint" that ties it to human activities, especially the

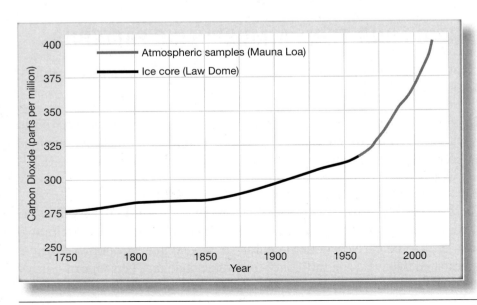

Figure 24-1: Change in atmospheric carbon dioxide concentration from 1750–2015. The dark line shows values derived from ice cores at Law Dome, Antarctica; the light line shows measured values from Mauna Loa, Hawai'i. (From Hess, *McKnight's Physical Geography*, 12th ed.)

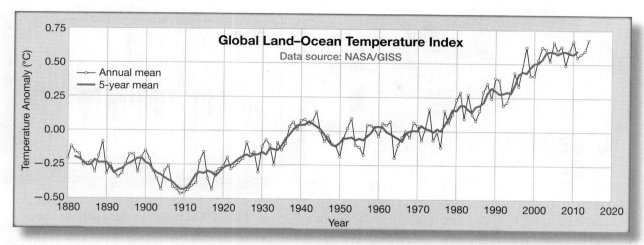

Figure 24-2: Average global temperature over land and ocean, 1880–2014, showing temperature difference relative to the 1951–1980 average. (From Hess, *McKnight's Physical Geography*, 12th ed.)

burning of fossil fuels. The vast majority of atmospheric scientists think that most, if not all, of the slight increase in average global temperature observed over the last century is a result of human-enhanced greenhouse effect associated with higher amounts of atmospheric CO_2 (Figure 24-2).

The effect of higher temperatures (such as that brought by human-enhanced greenhouse effect) on precipitation patterns is complex. For example, higher air temperatures can increase sea surface temperatures (thus promoting evaporation) as well as increase the water vapor capacity of air—and so increasing the potential for clouds and precipitation in some areas. On the other hand, changing global temperatures may alter general circulation patterns of the atmosphere that could lead to reduced precipitation in other areas.

Multiyear Atmospheric and Oceanic Cycles: A number of multiyear atmospheric and oceanic cycles are known to influence weather. For example, a strong **El Niño** event—marked by a weakening or reversal of the trade winds and warmer-than-usual water in the equatorial eastern Pacific Ocean—is typically associated with higher-than-average precipitation in some parts of the world (such as the southwestern United States), but is associated with lower-than-average precipitation in other parts of the world (Figure 24-3).

The **Pacific Decadal Oscillation** (PDO) leaves its mark in the temperature record of high-latitude Northern Hemisphere locations (Figure 24-4): During the "positive" (or "warm") phase of the PDO, warmer ocean water prevails in the eastern tropical Pacific Ocean and a stronger Aleutian Low pressure cell causes more frequent warm, southerly wind flows into interior Alaska and so increasing winter temperatures there; during the "negative" (or "cool") phase of the PDO, cooler ocean water prevails in the eastern tropical Pacific, and a weaker Aleutian Low reduces warm air flow into Alaska.

Urban Heat Islands: Urban areas tend to be warmer than rural areas. This often-observed increase in temperature associated with cities is known as the **urban heat island (UHI)** effect. The UHI effect is thought to be caused mostly from reduced nighttime cooling because buildings inhibit the loss of longwave radiation to space and reduce the mixing of warmer surface air with cooler air above. The United States Environmental Protection Agency (EPA) estimates that

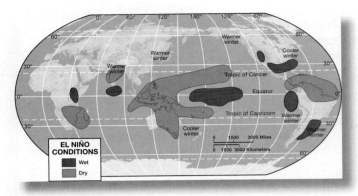

Figure 24-3: Generalized precipitation and temperature patterns during a strong El Niño event. (From Hess, *McKnight's Physical Geography*, 12th ed.)

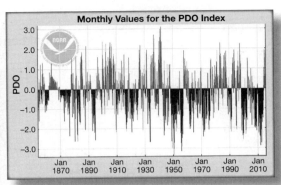

Figure 24-4: Pacific Decadal Oscillation index from 1860–2015. Positive values indicate warm water in the eastern tropical Pacific and the Gulf of Alaska. (From Hess, *McKnight's Physical Geography*, 12th ed.)

temperatures within a city of 1 million people are typically 1 to 3°C (1.8 to 5.4°F) warmer than nearby rural areas, with nighttime temperature differences much greater than this. However, some recent research suggests that the UHI effect may be smaller than previously thought—perhaps as little as a 0.9°C (1.6°F) difference between cities and the surrounding rural areas.

In addition to temperature, precipitation in urban areas (and downwind of urban areas) may be affected by the higher concentrations of aerosols released by industries or other human activity.

DETERMINING CLIMATES OF THE PAST

Detailed instrument records of weather go back only a few hundred years, so *proxy* ("substitute") measures of climate are used to reconstruct climates of the distant past—a field of study known as *paleoclimatology*. Many proxy measures can be used, such as tree rings (*dendrochronology*), ice cores, oceanic sediments, coral reefs, relic soils, and pollen, among many others. By correlating the results obtained from one method with those from another, as well as comparing proxy measures to the recent instrument record, a detailed history of Earth's climate can be constructed.

Oxygen Isotope Analysis: One of the most useful proxy measures of Earth's past climates comes from **oxygen isotope analysis**. Most oxygen atoms have 8 protons and 8 neutrons, giving them an atomic weight of 16 (^{16}O or "oxygen 16"). However, a small number of oxygen atoms have 2 extra neutrons, giving these atoms an atomic weight of 18 (^{18}O). ^{16}O and ^{18}O are known as *isotopes* of oxygen. The ratio of $^{18}O/^{16}O$ in common molecules such as water (H_2O) and calcium carbonate ($CaCO_3$) can tell us something about the environment in which those molecules formed.

Because they contain the "lighter" oxygen isotope, water molecules with ^{16}O evaporate more easily than those with ^{18}O. Thus, precipitation such as rain and snow tends to be relatively rich in ^{16}O. During an ice age, great quantities of ^{16}O are locked up in glacial ice on the continents, leaving a greater concentration of ^{18}O in the oceans; during a warmer, interglacial period, the glacial ice melts, returning ^{16}O to the oceans. Such variations in the concentration of ^{18}O in the oceans are reflected in the $^{18}O/^{16}O$ ratio of calcium carbonate found in sea shells and coral forming at the time.

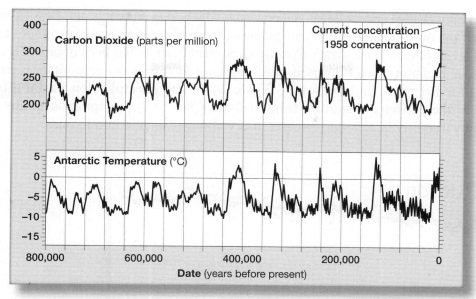

Figure 24-5: Carbon dioxide and temperature record for Antarctica over the last 800,000 years. CO_2 record derived from the EPICA Dome C ice core and other ice cores in Antarctica. Temperature record derived from Dome C data; temperature scale is relative to average temperature over the last 1000 years. (Adapted from Hess, *McKnight's Physical Geography*, 12th ed.)

Ice Cores: By drilling down into glaciers, a record of snowfall going back hundreds of thousands of years has been acquired in Antarctica. The ice core from "Dome C" (where the Antarctic ice cap is thickest) obtained by the multinational *European Project for Ice Coring in Antarctica* (EPICA) extracted a continuous 10-centimeter (4-inch) diameter ice core more than 3 kilometers (2 miles) long, providing climate and atmospheric composition data going back 800,000 years (Figure 24-5).

Because a greater number of water molecules with "heavier" ^{18}O evaporate from the oceans when temperatures are high than when temperatures are low, the $^{18}O/^{16}O$ ratio in a layer of ice serves as a "thermometer" for the climate at the time that snow fell. In addition, tiny air bubbles trapped deep in glacier ice are preserved samples of the ancient atmosphere, allowing direct measurements of the concentration of CO_2 and other gases.

THE INSTRUMENT RECORD OF RECENT CLIMATE

In most parts of the world it is difficult to find weather stations with uninterrupted records going back more than about 120 years. Gathering data from many sites helps cancel out local variations and measurement inconsistencies, and allows a composite picture of recent climate trends to be obtained. For example, by plotting the **average annual temperature** for different locations over many decades, long-term temperature trends may be seen (Figure 24-6).

Natural year-to-year weather variability is often great enough to mask long-term climate change, so various statistical and charting techniques may be employed to uncover trends. One such technique is to graph *5-year moving averages*, in which each year's annual temperature is averaged with the previous four years. The result is a smoother line that may uncover cycles of temperature change operating over periods of just a few years or decades.

It is also possible to plot a straight "best fit" line, calculated to show the most generalized temperature trend over a long period of time. However, such a line can be misleading because it can make temperature change appear linear and constant, which frequently it is not. Best-fit lines are often superimposed over the annual- and 5-year running averages, creating a useful composite "snapshot" with several levels of data generalization on the same chart (such as has been done in Figure 24-6).

Similar techniques can be used when looking at long-term precipitation patterns, such as is shown for Bozeman and Tucson in Figure 24-7. The **precipitation variability** of a location describes the expected departure from average ("normal") precipitation in any given year. Precipitation variability is generally greater in dry climates than in humid ones.

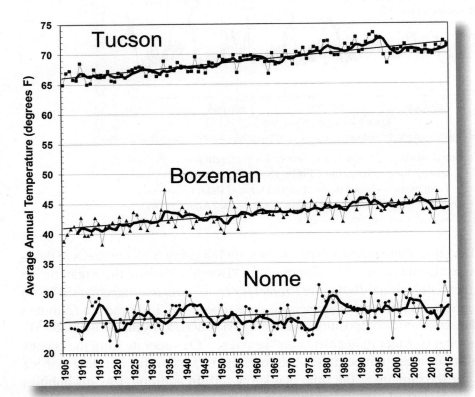

Figure 24-6: Average annual temperature in Tucson, Arizona; Bozeman, Montana; and Nome, Alaska, from 1905 to 2015. Five-year moving average shown with thick trend line; "best fit" shown with thin straight line. (Data sources: National Climate Data Center and Alaska Climate Research Center, University of Alaska, Fairbanks)

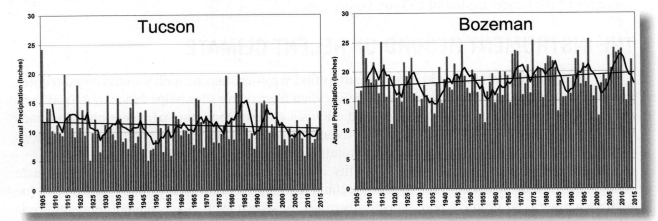

Figure 24-7: Annual precipitation in Tucson, Arizona, and Bozeman, Montana, from 1905 to 2015. Five-year moving average shown with thick trend line; "best-fit" line with thin straight line. (Data source: National Climate Data Center)

Name _____   Section _____

EXERCISE 24 PROBLEMS—PART I

The following questions are based on Figure 24-5, showing variations in CO_2 and temperature anomalies in Antarctica going back 800,000 years. This temperature record shows the major glacial (cold) and interglacial (warm) periods of the later **Pleistocene Epoch** (2.58 million to 11,700 years ago).

1.   Was there any time in the last 800,000 years when the concentration of carbon dioxide was higher than it is today?

2.   Was there any time in the last 800,000 years when temperature in the Antarctic was higher than it is today? If so, when?

3.   Were there times over the last 800,000 years when temperature in the Antarctic was lower than today? If so, how much colder?

4.   Using the period of peak of temperature associated with an interglacial period for reference, what was the approximate time interval between major periods of glaciation over the last 450,000 years?

5.   What appears to happen more abruptly, the onset of a glacial period or the onset of an interglacial period? Why do you say this?

6.   What is the general correlation between the concentration of CO_2 and temperature in the Antarctic over the last 800,000 years, as shown in Figure 24-5?

7.   Research suggests that over the last few hundreds of thousands of years, changes in atmospheric CO_2 concentration sometimes lagged behind a temperature increase by perhaps 1000 years—indicating that "feedback" loops associated with a warmer climate might lead to increasing CO_2 in the atmosphere rather than the other way around.

   (a)   Are any such "lags" visible in Figure 24-5?

   (b)   Looking at Figures 24-1 and 24-2, does the recent increase in global temperature exhibit such a lag?

Name _____   Section _____

EXERCISE 24 PROBLEMS—PART II

The following questions are based on Figures 24-8 and 24-9, charts showing the change in atmospheric carbon dioxide measured at the Mauna Loa Observatory in Hawai'i, from 1958 to April 2016.

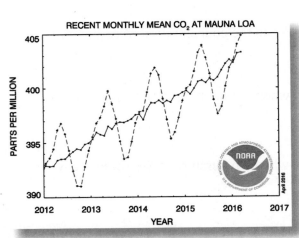

Figure 24-8: Atmospheric carbon dioxide measured at Mauna Loa Observatory, Hawai'i, from January 2012 to April 2016. The dark line is the seasonally corrected value. (Adapted from NOAA)

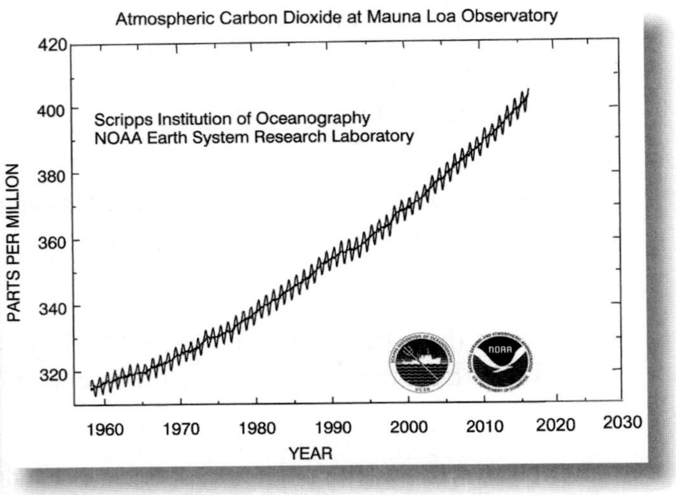

Figure 24-9: Atmospheric carbon dioxide measured at Mauna Loa Observatory, Hawai'i, from 1958 to April 2016. The dark line is the seasonally corrected value. (Adapted from NOAA)

1.   What is the approximate variation in CO_2 concentration each year? _____ ppm

2.   Notice that the concentration of CO_2 in the atmosphere reaches its highest point at the same time each year.

   (a)   During which time of year does CO_2 reach its highest point?

   (b)   What explains the timing of the annual fluctuation in CO_2? (Hint: Think about the process of photosynthesis and plant respiration, and the locations of large forests on Earth.)

3.   If the current trend continues, estimate when the seasonally adjusted atmospheric CO_2 levels will reach 420 ppm.

Name _____   Section _____

EXERCISE 24 PROBLEMS—PART III

The following questions are based on Figure 24-6, a chart showing average annual temperature from 1905 to 2015 in three U.S. cities: Nome, Alaska; Bozeman, Montana; and Tucson, Arizona.

1.   Generally, what has happened to average annual
      temperature in these three cities over the last century?   _____

2.   Using the straight best-fit line as an indicator, estimate the approximate observed tempera-
      ture change over the last century in each city; indicate if the temperature has increased (+)
      or decreased (–). (Note: these answers are based on simplistic assumptions.)

            Tucson: _____ °F   Bozeman: _____°F   Nome: _____°F

3.   (a)   Looking at the annual temperature value dots (not the 5-year moving average),
             what is the warmest year shown on the chart for each of the cities?

             Tucson: _____   Bozeman: _____   Nome: _____

      (b)   What is the coldest year shown on the chart for each of the cities?

             Tucson: _____   Bozeman: _____   Nome: _____

4.   (a)   Using the 5-year moving average lines for reference, in which of the cities does
             average annual temperature fluctuate the most from one year to the next or from
             one decade to the next?

      (b)   Do all of the major warm and cold periods in Nome
             correspond to warm and cold periods in the other two cities?   _____

5.   Although greenhouse gas concentration has increased steadily over the last century (see
      Figure 24-1), based on your answers for questions 3 and 4, is increased greenhouse effect
      likely to be the *only* factor influencing temperature change in these cities? Why?

6.   (a)   Looking at the annual temperature value dots and the 5-year moving averages, do
             you see any evidence of temperature changes following an 11-year sunspot cycle?

      (b)   If not, why might the influence of the sunspot cycle not be evident on these charts?

---

7. (a) Look at the temperatures from 1990 to 1992. Describe any possible effect of the 1991 eruption of Mount Pinatubo on the temperature record of these cities.

   (b) Based on your observations in question 7a, how important were volcanic eruptions in the overall temperature patterns of these cities over the last century?

8. (a) What cyclical factor helps explain the cool period in Nome between about 1945 and 1975? _____

   (b) What cyclical factor helps explain the warm period between about 1975 and 2005? _____

9. Tucson's population grew from about 7500 in the year 1900 to about 520,000 by 2015 (with about 1,000,000 in the surrounding county), and so a portion of the observed temperature increase here may be due to the urban heat island effect. Using your answer in question 2 as a starting point, use the EPA's upper-end estimate of the UHI effect (5.4°F) to calculate the approximate amount of temperature increase that probably *cannot* be explained by urbanization. (Note: This answer is based on simplistic assumptions.)

   _____ °F

The following questions are based on Figure 24-8, charts showing annual precipitation in Tucson, Arizona, and Bozeman, Montana, from 1905 to 2015.

10. Using the best-fit line for reference, what generally happened to annual precipitation in these cities over the last century?

   Tucson: _____   Bozeman: _____

11. (a) Which city exhibits the greatest precipitation variability? _____

    (b) Why?

12. Using the 5-year moving average lines for reference, did all of the major wet and dry periods of the last century occur at the same time in Bozeman and Tucson?

13. (a) Which city shows a greater increase in precipitation during the 1982-83 El Niño event? _____

    (b) Why might this be the case?

Groundwater

Objective:	To study patterns of groundwater flow.
Reference:	Hess, Darrel. *McKnight's Physical Geography*, 12th ed., pp. 269–275.

GROUNDWATER

Water below the surface of Earth is generally referred to as **groundwater**. Although the source of almost all groundwater is the infiltration of moisture from the surface, the depth of groundwater and its pattern of movement are influenced by a number of characteristics of the material below the surface.

Porosity and Permeability: The total quantity of water that can be contained in the rock and soil below the surface depends on the **porosity** of the material—the total percentage of voids, such as pore spaces and cracks. **Permeability** describes the ability of this subsurface material to transmit groundwater, and is a consequence of not only the size of openings, or *interstices*, but of their interconnectedness. Material that has high porosity—and so is able hold a lot of groundwater—is not necessarily permeable. For example, clay frequently has high porosity because of the many tiny openings between the flakes of clay, but it is often relatively impermeable because the interstices are so small that water is bound by molecular attraction to the flakes of clay.

Aquifers: Groundwater moves through, and is stored in, layers of permeable rock called **aquifers**. From the surface, water percolates down through permeable rock in the **zone of aeration,** where the voids are partially or temporarily filled with water. Below the zone of aeration is the **zone of saturation,** in which all pore spaces are filled with water. The top of the zone of saturation is called the **water table** (Figure 25-1). An aquifer may also be confined between impermeable layers of rock called *aquicludes* and so can become a *confined aquifer*.

Water Table Depth Contours: Although the orientation and slope of the water table often roughly conforms to the surface topography, it can vary considerably from this for a number of natural and human-caused reasons—becoming deeper in some places, closer to the surface in others. For example, if water is drawn from a well faster than it is replenished, the local water table will drop locally around the well, forming a cone of depression (Figure 25-2). In such cases, nearby wells may go dry.

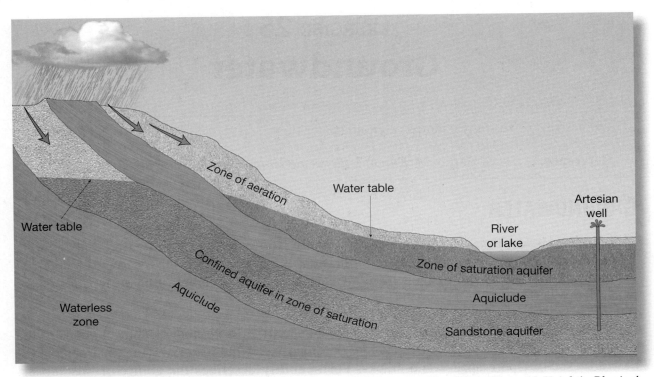

Figure 25-1: The water table is the top of the zone of saturation in an aquifer. (From Hess, *McKnight's Physical Geography*, 12th ed.)

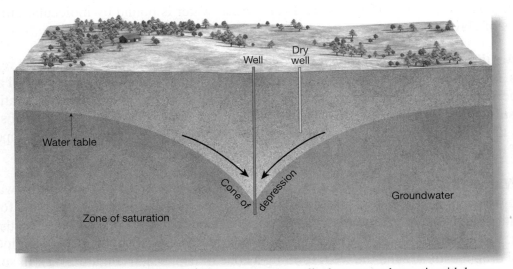

Figure 25-2: A cone of depression forms around a well where groundwater is withdrawn faster than it is recharged. (From Hess, *McKnight's Physical Geography*, 12th ed.)

The depth of the water table below the surface can be mapped with isolines known as *water table contours*. The data to draw water table contours can come from a number of sources, such as the height of water in wells and the elevation of lake and stream surfaces (where the local water table is at the surface).

Groundwater Flow Lines: Groundwater moves slowly through an aquifer in response to the pull of gravity. The path of moving groundwater is mapped with *flow lines*. Flow lines follow the slope of the water table, crossing the water table contours at 90° until they reach the surface at a stream or lake (Figure 25-3). Flow lines don't cross, but depending on the contours of the water table they may diverge or converge.

If a stream is fed by groundwater (an *effluent stream*), groundwater coming in from one side won't cross the stream and mix with groundwater coming in from the other side—instead groundwater from both sides enters the stream and is carried away. In such cases, any groundwater contaminant will follow the flow lines until it reaches a lake or stream. For the same reasons, if a contaminant is released into an effluent stream, it generally won't enter the water table, but will be carried downstream. However, if groundwater flows *away* from a stream (an *influent stream* such as most *intermittent streams*), contamination introduced into the stream can enter the zone of saturation of the groundwater.

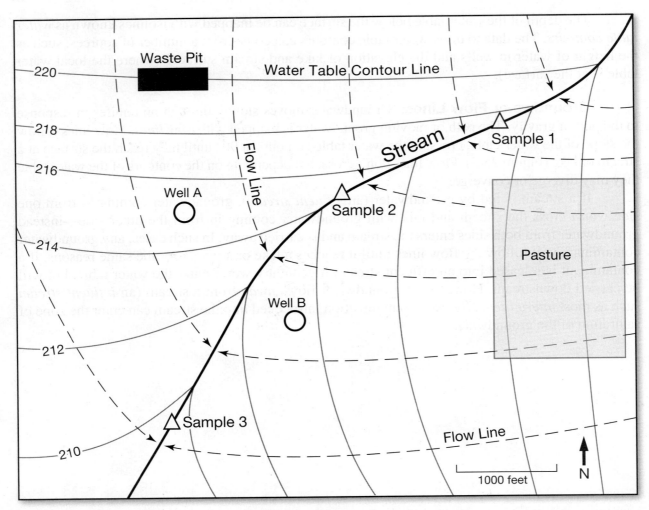

Figure 25-3: Map of a hypothetical landscape with an effluent stream flowing northeast to southwest across the area. Water table contours are drawn at 2-foot intervals, and flow lines show the direction of groundwater movement at the top of the zone of saturation. A waste pit in the northwest was dug down to a depth just slightly above the water table; a fenced pasture for grazing livestock is in the east; two wells and three water quality sample sites along the stream are also shown.

Name _____   Section _____

EXERCISE 25 PROBLEMS—PART I

The following questions are based on Figure 25-3, a map of a hypothetical landscape with an effluent stream flowing from northeast to southwest. Water table contours are drawn at 2-foot intervals and flow lines show the direction of groundwater movement at the top of the zone of saturation. The waste pit in the north was dug down to a depth just slightly above the water table in that location. A fenced pasture for grazing livestock is in the east. Two wells are shown, and the locations of three water quality sampling sites along the stream are indicated.

1.   (a)   If contaminated water is found in Well A, is it more likely to have come from the waste pit or the pasture?

    (b)   Why?

2.   (a)   If the water in Well B is contaminated, can groundwater pollution from the waste pit be blamed?

    (b)   Why or why not?

3.   (a)   Is it likely that the surface stream water in Sample Site 1 will be contaminated by groundwater from either the waste pit or the pasture?

    (b)   Why or why not?

4.   (a)   If the surface stream water in Sample Site 2 is contaminated by the entrance of groundwater, what is a likely source of the contamination?

    (b)   Why?

5.   (a)   If the surface stream water at Sample Site 3 is contaminated by the entrance of groundwater, could the pollution have come from either the waste pit or the pasture?

    (b)   How might you tell which is the case?

Name _____   Section _____

EXERCISE 25 PROBLEMS—PART II

The following questions are based on Figure 25-4, showing lakes in the "Alkali Lake, Nebraska" quadrangle (map reduced to approximate scale 1:75,000). These lakes are found in the Sand Hills area of Nebraska where the water table (reflected in the elevations of lake surfaces) is close to the surface.

1. Using the elevations of lake levels to represent the top of the water table, draw in water table contours using an interval of 10 feet. First, draw the 3830-foot contour in the southeast corner, then work toward the top of the map. Label each contour. Next, show two or three short flow line arrows crossing each contour to indicate the general direction of groundwater movement.

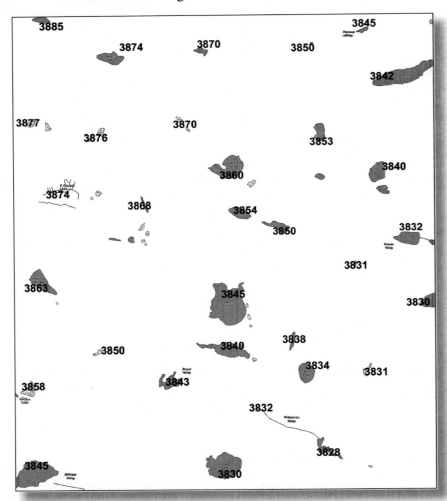

Figure 25-4: Lake elevations (in feet above sea level) in the USGS "Alkali Lake, Nebraska" quadrangle. (Adapted from *US Topo* topographic map; scale reduced to approximately 1:75,000)

2. Based on the pattern of water table contours, what should be the general direction of groundwater flow in this area?

EXERCISE 26
Biomes and Ecological Land Units

Lab Exercise
26

https://goo.gl/zPG2s2

Objective:	To analyze the relationships between the distribution of biomes, ecological land units, and the physical environment.
Resources:	Internet access or mobile device QR (Quick Response) code reader app (optional).
Reference:	Hess, Darrel. *McKnight's Physical Geography*, 12th ed., pp. 323–333.

ECOSYSTEMS AND BIOMES

Of the many interrelationships in the natural world, among the most striking is the relationship between the local physical environment—especially climate—and the local biological community.

Biogeographers have long described biological communities using two important concepts: the *ecosystem* and the *biome*. An **ecosystem** is a fundamental biological community of plants and animals that is tied together through *biogeochemical cycles* by the flow of energy and nutrients with the surrounding nonliving environment.

The difficulty of studying the distribution of ecosystems is that this concept can be used to describe biological communities at many different scales, from the microscopic world in a drop of water to the planetary scale. In order to study biological communities in a meaningful way at the global scale, we usually describe patterns of ecosystems in terms of biomes. A **biome** is a large, recognizable assemblage of plants and animals in functional interaction with its environment.

THE MAJOR BIOMES

The 10 major biomes of the world are shown in Map T-30 in the back of the Lab Manual. These biomes are primarily distinguished by the dominant types of vegetation they contain, although each biome generally includes characteristic animals and soil types, and is usually associated with characteristic climatic regions as well.

The map of world biomes is highly generalized. In reality, the natural patterns of vegetation and animal life have been significantly altered in many locations. Further, the boundaries between these biomes should be viewed as transition zones rather than abrupt borders.

The following brief descriptions highlight several important or distinguishing characteristics of each biome—see your textbook for additional information about each.

Tropical Rainforest: Tropical rainforests are characterized by very high species diversity (many different species present, but with just a few individuals of each species in a given area), and by relatively infertile soils. Tropical rainforests typically include a thick, continuous canopy of leaves produced by **broadleaf evergreen trees**, along with **epiphytes** (plants living above ground level without direct contact with the soil), vines, and a dense understory of vegetation on the surface.

181

The upper layers of the rainforest tree canopy are highly productive, and overall, most of the nutrients in these ecosystems are held in the living vegetation.

Tropical Deciduous Forest: Unlike the tropical rainforest biome, in the tropical deciduous forests many broadleaf trees shed their leaves during part of the year. The structure of tropical deciduous forests is similar to that of tropical rainforests, although the tree canopy in the deciduous forests tends to be less dense because of lower overall precipitation.

The usual transition between the tropical deciduous forest biome and the desert biome is either the tropical scrub or the tropical savanna biome.

Tropical Scrub: Tropical scrub biome typically contains thorny or spiny **shrubs**. The density of vegetation cover varies from areas of thick undergrowth to open growth of trees such as acacias.

Tropical Savanna: These tropical **grasslands** typically contain a wide variety of tall grasses, interspersed with scattered trees. Wildfires, both natural and those set by humans, are a regular occurrence in tropical savannas.

Desert: This biome includes both hot and relatively cool arid areas of the world, so the appearance of the vegetation cover varies greatly. In many deserts, the cover of vegetation is sparse; however, a rare but intense high rainfall event can trigger a sudden wildflower bloom that can cover the normally bare ground. Hot deserts often include dry shrubs and varieties of cactus and **succulents**, whereas cold deserts typically include short dry grasses and shrubs.

Mediterranean Woodland and Shrub: The mediterranean woodland is comprised of open grassland with interspersed trees (such as the oak in California), whereas the mediterranean shrub (also known as **chaparral**) consists of low, dense stands of shrubs and small trees.

Midlatitude Grassland: This biome includes grasslands such as the *pampa* of Argentina and the prairies of North America. The natural vegetation varies from the short grasses of the Asiatic steppes to the tall grasses of the North American prairie.

Midlatitude Deciduous Forest: These forests consist primarily of **broadleaf deciduous trees**, although many midlatitude deciduous forests also include **evergreen conifers** such as pines. Although once widespread, this is one of the most human-altered biomes in the midlatitudes, having been cleared for agriculture and other human activities.

Boreal Forest: The boreal forest forms an almost unbroken belt across the whole of North America and Eurasia between about 50° and 60° North, forming one of the most widespread biomes in the world. This biome typically consists of forests composed of just a few species of evergreen conifers such as pine, fir, larch, and spruce.

Tundra: The tundra biome consists of treeless plains with a low-lying and sometimes dense mat of shrubs, mosses, and sedges. Tundra is found in high latitude regions as well as in high mountain areas.

ECOLOGICAL LAND UNITS

Maps showing the distribution of ecosystems, as well as those showing the distribution of biomes (such as Map T-30 in the back of the Lab Manual), tend to emphasize the biological component of the environment—especially maps that use the dominant vegetation cover as the key defining characteristic of an ecological community. However, the ecosystem of a locality encompasses a much wider range of components than just vegetation.

An ecosystem can be conceptualized as beginning below the surface with the bedrock, and then extending upward through the soil and groundwater to the surface. On the surface, the local topography, surface water, plants, and animals come into play. The "top" of an ecosystem is the atmosphere—both the local conditions, or "microclimate," and the broader regional climate.

In order to more fully appreciate the role of the physical environment in the distribution of ecosystems, *ecological land classification* can be used. Whereas ecosystem and biome maps emphasize the biological components of an environment, ecological land classification maps put equal emphasis on the physical components of an environment that are influencing the biological components. In this case, the unit of study is known as an *ecological land unit*.

An **ecological land unit** (ELU) is an area with a distinct combination of local climate, topography, rock type, and land cover. Contemporary geospatial technology—such as geographic information systems (GIS) and remote sensing (see Exercise 8)—make it possible to map ELUs at the global scale in remarkable detail. For example, the U.S. Geological Survey and Esri® developed an interactive global map of ELUs (**https://ecoexplorer.arcgis.com/**).

This ecological land unit map shows the numerous possible combinations of four layers of data:

Bioclimates: A total of 37 classes of climate are defined, primarily by incorporating a measure of a location's temperature regime (*growing degree days*, which describe the seasonal accumulation of warmth) and moisture regime (the *aridity index*, which compares the annual precipitation to the potential evapotranspiration).

Landforms: The topography of a location is generalized into 16 classes based on the steepness of slope and the relative relief of the region.

Lithology: The lithological component of the landscape is generalized into 15 classes based on the dominant rock type found at the surface.

Land cover: The 36 classes of land cover include not only the dominant types of natural vegetation, but also agricultural areas and artificial surfaces produced by humans.

Each possible combination of the four environmental components is called an *ecological facet*. After unreasonable combinations are eliminated (such as a hot environment with a cover of permanent snow) more than 47,000 combinations are possible. Figure 26-1 shows in tabular format these four components aggregated and simplified into just 17 classes of bioclimate, 3 classes of landforms, 11 classes of lithology, and 9 classes of land cover. Even with this simplification, more than 3900 possible ecological land units are possible!

Bioclimate	Landforms	Lithology	Land Cover
Arctic	Plains	Pyroclastics	Swampy/Often Flooded
Cold Wet	Hills	Unconsolidated Sediment	Sparse Vegetation
Cold Moist	Mountains	Non-Carbonate Sedimentary Rock	Mostly Needleleaf/Evergreen Forest
Cold Semi-Dry		Carbonate Sedimentary Rock	Mostly Deciduous Forest
Cold Dry		Mixed Sedimentary Rock	Mostly Cropland
Cool Wet		Metamorphics	Grassland, Scrub, or Shrub
Cool Moist		Evaporites	Bare Area
Cool Semi-Dry		Acidic (felsic) Volcanics	Artificial Surface/Urban Area
Cool Dry		Acidic (felsic) Plutonics	Surface Water
Warm Wet		Non-Acidic Volcanics	
Warm Moist		Non-Acidic Plutonics	
Warm Semi-Dry			
Warm Dry			
Hot Wet			
Hot Moist			
Hot Semi-Dry			
Hot Dry			

Figure 26-1: Simplified aggregate categories of bioclimate, landforms, lithology, and land cover used in the USGS/Esri® interactive ecological land units map.

Ecophysiographic Diversity: One of the results of mapping ELUs is that regions of high *ecophysiographic diversity* can be identified. By comparing the number of distinct ecological facets found in a given 5-square-kilometer area to the average number of EFs found worldwide, "hot spots" of environmental diversity can be identified. Although areas of high ecophysiographic diversity may not correlate directly with the actual *biodiversity* of that location (the variety of different organisms present), such areas may suggest places of biogeographic interest.

The location of highest ecophysiographic diversity in the world was found to be in the Sweetwater Mountains, just north of Mono Lake in California (38°25'30" N, 119°17'41" W).

Name _____   Section _____

EXERCISE 26 PROBLEMS—PART I

Using the global map of biome distribution (Map T-30 in the back of the Lab Manual), the map of climate distribution (Map T-29), the maps of global temperature (Maps T-28a and T-28b), the map of global precipitation (on the inside front cover of the Lab Manual), as well as information about Köppen climate types from Exercise 23, answer the following questions.

1. (a) Which Köppen climate type (or types) is most closely associated with the tropical rainforest biome?

   (b) Describe the general temperature and precipitation characteristics of the climate in regions of the tropical rainforest biome.

2. (a) Describe the general locations of the tropical deciduous forest biome relative to the tropical rainforest biome (especially note the pattern in South America and Africa).

   (b) Based on the differences in location noted in problem 2a, in what way(s) will the climate in regions of tropical deciduous forest differ from the climate in regions of tropical rainforest?

   (c) How do the climate differences noted in problem 2b help explain the presence of large numbers of deciduous trees in the tropical deciduous forest biome?

3. (a) Which Köppen climate type (or types) is most closely associated with regions of the tropical scrub and the tropical savanna biomes?

   (b) What characteristics of the climate in these regions would tend to limit the growth of extensive forest (such as those found in the tropical rainforest and tropical deciduous forest biomes)?

4. Which Köppen climate type (or types) is most closely associated with the desert biome?

5. (a) Which Köppen climate type (or types) is most closely associated with the mediterranean woodland and shrub biome?

   (b) Describe the general seasonal patterns of temperature and precipitation in these locations.

---

6. (a) Which Köppen climate type (or types) is most closely associated with the midlatitude grassland biome?

    (b) What characteristics of the climate in these regions would tend to limit the growth of forest?

7. Describe the general climate characteristics of regions of the midlatitude deciduous forest biome (you may specifically note the Köppen climate type or types, but this isn't necessary).

8. (a) Which Köppen climate type (or types) is most closely associated with the distribution of the boreal forest biome?

    (b) Describe the general characteristics of temperature and precipitation in these regions.

9. (a) Which Köppen climate type (or types) is most closely associated with the distribution of the tundra biome?

    (b) Describe the temperature and precipitation characteristics of the climate in these regions.

EXERCISE 26 PROBLEMS—PART II—GOOGLE EARTH™

Go to the Hess *Physical Geography Laboratory Manual*, 12th edition, website at **www.MasteringGeography.com** and open the KMZ file for Exercise 26, or scan the QR (Quick Response) code for this exercise to open a Google Earth™ video. After flying to Points 1 through 4, answer the following questions.

1. Based on the location and the general appearance of vegetation in the area, which biome is most likely found around:

    (a) Point 1? _____   (b) Point 2? _____

    (c) Point 3? _____   (d) Point 4? _____

2. Why is the vegetation generally denser on the north sides of the slopes around Point 1?

3. Explain the meandering pattern made by the lines of green trees near Point 3.

Name _____   Section _____

EXERCISE 26 PROBLEMS—PART III—INTERNET

In this exercise, you will use the USGS/Esri® interactive map to study ecological land units (ELUs) around the world. Go to the Hess *Physical Geography Laboratory Manual,* 12th edition, website at **www.MasteringGeography.com**. Select Exercise 26, and go to the USGS/Esri® interactive map of ecological land units (**https://ecoexplorer.arcgis.com/**).

• When you click on the main map of the world, an "x" marks your location. The type of local bioclimate, landforms, rock type, and land cover appears across the top of the map; maps in the left-side panels show the other places in the world with the same individual attribute as the location selected on the main map. On the main map, red dots show all of the other places in the world with the same combination of attributes—and so the same kind of ELU.

• Navigate by zooming in or out, holding and dragging your mouse across the world map, or by entering place names in the "Search" box. You can also use the "Search" box to navigate to a specific latitude and longitude: Enter coordinates using decimal degrees with longitude first (negative values for W longitude and S latitude; positive values for N latitude and E longitude), so 35°15' N, 125°30' W would be entered as –125.50 35.25. To zoom back out to the world map, click on the "globe" in the upper right part of the screen.

• Use the dropdown "Map" menu in the upper right to select the background map. Begin with "Imagery."

1.   Enter the name of your city and state in the "Search" box; the map will zoom in on your location. Click on your city or neighborhood. What are the environmental attributes of the ELU in this location?

    (a)   Bioclimate: _____   (b)   Landforms:_____

    (c)   Rock Type: _____   (d)   Land Cover:_____

    (e)   Is this ELU (shown in red on the map) widespread? Why or why not?

    (f)   Click on a few other locations around your city. Which of the four environmental attributes vary the most in your area? Why?

2.   Enter "Mulalo Ecuador" into the "Search" box; you will zoom in on a region in the Andes Mountains of South America. Click on the city of Mulaló. What are the environmental attributes of the ELU in this location?

    (a)   Bioclimate: _____   (b)   Landforms:_____

    (c)   Rock Type: _____   (d)   Land Cover:_____

    (e)   Is the ELU of Mulaló widespread or localized? _____

(f)   Click on a few other locations or towns around Mulaló. Does the ELU change? Why?

3.   Click on Mount Cotopaxi (the high volcano a short distance to the east of Mulaló). What are the environmental attributes of this ELU?

(a)   Bioclimate: _____   (b)   Landforms:_____

(c)   Rock Type: _____   (d)   Land Cover:_____

(e)   What other places in this region are shown in red as the same ELU?

(f)   What helps explain this?

4.   Enter "Puerto Napo Ecuador" into the "Search" box; you will zoom in on a region in the upper Amazon Basin rainforest in eastern Ecuador. Click on the city name. What are the environmental attributes of the ELU in this area?

(a)   Bioclimate: _____   (b)   Landforms:_____

(c)   Rock Type: _____   (d)   Land Cover:_____

(e)   Compared with the ELU around Mulaló, how widespread is this ELU? Why?

(f)   Click on a few other locations or towns in this area. Which environmental attributes vary most? Why?

5.   Enter "Mono Lake California" into the "Search" box; zoom in as far as you can on the lake. Next, enter the coordinates –119.27 38.38 into the "Search" box (38.38° N, 119.27° W); your view is now centered over the Sweetwater Mountains—an area of very high eco-physiographic diversity. Click on the mountains. What are the environmental attributes of the ELU in this location?

(a)   Bioclimate: _____   (b)   Landforms:_____

(c)   Rock Type: _____   (d)   Land Cover:_____

(e)   Click on several other locations in this immediate area. How do the ELUs change?

(f)   What environmental factors might help explain the great diversity of ELUs in this small area?

Lab Exercise
27

https://goo.gl/CW8GVw

EXERCISE 27
Soils

Objective:	To study the characteristics of soils and the factors influencing the development of soils.
Resources:	Internet access or mobile device QR (Quick Response) code reader app (optional).
Reference:	Hess, Darrel. *McKnight's Physical Geography*, 12th ed., pp. 344–358.

SOIL-FORMING FACTORS

Soil represents the outcome of an ever-continuing set of soil-forming processes. The soil presently found in an area should not be thought of as a final product, but rather as a stage in an ongoing evolution. Soil takes in inputs from its surrounding environment and is acted upon by that environment. Five principal factors influence the development of soil.

Parent Material: The development of soil begins with the physical and chemical breakdown of the bedrock. In addition to the chemical composition of this *parent material*, the texture that results as rock disintegrates also influences the characteristics of soil. For example, bedrock such as sandstone tends to weather into relatively coarse particles, leading to soils easily penetrated by water and air; in contrast, bedrock such as shale tends to weather into minute particles, leading to soil with many tiny openings that can hold a lot of water, but one that may not allow rapid infiltration of water from the surface.

Climate: Temperature and moisture availability are the two most significant climatic factors acting upon soil. In general, high temperatures and ample moisture tend to increase chemical and biological processes, and commonly will produce deeper soils.

Topography: In addition to the topographic influence of water drainage to soil development, the slope of the land also affects soil formation. Soil generally becomes deeper in flat areas because soil formed on steep slopes tends to be removed by erosion.

Biological Factors: Even though only a small portion of soil typically consists of organic material, this small fraction is extremely important. Biological factors in soil development include both the burrowing activity of animals such as earthworms, as well as the decomposing effects of microorganisms that produce *humus*, a dark, organic material.

Time: Because all soils are in various stages of ongoing evolution, time becomes a key factor in determining soil characteristics. Soil-forming processes usually require long periods of time to operate, and over time the characteristics of the original parent material become less important that other soil-forming factors.

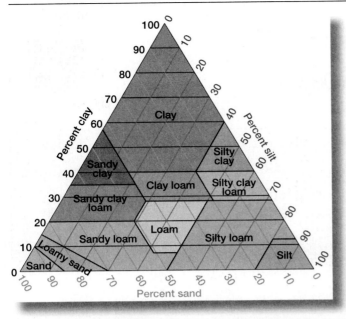

Figure 27-1: The standard soil-texture triangle. The texture of a soil is determined by the relative proportions of sand, silt, and clay particles. (From Hess, *McKnight's Physical Geography*, 12th ed.)

SOIL COMPONENTS AND PROPERTIES

A typical soil will contain about half mineral matter (disintegrated rock), one quarter pore space filled with air, one quarter pore space filled with water, along with a small amount of organic matter. It is the numerous combinations of these components that determine many important properties of soil.

Soil Texture: Soils are composed of a variety of different sizes of particles, or **separates**, each classified by diameter size. For example, common separates include (from larger to smaller): *gravel* (greater than 2 mm [0.08 in.]), *medium sand* (0.25–0.5 mm [0.01–0.02 in.]), *medium silt* (0.006–0.02 mm [0.00024–0.0008 in.]), and *clay* (less than 0.002 mm [0.00008 in.]). Figure 27-1 shows the standard classification scheme for soil texture based on the combination of *sand-*, *silt-*, and *clay*-sized particles. A *loam* is the soil texture in which there is a fairly even mix of all three separate sizes.

Structure: Soil structure is largely determined by the way in which soil particles tend to aggregate into clumps, or *peds*. The shape, size, and stability of peds influence how easily water, air, and organisms (including plant roots) are able to move through the soil. For example, soils with rounded *granular* and *crumb* structures tend to have more pore space than soils with thin and flat *platy* structures.

Soil Water: Of the many functions of water in soil, its ability to dissolve, transport, and deliver nutrients to plant roots is among the most significant. *Leaching* is a process in which water flushes nutrients out of the upper layers of soil. Mineral matter in general may be carried to deeper soil layers by percolating groundwater in the process of *eluviation*. These soil particles are then deposited in deeper soil layers in the process of *illuviation*.

Two soil structure factors are especially important with regard to soil moisture: *porosity* and *permeability*. **Porosity** is the amount of pore space in a soil, typically defined as the volume of voids divided by the total volume. Porosity is usually expressed as a percentage of total volume, and describes a soil's ability to hold water and air.

Permeability describes how easily water can move through the soil. Some soils, such as sandy soils, have both high porosity and high permeability; such soils may dry out quickly because water percolates through rapidly. Other soils, such as those with high clay content, have high porosity but low permeability, meaning that they can hold a great deal of water but that this water cannot infiltrate quickly from the surface or move through the soil readily.

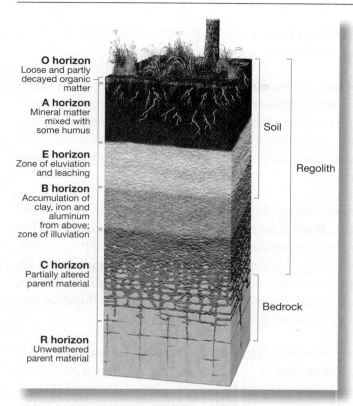

O horizon
Loose and partly decayed organic matter

A horizon
Mineral matter mixed with some humus

E horizon
Zone of eluviation and leaching

B horizon
Accumulation of clay, iron and aluminum from above; zone of illuviation

C horizon
Partially altered parent material

R horizon
Unweathered parent material

Soil

Regolith

Bedrock

Figure 27-2: Idealized soil profile. The O horizon consists mostly of organic matter. The A horizon, or "topsoil," contains both mineral and organic matter. The E horizon is primarily where eluviation takes place. The B horizon, or subsoil, is a layer of illuviation. The C horizon is unconsolidated but weathered parent material. The R horizon is bedrock. The true soil, or solum, consists of the O, A, E, and B horizons. (From Hess, *McKnight's Physical Geography*, 11th ed.)

SOIL HORIZONS AND PROFILES

The five soil-forming factors (parent material, climate, topography, biological factors, and time) influence four sets of processes that deepen and "age" a soil: *addition* (the ingredients added to the soil), *loss* (the ingredients removed or lost from the soil), *transformation* (the ingredients that are altered within the soil), and *translocation* (the ingredients that are moved within the soil). One of the outcomes of these processes is the development of *soil horizons*.

Soil horizons are the different vertical layers or zones that can develop in a soil. These layers typically develop roughly parallel with the surface. Figure 27-2 shows the different horizons that can develop in soil. The *O horizon* (if present) consists of decaying organic matter. The *A horizon* or "topsoil" contains both mineral and organic material. The *E horizon* is an eluvial layer from which minerals have been removed. The *B horizon* or "subsoil" is a layer of illuviation where material from the E horizon above has been deposited. The *C horizon* is a layer of unconsolidated parent rock, and the *R horizon* is largely unweathered bedrock. A true soil, or *solum*, consists of the O, A, E, and B horizons.

Soil classification is largely based on the extent of development and characteristics of the horizons in a **soil profile**—a vertical cross section down through a soil (see Figure 27-2). For example: The lack of soil development below the A horizon usually indicates immature soils; the formation of an illuvial B horizon usually indicates a "mature" soil; "fossil" horizons may remain from a period of different climate in the past; and in some cases, horizons may be missing entirely.

PEDOGENIC REGIMES

Pedogenic regimes ("soil-forming" regimes) are the environmental settings in which certain physical, chemical, and biological processes prevail to produce distinctive soil characteristics. Especially important in determining these pedogenic regimes are differences in climate and the dominant kinds of vegetation present—both of these factors are frequently reflected in the local *biome* (see Exercise 26). Five major pedogenic regimes are generally recognized.

Laterization: *Laterization* occurs in hot and humid areas, and is characterized by rapid weathering and the leaching of most minerals, leaving primarily iron and aluminum oxides. Laterization is common in the wet tropics where high temperature and high rainfall lead to rapid chemical weathering. The resulting soils (sometimes called *latosols*) frequently appear brick-red from the prevalence of iron oxides, and usually perform poorly when cultivated for agriculture because little organic matter accumulates in the A horizon.

Podzolization: *Podzolization* occurs in cool but fairly humid areas, especially in regions with long, cold winters and short, cool summers. Podzolization is especially common in the higher midlatitude coniferous forests of the Northern Hemisphere (and to a lesser extent in milder environments where pine forests are dominant). Chemical weathering is generally slow, but acidic plant litter leads to high soil acidity, and so leaching in the A horizon is very effective. Podzolization often leaves ashy, gray-colored soils (sometimes called *podzols*) with light-colored, highly eluviated and sandy A horizons, and darker, orange or yellow illuviated B horizons below.

Gleization: *Gleization* is prominent in cool or cold, wet, waterlogged regions and tends to produce slowly decomposing, acidic soils. Gleization is common in poorly drained locations, such as those where deposition from Pleistocene glaciers disturbed the preglacial drainage system. Because of the slow bacterial activity, poorly decomposed vegetation or *peat* is common in the upper parts of such *gley soils*, which often include a layer of gray or bluish waterlogged clay.

Calcification: *Calcification* occurs in arid or semiarid climates, including the drier grasslands and prairies of the midlatitudes, as well as in some areas of steppe and desert climate in the subtropics. Because of limited amounts of percolating water, both leaching and eluviation are limited, and so such soils are often rich in organic material (soils developed under undisturbed midlatitude grasslands are frequently very fertile when cultivated). With limited soil moisture, chemical compounds such as calcium carbonate are carried downward a short distance into the B horizon; these compounds are also carried upward from below by capillary water and grass roots (and then returned to the soil when the grass dies). Under such conditions, calcium carbonate ($CaCO_3$) may become so concentrated in the B horizon that a *calcic hardpan* develops.

Salinization: *Salinization* occurs in hot, arid areas—especially in dry, enclosed valleys and basins where drainage is poor or the groundwater level is high. In such environments, moisture is drawn up from the soil through capillary action by the rapid evaporation, leaving behind salts on or just below the surface. Salinization may also develop in regions where extensive irrigation takes place in an arid environment—without adequate precipitation or artificial drainage to flush away salts, soil may be left too salty to cultivate after just a few years.

Pedogenic Regimes and Soil Type: Although pedogenic regimes are a useful starting point for understanding the dominant soil-forming process operating in an area, the actual soil type found in any given location often reflects the outcome of a number of other factors. Thus, the classification of a soil (for example, into one of the 12 *soil orders* of the U.S. *Soil Taxonomy*) depends on specific observed characteristics in the horizons of a soil profile, and cannot always be anticipated simply based on location.

Name _____   Section _____

EXERCISE 27 PROBLEMS—PART I

Using Figure 27-1, the standard soil texture triangle, answer the following questions.

1.   Determine the soil texture for the following combinations of sand, silt, and clay.

    (a)   Sand 20 percent; Silt 20 percent; Clay 60 percent: _____

    (b)   Sand 60 percent; Silt 30 percent; Clay 10 percent: _____

    (c)   Sand 50 percent; Silt 10 percent; Clay 40 percent: _____

2.   Determine a representative combination of sand, silt, and clay for the following soil textures.

    (a)   Silty Clay: Sand _____%; Silt _____%; Clay _____%

    (b)   Silty Loam: Sand _____%; Silt _____%; Clay _____%

    (c)   Loam: Sand _____%; Silt _____%; Clay _____%

3.   (a)   Will a soil comprised of a combination of sand 90 percent,
          silt 5 percent, and clay 5 percent tend to have high or
          low porosity?   _____

    (b)   Will this soil tend to have high or low permeability?   _____

    (c)   Is such a soil likely to dry out quickly?   _____

    (d)   Why?

4.   (a)   Will a soil comprised of a combination of sand 10 percent,
          silt 20 percent, and clay 70 percent tend to have high or
          low porosity?   _____

    (b)   Will this soil tend to have high or low permeability?   _____

    (c)   Is such a soil likely to allow the rapid infiltration of
          water from the surface?   _____

    (d)   Why?

5.   In terms of porosity and permeability, why is loam often considered the ideal soil texture
for agriculture?

Name _____   Section _____

EXERCISE 27 PROBLEMS—PART II

The following questions are based on Figure 27-3, the photograph of a soil profile. You may view this image in color by going to the Lab Manual website at **www.MasteringGeography.com**, or by scanning the QR code for this exercise. The depth scale is in inches.

1. Using a colored pencil, draw in lines marking the approximate boundaries between the three most obvious soil horizons in this profile.

2. Based on their general appearance and position, name the most likely soil horizons visible in this profile.

    (a)   Top zone: _____ horizon

    (b)   Middle zone: _____ horizon

    (c)   Bottom zone: _____ horizon

3. What suggests that eluviation has taken place within the middle horizon?

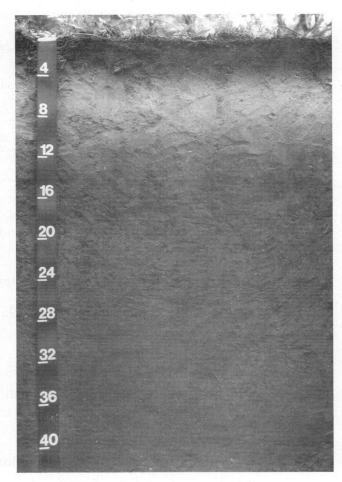

Figure 27-3: Soil profile example. The depth scale is in inches. (From USDA, Natural Resources Conservation Service)

4. (a)   Is unaltered parent material (the R horizon) visible in this soil profile?

    (b)   Is this soil more likely to have developed in a fairly flat area or on a steep slope?

    (c)   How do you know?

Name _____   Section _____

EXERCISE 27 PROBLEMS—PART III—GOOGLE EARTH™

To answer the following questions, go to the Hess *Physical Geography Laboratory Manual*, 12th edition, website at **www.MasteringGeography.com**, then Exercise 27 and select "Exercise 27 Part III Google Earth™" to open a KMZ file in Google Earth™, or scan the QR code for this exercise and view "Exercise 27 Part III Google Earth™ video."

In this exercise you will fly to different locations around the world, identify the most likely climate and biome of the region (see Map T-29, showing world climates, and map T-30, the map of world biomes), and then predict the most likely pedogenic regime present.

Point 1: (a)   Climate: _____

          (b)   Biome: _____

          (c)   Likely pedogenic regime: _____

          (d)   What makes this the most likely pedogenic regime?

          (e)   What soil characteristics would you expect to see in this area?

Point 2: (a)   Climate: _____

          (b)   Biome: _____

          (c)   Likely pedogenic regime: _____

          (d)   What makes this the most likely pedogenic regime?

          (e)   What soil characteristics would you expect to see in this area?

Point 3: (a)   Climate: _____

          (b)   Biome: _____

          (c)   Likely pedogenic regime: _____

          (d)   What makes this the most likely pedogenic regime?

          (e)   What soil characteristics would you expect to see in this area?

Point 4: (a)   Climate: _____

          (b)   Biome: _____

          (c)   Likely pedogenic regime: _____

          (d)   What makes this the most likely pedogenic regime?

          (e)   What soil characteristics would you expect to see in this area?

Point 5: (a)   Climate: _____

          (b)   Biome: _____

          (c)   Likely pedogenic regime: _____

          (d)   What makes this the most likely pedogenic regime?

          (e)   What soil characteristics would you expect to see in this area?

EXERCISE 28
Contour Lines

Objective:	To learn to interpret elevation contour lines.
Reference:	Hess, Darrel. *McKnight's Physical Geography*, 12th ed., pp. 37–39 and A3–A7.

CONTOUR LINES

As we saw in Exercise 6, **isolines** are used to illustrate the distribution of various phenomena. For example, isotherms are used to show patterns of temperature and isobars to show patterns of pressure. In the study of landforms, we often use maps showing elevation with isolines known as **contour lines**.

Contour lines are lines that connect points of equal elevation. Contour lines enable us to study the topography of a region from a two-dimensional map. Figure 28-1 shows a simple contour line map and a profile cross section through the landscape.

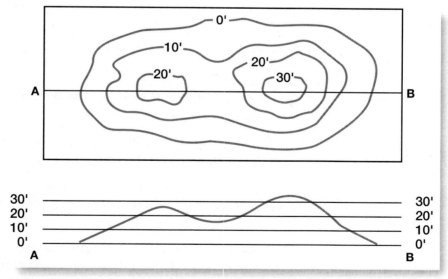

Figure 28-1: Simple contour line map and profile.

SAMPLE CONTOUR LINE MAP

Figure 28-2 shows a fictitious landscape and a contour line map of the same landscape with various elevations and features labeled.

CONTOUR LINE RULES

The following rules will help you interpret contour lines:

1. A contour line connects points of equal elevation.

2. The difference in elevation between two contour lines is known as the **contour interval**.

3. Usually every fifth contour line is a wider, darker **index contour**. (On some maps, every fourth line is an index contour.) You will usually find the elevation labeled somewhere along an index contour.

4. Elevations on one side of a contour line are higher than on the other side.

5. Contour lines never cross one another, although they may touch at a vertical cliff.

6. Contour lines have no beginning or end. Every line closes on itself, either on or off the map.

7. Uniformly spaced contours indicate a uniform slope.

8. If spaced far apart, contour lines indicate a gentle slope. If spaced close together, they represent a steep slope.

9. When crossing a valley, gully, or "draw," a contour line makes a "V" pointing uphill.

10. When crossing a spur or a ridge running down the side of a hill, a contour line makes a "V" pointing downhill.

11. A contour line that closes within the limits of the map represents a hill or rise. The land within the closed contour is higher than the land outside the closed contour.

12. The top of a hill shown with closed contour lines is higher than the uppermost closed contour, but lower than the next highest contour that hasn't been shown on the map.

13. A small depression is represented by a closed contour line that is hachured on the side leading into the depression. Hachured contours are called **depression contours**.

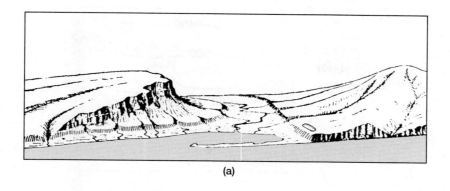

(a)

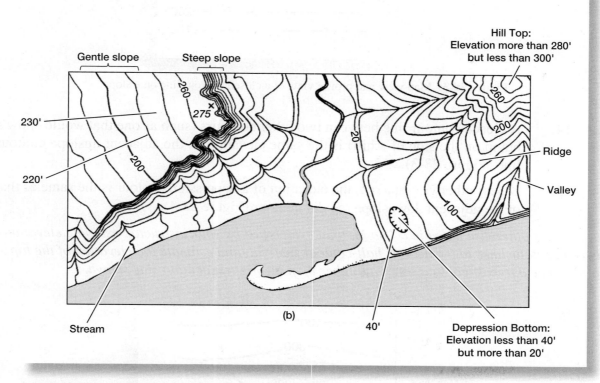

Gentle slope   Steep slope

Hill Top:
Elevation more than 280'
but less than 300'

230'

220'

Ridge

Valley

(b)

Stream

40'

Depression Bottom:
Elevation less than 40'
but more than 20'

Figure 28-2: (a) Fictitious landscape; (b) Sample contour line map (contour interval 20 feet; adapted from U.S. Geological Survey).

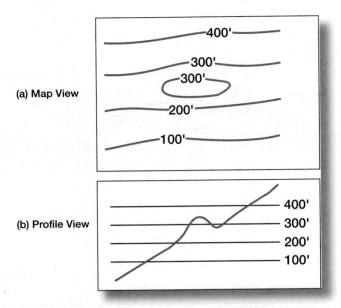

Figure 28-3: Map view and profile view of a closed contour line on a slope.

14. A closed contour line between two other contours (such as one that would show a bump on the side of a hill) is the same elevation as the adjacent upslope contour line (Figure 28-3).

15. Unless otherwise marked, the elevation of a depression contour is the same as that of the adjacent downslope regular contour (Figure 28-4).

Note: Unless otherwise noted in a Lab Manual exercise or by your instructor, estimate elevations between contour lines to the nearest half-contour interval, and estimate the elevation of the top of a hill to be one-half-contour interval higher than the highest contour line shown.

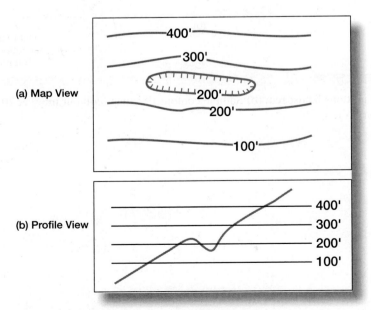

Figure 28-4: Map view and profile view of depression contour line on a slope.

Name _____   Section _____

EXERCISE 28 PROBLEMS

The questions in this exercise are based on this contour line map with elevations shown in feet.

- North is to the top of the map.

- Streams are shown with dashed lines.

- A graphic scale for measuring horizontal distances is shown below the map.

- Estimate elevations between contour lines to the nearest half-contour interval; assume that the top of a hill is one-half-contour interval higher than the highest contour line shown.

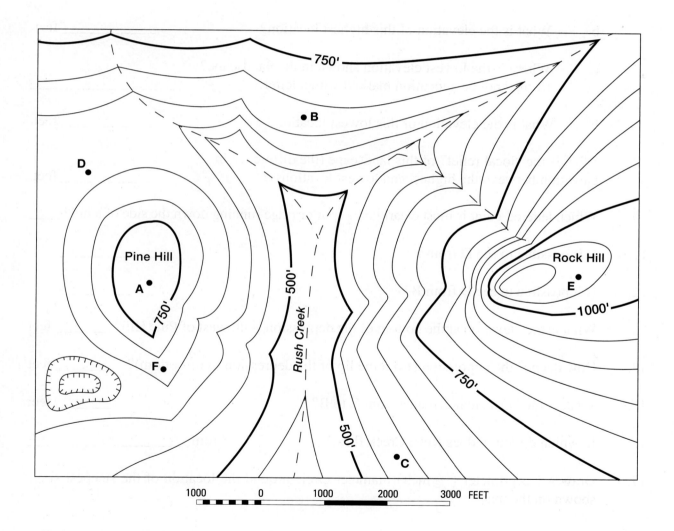

1.   What is the contour interval?   _____ feet

2.   What is the elevation of Point A?   _____ feet

3.   What is the elevation of Point B?   _____ feet

4.   What is the elevation of Point C?   _____ feet

5.   Which lettered point has the highest elevation?   _____

6.   Which lettered point has the lowest elevation?   _____

7.   a.   Where is the highest elevation shown in this landscape?
             (It may not be a location marked with a letter.)   _____

      b.   What is the elevation of this highest location?   _____ feet

8.   a.   Where is the lowest elevation shown in this landscape?
             (It may not be a location marked with a letter.)   _____

      b.   What is the elevation of this lowest location?   _____ feet

9.   What is the "local relief" of this landscape (the difference in
      elevation between the highest and lowest locations)?   _____ feet

10.   Which lettered point is most clearly on a spur or ridge running down the side of a hill? _____

11.   Is it possible to see D from F?   _____

12.   Is it possible to see D from B?   _____

13.   What is the elevation at the bottom of the depression southwest of Pine Hill? _____ feet

14.   How deep is the depression (from the lip of the depression to its bottom)?   _____ feet

15.   What is the horizontal distance from C to B?   _____ feet

16.   In which direction does Rush Creek flow?   From _____ to _____

17.   Draw a 1 centimeter ($^1/_2$ inch) diameter circle around the location of the steepest slope
      shown on the map.

18.   Draw in three more streams as indicated by the contour pattern (but not shown on the map
      with dashed lines).

EXERCISE 29
U.S. Geological Survey Topographic Maps

Objective:	To learn the features of standard U.S. Geological Survey topographic maps.
Materials:	A complete USGS topographic quadrangle is helpful, but not required.
Resources:	Internet access or mobile device QR (Quick Response) code reader app (optional).
Reference:	Hess, Darrel. *McKnight's Physical Geography*, 12th ed., pp. A3–A7.

TOPOGRAPHIC MAPS

Topographic maps are large-scale maps that use contour lines to portray the elevation and shape of the topography. Topographic maps show and name both natural and human-made features. The U.S. Geological Survey (USGS) is the principal government agency that provides topographic maps of the United States. USGS topographic maps cover the entire United States at several different scales.

The largest scale printed maps are those at a scale of 1:24,000 (1 inch represents 2000 feet; 1 cm represents 0.24 km). The long-established 1:62,500 scale maps (1 inch represents about 1 mile; 1 cm represents about 0.6 km) have been replaced by 1:100,000 scale maps (1 inch represents about 1.6 miles; 1 cm represents 1 km). The entire country is also mapped at a scale of 1:250,000 (1 inch represents about 4 miles; 1 cm represents 2.5 km). The primary scale for mapping Alaska has been 1:63,360 (1 inch represents 1 mile; 1 cm represents 0.63 km), although larger scale maps cover part of this state as well. Printed versions of topographic maps are available for purchase through the USGS; however, *US Topo* (**http://nationalmap.gov/ustopo/**) lets you download free electronic versions of USGS topographic maps in an easy-to-use PDF format. The most recent *US Topo* maps let you turn on and off data layers such as contour lines and satellite imagery. In addition, *The National Map* (**http://nationalmap.gov**) lets you select and download a wide variety of topographic maps, aerial photographs, and satellite imagery of most parts of the United States (see Exercise 7). Electronic copies of old maps (going back to the 1880s) are obtained from the Historical Topographic Map Collection at **http://ngmdb.usgs.gov/maps/TopoView/**.

MARGINAL DATA ON USGS MAPS

In this exercise, we will focus primarily on the information found in the margins of these maps (this information has been omitted from most of the maps reproduced for future exercises). Maps T-18 and T-19, reproduced in color in the back of the Lab Manual, show the lower right corner of a standard USGS topographic map with a scale of 1:24,000 (one of the USGS "$7\frac{1}{2}$ minute" maps).

The name of this map, or **quadrangle**, is "Greasewood Spring, Arizona." Below the name is the date of publication, in this case, 2014. To the left of the name is a small map showing the quadrangle location in Arizona.

The latitude and longitude are printed at each corner of the quadrangle. The coordinates of the lower right (southeast) corner of this map are 35°22′30″ North, 109°52″30″ West (notice that on these maps, "north" latitude and "west" longitude are understood).

The eight quadrangles that surround this map are shown in a chart below the index map of Arizona. On older USGS maps, adjoining quadrangles are named in parentheses around the margins of the map. The quadrangle to the southeast of Greasewood Spring is called "Betty Well."

The scales are shown at the bottom center of the map. Below the fractional scale (1:24,000), three graphic scales are shown, in kilometers, miles, and feet. Note that "0" is not at the far left edge of the graphic scales. Below the scales the contour interval is given. The **datum** is the reference point from which elevations are measured. On USGS topographic maps the datum is usually mean (average) sea level.

The **declination arrow** (or *declination diagram*) is found at the lower left corner of the map. True north is shown with the tallest arrow, labeled with either a star or a large "N." The "MN" arrow shows the direction of **magnetic north**. The location of the north magnetic pole is not the same as the true geographical North Pole, so it is necessary to adjust for this difference when using a magnetic compass. The position of the north magnetic pole changes with time, and so the compass correction indicated on the map may not be exact some years after the original survey.

The "GN" arrow shows **grid north**. In addition to the grid system of latitude and longitude, other kinds of grids are also marked on many topographic maps. Grid north refers to the orientation of the **Universal Transverse Mercator grid** (UTM) used by the military. Abbreviated numbers for the UTM grid are marked every 1000 or 10,000 meters around the margins of the map. North–south locations are indicated in meters north or south of the equator, while east–west locations are indicated in meters from a standard meridian. Similar state grids are often marked every 1000 or 10,000 feet.

UTM grid numbers are shown along the right and bottom margins of Maps T-18 and T-19. Along the right margin of the map, the number $^{39}16^{000\,m.}$ N indicates a location 3,916,000 meters north of the equator, while along the bottom margin of the map, the number $^{6}01^{000\,m.}$ E indicates a location 601,000 meters east of a standard meridian.

In addition to the UTM numbers along the margins of the map, additional indications of latitude and longitude are also provided. Notice along the right margin of the map, near the top of Map T-19, the number 25′ appears—this marks the location where the latitude is 35°25′ N. Along the bottom margin of the map, near the left side of Map T-18, you see the number 57′30′ —this marks the location where the longitude is 109°57′30″ W. Most topographic maps will have these supplementary marks for latitude and longitude at regular intervals along the margins.

TOPOGRAPHIC MAP SYMBOLS

Standard symbols and colors are used on USGS topographic maps. Brown lines are elevation contours. Spot elevations are shown by black numbers next to an "X," whereas more precisely surveyed points known as **benchmarks** are shown as numbers next to the letters "BM" (or "VABM"). Blue lines and numbers are used to show water features (blue contour lines on a white background indicate glaciers). Green is used for various kinds of vegetation or forest cover. Human-built features, such as roads, are shown in black and red, while urbanized areas are shown with either red or gray shading. "Photorevised" features are shown in purple on older USGS maps that were updated with aerial photographs. A chart showing standard symbols used on USGS topographic maps is found on the inside of the back cover of the Lab Manual.

Name _____ Section _____

EXERCISE 29 PROBLEMS—PART I

The following questions are based on the lower right (southeast) corner of the "Greasewood Spring, Arizona," quadrangle (Maps T-18 and T-19 shown in color in the back of the Lab Manual):

1.   What is the contour interval of the map?   _____ feet

2.   What is the difference in elevation between index contours?   _____ feet

3.   What is the elevation of the highest contour line at the top of the
        eastern Twin Butte?   _____ feet

4.   What do the dashed blue lines represent?   _____

5.   What do the gray lines around and between the Twin Buttes represent?   _____

6.   What is the name of the adjoining quadrangle to the south?

7.   The latitude of the upper right (northeast) corner of the map is 35°30′ N, while the longitude of the lower left (southwest) corner of the map is 110°00′ W. Why is this called a "7.5 minute" topographic map? (Hint: What is "minute" referring to?)

8.   At the time this quadrangle was printed, what was the difference
        (in degrees) between true north and magnetic north for this map?   _____ degrees

9.   On all three graphic scales for the map, "0" is not at the far left. Explain the reason for this.

10.   Using the graphic map scales, determine the maximum
        width of the eastern Twin Butte (use the 6100′ contour
        to be the outer edge of the butte).   _____ feet

                                                                                                                            _____ kilometers

Name _____   Section _____

EXERCISE 29 PROBLEMS—PART II—INTERNET

Scan the QR (Quick Response) code for this exercise, or go to the Hess *Physical Geography Laboratory Manual*, 12th edition, website at **www.MasteringGeography.com**, to view Figure 29-1 and Figure 29-2. Figure 29-1 shows the *US Topo* imagery layer of the same part of the "Greasewood Spring, Arizona" quadrangle shown in Maps T-18 and T-19. For reference, Figure 29-2 shows an older "Greasewood Spring" quadrangle in color.

1.   What kinds of details of the landscape are visible in the imagery (Figure 29-1) that are not visible in the topographic map using contour lines (Maps T-18 and T-19)?

2.   What kinds of features are harder to recognize in the imagery than on the topographic map?

3.   (a)   Do the roads in this area appear to be paved or unpaved?

    (b)   Why do you say this?

    (c)   What kind of symbol is used to show these roads on the older topographic map (Figure 29-2)?

4.   How did the magnetic declination change from 1972 (Figure 29-2) to 2014 (Map T-18)?

EXERCISE 30
Topographic Profiles

Objective:	To learn to draw and interpret topographic profiles.
Materials:	Ruler.

TOPOGRAPHIC PROFILES

Features on topographic maps are shown in "plan view," looking down on the surface. A **topographic profile** is a diagram showing the changes in elevation of the landscape along a line. It creates a "side view" of the landscape, much like the silhouette of a skyline. Topographic profiles are used in the study of landforms to help emphasize patterns in the topography.

In this example, we will draw a topographic profile of a simple contour line map showing a hill (Figure 30-1). We will draw the profile along line AB, using a profile graph to show vertical and horizontal distance.

We begin by laying the edge of a piece of paper along line AB (Figure 30-2). At each place where a contour line meets the edge of the paper, make a short mark and write down the elevation. Continue across the map, also marking the positions of mountain peaks, passes, streams, and any important cultural features along the line of the profile. Next, line up the edge of paper along the bottom of the graph (Figure 30-3). Carefully mark the elevations of each point along the profile on the graph. Finally, connect these points with a smooth line (Figure 30-4).

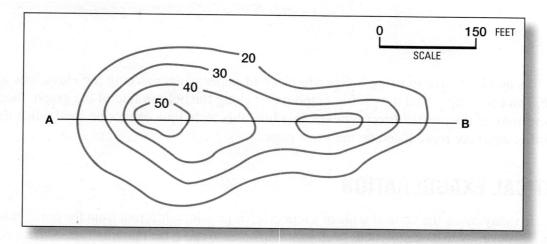

Figure 30-1: Sample contour line map (contour interval 10 feet).

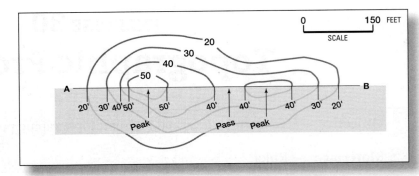

Figure 30-2: Marking elevations along profile line.

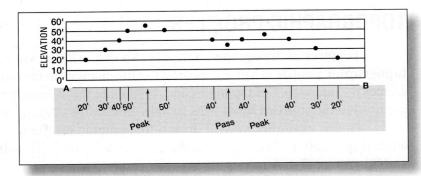

Figure 30-3: Transferring elevations to profile graph.

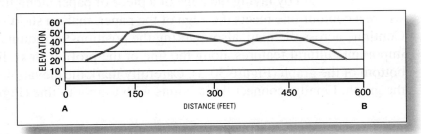

Figure 30-4: Completed topographic profile.

As an alternative to using a separate piece of paper when marking the elevations along a profile, you can simply fold the profile graph paper along the bottom line of the graph, then mark the elevations of the profile directly below this line; this technique cannot be used when the map and profile chart are reproduced on the same page.

VERTICAL EXAGGERATION

In many cases, the vertical scale of a topographic profile is different from the horizontal scale. This is done to emphasize differences in elevation in order to make the pattern of relief more obvious. We say that such profiles have been drawn with **vertical exaggeration**. If great vertical exaggeration is being used, small hills can begin to appear as tall peaks. To determine the amount of vertical exaggeration, compare the horizontal and vertical scales. In the example above, the horizontal scale is 1 inch represents 150 feet, while the vertical scale is 1 inch represents 60 feet. Therefore:

$$\text{Vertical Exaggeration} = \frac{\text{Horizontal}}{\text{Vertical}} = \frac{150 \text{ ft}}{60 \text{ ft}} = 2.5 \times$$

Name _____   Section _____

EXERCISE 30 PROBLEMS—PART I

1. Construct a topographic profile of the map below along line AB. Draw your profile in the graph provided.

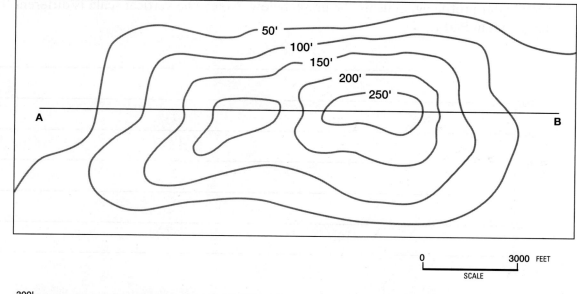

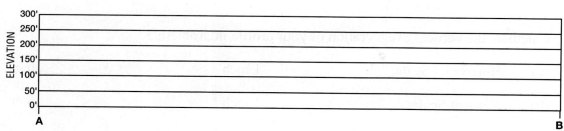

2. Calculate the vertical exaggeration of your profile in problem. Use a ruler to compare the horizontal (map) scale with the vertical scale of the graph.

    (a)   Horizontal Scale: 1 inch   =   _____ feet

    (b)   Vertical Scale: 1 inch   =   _____ feet

    (c)   Vertical Exaggeration   =   _____ ✕

Name _____   Section _____

EXERCISE 30 PROBLEMS—PART II

Complete these problems after you have finished drawing your profile in Part I.

3. Draw your profile again using the graph below. Note: The vertical scale is different from the graph in Part I.

4. Calculate the vertical exaggeration of your profile in problem 3:

    (a)   Horizontal Scale: 1 inch   =   _____ feet

    (b)   Vertical Scale: 1 inch   =   _____ feet

    (c)   Vertical Exaggeration   =   _____ ✕

5. What are the advantages and disadvantages of the amount of vertical exaggeration in your second profile?

EXERCISE **31**
U.S. Public Land Survey System

Objective:	To learn to use the Public Land Survey System.

PUBLIC LAND SURVEY SYSTEM

The **Public Land Survey**, or **township grid**, was established by the U.S. federal government in 1785 in order to keep track of land ownership in the American frontier. This grid covers most of the continental United States west of the Mississippi and Ohio rivers, with the exception of some regions such as those under old Spanish land grants.

The starting point for the grid is a series of parallels known as **base lines**, and a series of **principal meridians** (Figure 31-1). Note that most sets of base lines and principal meridians are named after the same reference point, such as the *Boise Base Line* and the *Boise Principal Meridian* in Idaho. Beginning at the intersection of a base line and a principal meridian, rows of 36-square-mile tracts of land known as **townships** were established (Figure 31-2a).

TOWNSHIP AND RANGE

Each township is a square tract of land, 6 miles to a side, and is identified by its position north or south of a base line and east or west of a principal meridian. The first position north of a base line is called "Township 1 North" (T1N), the second position north is T2N, and so on. The first position south is "Township 1 South" (T1S).

The first position west of a principal meridian is called "Range 1 West" (R1W), and the first position east is "Range 1 East" (R1E). Each 36 square-mile township is identified by both a **township** and a **range**. For example, one of the townships would be designated "Township 3 North, Range 2 East" (see Figure 31-2a).

(Note: The term "township" has two meanings in the context of the Public Land Survey—a 36-square-mile tract of land, as well as the positions of these tracts north and south of a base line. It may help to think of "T1N" and "T2N" as referring to "tier" 1 North, "tier" 2 North, and so on.)

A township is divided into 36 **sections**. Each section is 1 square mile (640 acres) in area and is given a number, from 1 to 36. Notice the specific numbering pattern of sections within a township—beginning with "1" in the northeast corner, and then "zigzagging" down to "36" in the southeast corner (Figure 31-2b). Each section is subdivided into "quarter sections" (160 acres), and each quarter section is further divided into "quarters of quarter sections" (40 acres), or as shown in Figure 31-2c, into even smaller tracts of land. The shaded 10-acre plot shown in Figure 31-2c would be called the "Southeast Quarter of the Southwest Quarter of the Northeast Quarter, Section 24, Township 2 South, Range 3 West" (SE$^1/_4$, SW$^1/_4$, NE$^1/_4$, Sec. 24, T2S, R3W).

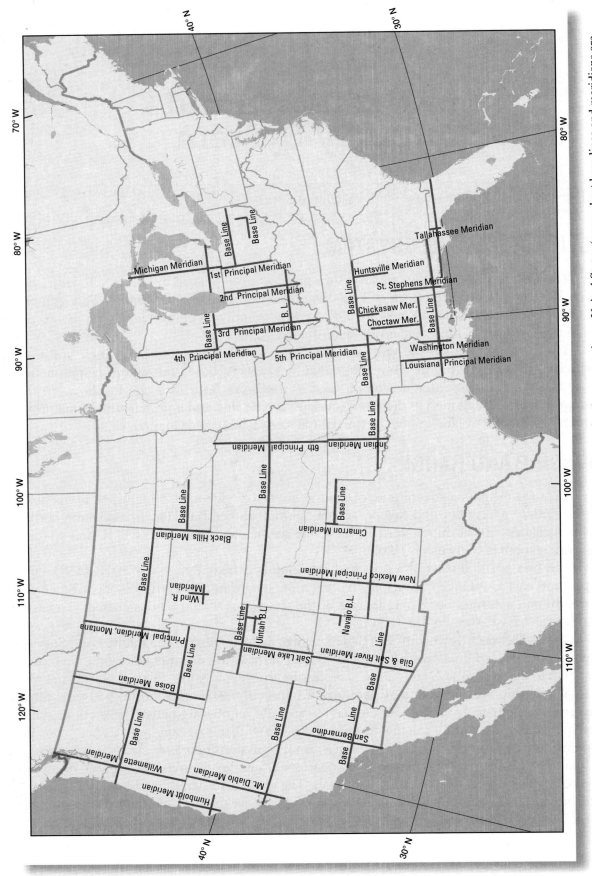

Figure 31-1: Base lines and principal meridians of the Public Land Survey System for the conterminous United States (some short base lines and meridians are not shown). (Adapted from U.S. Geological Survey)

Figure 31-2: Public Land Survey System.

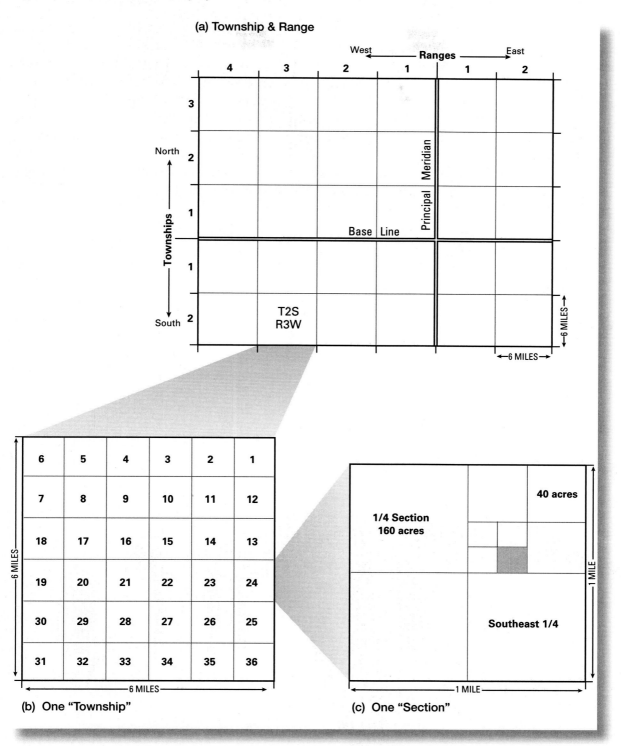

(a) Township & Range

(b) One "Township"

(c) One "Section"

TOWNSHIP GRID ON TOPOGRAPHIC MAPS

On USGS topographic maps, the Public Land Survey grid is usually shown with red lines and section numbers (for example, look at Map T-11, the "Whitewater, Wisconsin," quadrangle). The township and range numbers are shown around the margins of the map (see Map T-17, the "Antelope Peak, Arizona," quadrangle, shown in color in the back of the Lab Manual). The base line and principal meridian are often not identified. You will also notice that a row of townships is occasionally offset relative to the row to the north or south. This is to compensate for the constriction of a township that would result from the convergence of the meridians as latitude increases.

A number of human features in the American landscape are influenced by the Public Land Survey System. Agricultural fields, roads and political boundaries often follow the orientation of the grid.

Name _____   Section _____

EXERCISE 31 PROBLEMS

The following questions are based on Map T-17 in the back of the Lab Manual, showing a portion of the "Antelope Peak, Arizona," quadrangle (scale 1:62,500; contour interval 25 feet); a portion of this map is reproduced in black-and-white in Figure 31-3 as well. The map's marginal information is visible along the left and top margins ("T.5S," "R.2E," etc.).

1.   The word "Hidden" appears on the map within which township?

    Township _____, Range _____

2.   The words "Vekol Wash" appear within which township?

    Township _____, Range _____

3.   The "Booth Hills" are found within which section and township?

    Section _____, Township _____, Range _____

4.   Find the hill in Sec. 8, T6S, R2E. This hill covers approximately how many acres?   _____ acres

5.   Describe two kinds of human/cultural features that follow the Public Land Survey grid.

6.   Why might so many "spot elevations" appear on the map at the corners or intersections of sections?

7.   With a red pencil, carefully mark off and shade in:

(a)   The Northeast Quarter of Section 25, T5S, R1E.

(b)   The Northwest Quarter of the Southeast Quarter of Section 5, T6S, R2E.

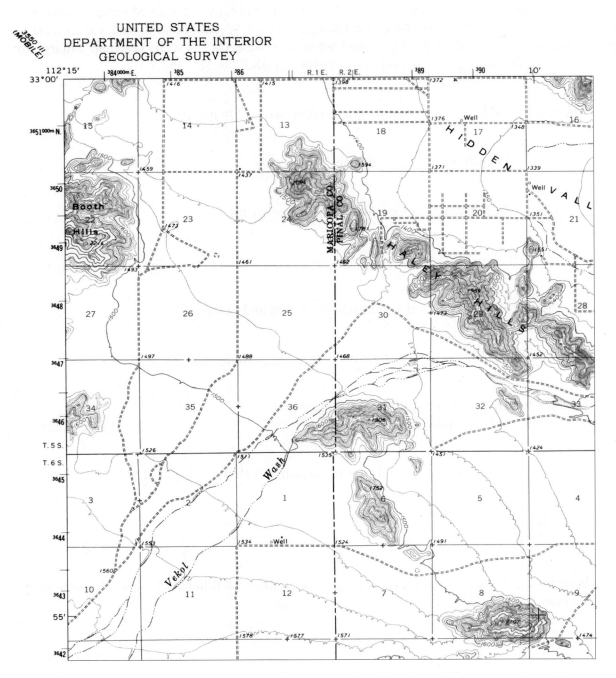

Figure 31-3: USGS "Antelope Peak, Arizona," quadrangle (scale 1:62,500; contour interval 25 feet; ↑N).

Aerial Photographs and Stereograms

Objective:	To learn to interpret aerial photographs and stereograms.
Materials:	Lens stereoscope.
Reference:	Hess, Darrel. *McKnight's Physical Geography*, 12th ed., pp. 41–43.

AERIAL PHOTOGRAPHS

Photographs taken from the air are valuable tools in landscape studies. Although it might seem that photographs would be easier to interpret than maps, in many cases the opposite is true. Maps show selected information and use symbols to distinguish different features that may look alike from the air. On the other hand, photographs show "everything"—and that can make interpretation difficult.

A number of characteristics can be used to help identify features shown in aerial photographs:

- **Shape:** The shapes of objects can offer clues to their nature. This is especially true of human-made features such as airports and stadiums.
- **Pattern:** Look for regularities in the landscape. Evenly spaced lines or dots often show human-built features such as cultivated fields or orchards. A perfectly straight line that cuts across the landscape may be a road or canal, whereas a nearly straight line can be a natural feature such as a fault or a major joint in the rock structure.
- **Size:** Judging the size of an object can be difficult. A house and a large apartment may look the same from the air and you'll need to use nearby objects, such as parked automobiles, for scale.
- **Shadows:** In some cases shadows can reveal the nature of objects that are otherwise hard to see—such as a vertical cliff or a tall rock spire. Shadows can also mislead us. For example, "shaded relief" maps (such as color Maps T-10 and T-13 in the back of the Lab Manual) show the landscape as if the light source was shining from the northwest, and so "shadows" appear on the southeastern side of a mountain peak; however, on aerial photographs, the *actual* shadows cast by the Sun may appear on the northwest side of a peak and this can produce an optical illusion that makes a mountain peak look like a depression and a valley like a high ridge!
- **Texture:** A coarse texture may show a jumble of boulders from a landslide, whereas a smoother texture may indicate a deposit of smaller material such as sand; a forest consisting of large trees may look "bumpy," whereas grassland may appear much smoother.
- **Tone and Color:** In natural-color imagery, green can reveal vegetation, and shades of brown and tan different types of surface rocks. Wet soil tends to be darker than dry soil; clear water tends to look darker than muddy or turbulent water.
- **Locational Context:** Knowing something of the general location shown in an aerial photograph will help in your analysis. In a volcanic area, a dark, black surface might represent a lava flow, whereas in another kind of landscape, it might represent dense forest cover.

STEREOGRAMS

A **stereogram** or **stereopair** is a set of carefully matched **vertical aerial photographs**. A sample stereogram is shown in Figure 32-3. At first glance, a stereogram may look like two identical photographs, but it actually consists of two slightly different photographs.

Stereograms come from photographs taken from an airplane with a camera pointing directly down toward the surface. As the plane flies over an area, it takes a sequence of photographs so that the area shown in one photograph overlaps the area shown in the next photograph by about 60 percent (Figure 32-1). Stereograms are produced from two of these photographs. Both photographs in the pair show the same area, but from slightly different angles.

Stereograms are useful in the study of landforms, because by viewing the stereograms with simple equipment we can see the landscape in "three dimensions." We see in stereo with depth perception because our two eyes see an object from slightly different angles. Because the two photographs in a stereogram also show objects from slightly different angles, if we view the right photograph with our right eye and the left photograph with our left eye, we can produce a stereovision view of the landscape.

THE LENS STEREOSCOPE

The images in stereograms are separated by a distance of 63 mm (about 2.5 inches)—this is the approximate distance between our eyes. They are designed to view with an instrument called a **lens stereoscope** (Figure 32-2). A lens stereoscope consists of a pair of magnifying lenses, usually supported with an adjustable stand.

To view a stereogram, set the spacing of an adjustable stereoscope to 63 mm (or approximately so). Set the stereoscope on the stereogram so that the right lens is directly over one image of the stereogram and the left lens is directly over the other image. Slowly rotate the stereoscope slightly clockwise and counterclockwise until the stereo image comes into view. (To use simple handheld stereoscopes that don't have a stand, rock the stereoscope back and forth slightly on the bridge of your nose until the image appears in stereo.)

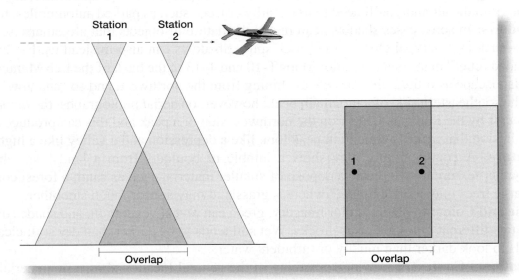

Figure 32-1: Overlapping vertical aerial photographs taken from an airplane. (Adapted from McKnight and Hess, *Physical Geography*, 9th ed.)

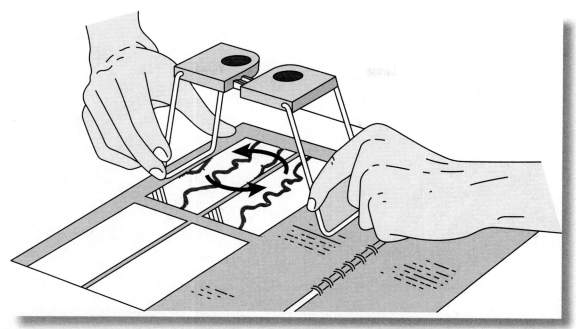

Figure 32-2: Using a lens stereoscope. (Adapted from W. Kenneth Hamblin and James D. Howard, *Exercises in Physical Geology*, 9th ed. Englewood Cliffs, NJ: Prentice Hall, 1995, p. 71.)

With practice, most people are able to see three-dimensional images from stereograms, but some people—even those with good vision—may not. If you cannot see the stereo image, you still can complete the exercises in this Lab Manual that use stereograms; it will require just a little more effort to interpret the photographs.

Even if you have no trouble seeing the stereo image, it isn't a good idea to view stereograms for extended periods of time—especially until you become proficient at lining up the images perfectly. Using the stereoscope may lead to eye fatigue, so **view a stereogram for no longer than about five seconds without looking away for a few seconds to relax your eyes**.

Although the stereograms in the Lab Manual have a nominal image separation of 63 mm, in some cases it is impossible to optimize the image separation for all parts of the stereogram. This is especially true in scenes with very high relief. This means that you may not see a clear stereo image over an entire stereogram. Also, some stereograms consist of three photographs instead of two. To view these stereograms, simply match the center photograph with one of the side photographs.

VERTICAL EXAGGERATION

Because the pair of photographs in a stereogram was taken from locations perhaps a kilometer or so apart, when viewed with a stereoscope, the vertical dimension in the landscape is exaggerated. When you view a stereogram, you are viewing the landscape below as if your eyes were separated by a kilometer! This **vertical exaggeration** means that slopes will look steeper, mountain peaks will look higher, and depressions will look deeper. Often this vertical exaggeration is helpful to us by accentuating the surface features, but you need to keep in mind that the topography isn't really as steep as it appears.

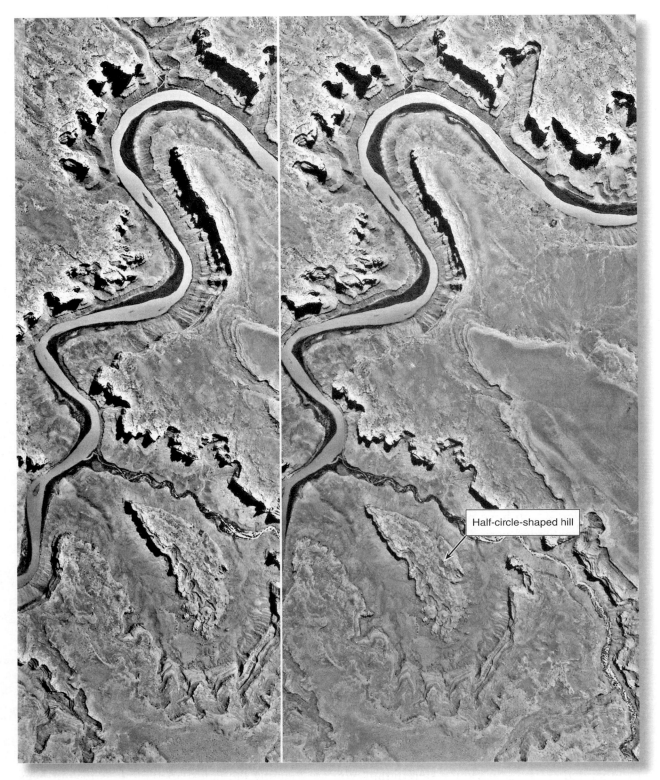

Figure 32-3: Stereogram of the entrenched meanders of the Green River in Utah. North is to the left side of the page (scale 1:40,000; USGS photographs, 1993; ← N).

Name _____   Section _____

EXERCISE 32 PROBLEMS—PART I

The following questions are based on Figure 32-3, a black-and-white stereogram, and color Maps T-20a and T-20b (in the back of the Lab Manual). These maps show the entrenched meanders of the Green River near Canyonlands National Park, Utah. A larger portion of this region is shown in color Map T-7 in the back of the Lab Manual.

Begin by looking at the flat-topped, half-circle-shaped hill about 4 centimeters (1.5 inches) from the bottom of the stereogram.

1.   When viewed without a stereoscope, the shadow on the left side (the north side) of this hill appears larger in the left image than in the right image. Why?

2.   Locate the dry gorge that circles around this half-circle-shaped hill. With the stereoscope, view this dry gorge on the left (north) side of the hill:

   (a)   This dry gorge is approximately 2000 feet (about 600 meters) wide. Viewing the stereogram, approximately how deep does the gorge **appear**? As a starting point for your estimate, consider if the gorge appears deeper than it is wide, wider than it is deep, or equally deep and wide. (There is no correct answer for this question; base your answer on how things look *to you*.)

                                        Apparent depth _____ feet

   (b)   Use the topographic map (Map T-20a) to determine the **actual** depth of the dry gorge (you can simply count the number of contour lines up the side of the gorge, and then multiply the number of contour lines by the contour interval).

                                        Actual depth _____ feet

   (c)   Was your estimate in problem 2a far off? If so, why?

3.   Using the stereogram, describe the location of an apparent overhanging cliff.

Name _____   Section _____

EXERCISE 32 PROBLEMS—PART II

The following questions are based on color Maps T-20a and T-20b (in the back of the Lab Manual) and Figure 32-3, a black-and-white stereogram, showing the entrenched meanders of the Green River near Canyonlands National Park, Utah. A larger portion of this region is shown in color Map T-7 in the back of the Lab Manual.

1. In the aerial imagery (Map T-20b), notice that the dry gorge around the half-circle-shaped hill looks like a ridge surrounding a depression. Why does it appear this way?

2. Describe one kind of detail on the surface of the cliffs along the Green River that is visible in the aerial photographs, but not on the topographic map.

3. How can different layers of rock be discerned from the aerial photographs?

4. Look at the vegetation shown as green in the color aerial photograph (Map T-20b).

    (a) Does all of the vegetation along the river course appear to be the same as that shown along the narrow wash extending to the southwest through Horseshoe Canyon?

    (b) Explain.

5. Look for faint, slightly sinuous lines of discontinuous vegetation on the plateau above the river in the southeastern part of the aerial photograph. In what kinds of topographic features might this vegetation be growing?

EXERCISE **33**
Plate Tectonics

Objective:	To study the tectonic processes and topographic features associated with plate boundaries and hot spots.
Reference:	Hess, Darrel. *McKnight's Physical Geography*, 12th ed., pp. 400–416.

PLATE TECTONICS

The model of **plate tectonics** is the starting point for understanding the distribution and formation of many collections of landforms around the world. Figure 33-1 is a map showing the principal plates and plate boundaries. These **lithospheric plates** are 65 to 100 kilometers (40 to 60 miles) thick and consist of the crust and upper mantle. The plates move over the layer of the mantle known as the **asthenosphere** at speeds averaging from 2.5 to 10 centimeters (1 to 4 inches) per year.

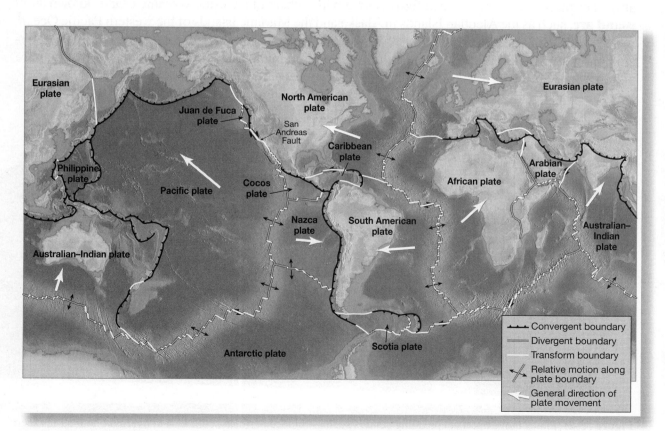

Figure 33-1: Major lithosphere plates. Barbed lines show collision; dark gray lines with offsets show spreading; single white lines show transform boundaries; arrows indicate generalized direction of plate movement. (From Hess, *McKnight's Physical Geography*, 12th ed.)

223

PLATE BOUNDARIES

The three different kinds of plate boundaries are associated with different kinds of topographic features and tectonic activity.

Divergent Boundaries: At divergent boundaries (also called "spreading centers"), plates are moving apart. The most common kind of spreading center is the **midocean ridge** where new basaltic ocean floor is created (Figure 33-2). Spreading may also take place within a continent. In this case, blocks of crust may drop down as the land is pulled apart, producing a **continental rift valley** (such as the Great East African Rift Valley).

Convergent Boundaries: At convergent boundaries, where plates collide, three circumstances are possible:

1. If the edge of an oceanic plate collides with the edge of a continental plate a **subduction zone** is formed. The denser oceanic plate is subducted below the continent, producing an **oceanic trench**. As the oceanic lithosphere descends, water and other volatile materials are driven out of the ocean rocks, leading to the partial melting of the mantle. The **magma** that is generated rises, producing intrusions of **plutonic rock** such as granite and a chain of andesitic **volcanoes**, such as the Andes in South America or the Cascades in North America (see Figure 33-2).

2. If the edge of an oceanic plate collides with the edge of another oceanic plate, subduction also takes place. An oceanic trench forms, along with a chain of andesitic volcanic islands known as an **island arc**, such as the Aleutian Islands in Alaska and the Mariana Islands of the western Pacific Ocean.

3. If the edge of a continent collides with the edge of another continent, the relatively buoyant continental material is not subducted. Instead, a mountain range is uplifted. The Himalayas are a dramatic example of this kind of plate boundary interaction.

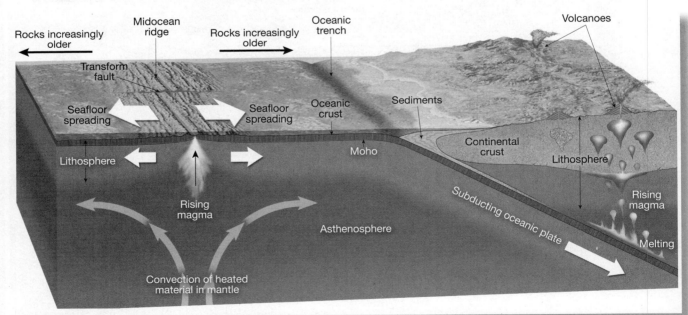

Figure 33-2: Plates move apart at spreading centers such as midocean ridges, collide at convergent boundaries such as subduction zones, and slide past each other along transform faults. (From Hess, *McKnight's Physical Geography*, 12th ed.)

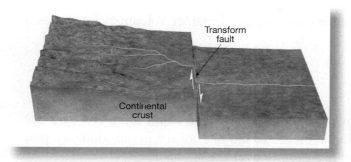

Figure 33-3: Along a transform plate boundary, such as the San Andreas Fault system of California, the movement is lateral. (From Hess, *McKnight's Physical Geography*, 12th ed.)

Transform Boundaries: Plates slide past each other at transform boundaries, such as along the San Andreas Fault system in California (Figure 33-3).

EVIDENCE SUPPORTING PLATE TECTONICS

Evidence supporting the theory of plate tectonics comes from global patterns of landforms and tectonic activity. In addition to the matching shape of the continental margins on both sides of the Atlantic Ocean (which spread apart from the Mid-Atlantic Ridge), the age of the ocean floor provides evidence of movement. The ocean floors are youngest at midocean ridges, where new lithosphere is being formed, and become progressively older away from a ridge in both directions. This was verified through ocean core samples, as well as **paleomagnetic** evidence (changes in Earth's magnetic field that have been recorded in the volcanic rocks of the ocean floor).

Plate boundaries are often the sites of significant volcanic activity. At spreading centers, magma is moving up to the surface, creating new lithosphere as the plates spread apart. Magma generated in subduction zones can produce a chain of continental volcanoes or a volcanic island arc.

The distribution of **earthquakes** also provides clues to plate activity. Most earthquakes around the world occur in association with plate boundaries. Shallow-focus earthquakes, within about 70 kilometers (45 miles) of the surface, occur at all plate boundaries. However, in subduction zones, bands of progressively deeper earthquakes are observed, produced when an oceanic plate is thrust deep down into the asthenosphere.

HOT SPOTS

One of the important modifications of basic plate tectonic theory is the concept of the **hot spot**. These are locations where a fairly narrow plume of magma is rising from the asthenosphere to the surface, producing volcanoes. Many hot spots apparently develop from **mantle plumes** that originate deep within the mantle.

Hot spots may occur well away from plate boundaries, often in the middle of a plate. It is not yet completely understood why these hot spots occur where they do, but the existence of hot spots has been helpful in verifying plate motion.

Evidently, hot spots can remain active in the same location for millions of years. While the hot spot remains in the same place, the plate above continues to move over it. Currently active volcanoes are found directly over the hot spot, while the moving plate carries older volcanoes off the plume, at which time they become inactive. Ongoing plate motion carries these old volcanoes farther and farther away from the hot spot, resulting in a chain of extinct volcanoes known as a "hot spot trail."[1]

THE HAWAIIAN HOT SPOT

The Hawaiian Islands are the best-known example of an island chain produced by a hot spot. The only currently active volcanoes are found on the island of Hawai'i in the southeast part of the island chain. It is believed that this island is currently over the hot spot.

Figure 33-4 is a map showing the ages of volcanic rocks in the Hawaiian chain. Notice that the age of the volcanic rocks becomes progressively older as we follow the islands to the northwest. The pattern of islands in the Hawaiian chain shows the general direction of movement of the Pacific Plate, and from the ages of the rocks, we can infer the rate of plate movement.

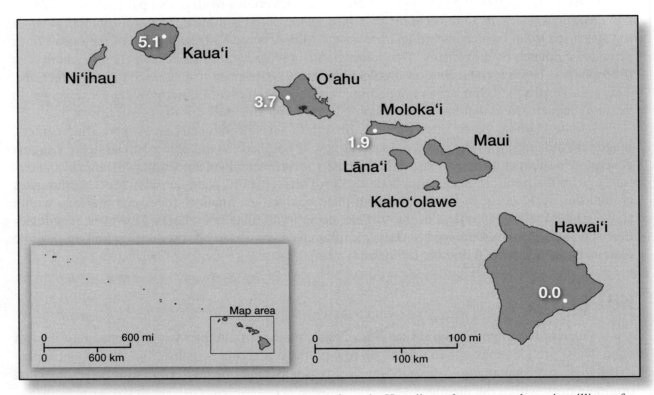

Figure 33-4: The Hawaiian Islands. The ages of the basalt from the Hawaiian volcanoes are shown in millions of years; map scale 1:4,200,000. (Adapted from McKnight, *Physical Geography*, 4th ed.)

[1]Recent geophysical evidence suggests that the locations of some mantle plumes may slowly change over time, making a complete explanation of some hot spots more complex than geologists once thought.

EXERCISE 33 PROBLEMS—Part I Introduction

In the Part I problems for this exercise, you will study the tectonic map of a hypothetical ocean basin (shown on page 229). The map shows the location of volcanoes, earthquakes, and the age of ocean floor rocks. From this map, you will determine the probable location of the plate boundaries and the locations of major topographic features in the region.

On the map, the edges of two continents are shown (in the upper right corner and the lower left corner). Six islands are also shown in the ocean basin.

The symbols used on the tectonic map are described below.

Earthquake Epicenter Location and Depth:

The locations of earthquake **epicenters** are shown with letters. The depth of an earthquake (the distance of the earthquake hypocenter or "focus" below the surface) is indicated with an "S" (shallow focus), "I" (intermediate focus), or "D" (deep focus):

S = Shallow Earthquakes		0–70 kilometers (0–45 miles) deep
I = Intermediate Earthquakes		70–200 kilometers (45–125 miles) deep
D = Deep Earthquakes		200–500 kilometers (125–310 miles) deep

Active Volcano:

Continent or Island:

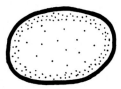

Age of Volcanic Ocean Floor Rocks:

The circled numbers represent the age of volcanic ocean floor rocks in millions of years.

For example, (20) indicates the location of 20-million-year-old rocks.

EXERCISE PROCEDURE:

The first step of the exercise is to draw in the approximate plate boundaries as indicated by the tectonic activity on the map.

Clues include:

(a)   The pattern of earthquakes. For example, subduction produces a pattern of deeper and deeper earthquakes as one plate plunges below the other.

(b)   The age pattern of volcanic ocean floor rocks suggests the location where new ocean floor is being created at a midocean ridge.

(c)   Volcanic activity may be associated with subduction, spreading centers, or hot spots.

Use the following symbols to indicate the extent of all plate boundaries. Both the map symbols, and a side view of the circumstance they represent, are shown below. Arrows indicate direction of plate movement.

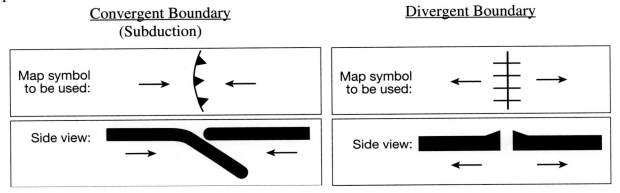

Note:

- No transform boundaries are found on the map.

- Assume that only one of the volcanoes on the map is associated with a hot spot.

Name _____   Section _____

EXERCISE 33 PROBLEMS—PART I

After drawing in the plate boundaries using the appropriate symbols on the tectonic map below, answer the questions on the following page. When asked to cite evidence to support your answers, only cite evidence *that you can see on this map*.

TECTONIC MAP OF HYPOTHETICAL OCEAN BASIN

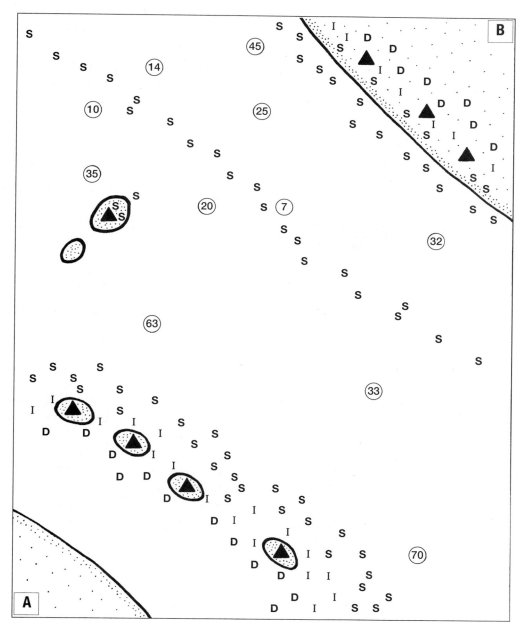

Scale: 1 cm = 300 km (1 inch = 500 miles)

1. (a)   How many different *plates* (not boundaries) are clearly shown on the map? _____

    (b)   How many of the plates on the map consist entirely of ocean floor
          (or ocean floor with islands)?   _____

2. (a)   With the number "2" indicate the most likely location on the map of a midocean
          ridge (such as the Mid-Atlantic Ridge).

    (b)   What type of plate boundary is this? _____

    (c)   What evidence *shown on the map* indicates that this type of boundary is present?

3. (a)   With the number "3" indicate the most likely location on the map of a major
          volcanic mountain range similar to the Andes in South America.

    (b)   What type of plate boundary is this? _____

    (c)   What evidence *shown on the map* indicates that this type of boundary is present?

4. (a)   With the number "4" indicate the most likely location on the map of a volcanic island arc.

    (b)   What type of plate boundary is this? _____

    (c)   What evidence *shown on the map* indicates that this type of boundary is present?

5.   With the number "5" label all plate boundaries where oceanic trenches should be found.

6. (a)   Assume that only one of the volcanoes on the map is associated with a hot spot.
          With the number "6" label this volcano.

    (b)   With a 2-centimeter (about 1-inch) long arrow extending from this volcano,
          indicate the direction in which you would expect to find progressively older
          extinct volcanoes left by the hot spot.

7.   In the space below, draw an approximate continuous cross section ("side view") of the
          ocean basin from Point "A" to Point "B" (from lower left to upper right). Use the "side
          view" drawings on page 228 and Figure 33-2 for reference, and use arrows to indicate the
          relative direction of plate motion. If subduction is taking place, clearly show which plate
          is subducting beneath the other.

A   **B**
Lower Left   Upper Right

Name _____   Section _____

EXERCISE 33 PROBLEMS—PART II

Using the map of the Hawaiian Islands and the ages of the basaltic lava (Figure 33-4), compute the approximate rate of movement of the Pacific Plate as it passes over the Hawaiian hot spot.

You will compare the age and distance between several different volcanoes on the islands. The ages of volcanic rocks on the islands are given in millions of years. White dots on the islands mark the location of a volcano. For example, on the island of Hawai'i, the dot next to "0.0" marks the location of the currently active volcano, Kīlauea.

In this exercise, you will compare Hawai'i (0.0 years—currently active volcanoes), Moloka'i (1.9 million years), O'ahu (3.7 million years), and Kaua'i (5.1 million years). For the purposes of this exercise, we will take the position of the Kīlauea volcano to represent the location of the Hawaiian hot spot (keep in mind that this is a simplistic assumption). We also assume that the Hawaiian hot spot is completely stationary over long periods of time.

1.   Complete the chart on the following page:

    (a)   First, determine the distance between each pair of locations listed on the chart. With a ruler, carefully measure the distance between locations on the map to the nearest millimeter if you use S.I. units and to the nearest 1/16 inch if you use English units. If you use English units, convert fractions of inches to decimals to make other calculations easier (for example, $1\frac{1}{8} = 1.125$). This figure is the "Measured Distance on Map." Then multiply this measured distance on the map by 4,200,000 (the denominator of the fractional map scale) to determine the "Actual Distance" in millimeters (or inches).

    (b)   Next, determine the "Age Difference" in years between each pair of locations. (Be sure to include the correct number of zeros in your figure.)

    (c)   Finally, divide the "Actual Distance" between locations by the "Age Difference" to estimate the rate of plate movement in millimeters (or inches) per year.

Locations	Measured Distance on Map (in mm or inches)	Actual Distance (in mm or inches)	Age Difference (in years)	Rate of Plate Movement (mm or inches per year)
Kauaʻi to Hawaiʻi				
Oʻahu to Hawaiʻi				
Molokaʻi to Hawaiʻi				
Kauaʻi to Oʻahu				
Kauaʻi to Molokaʻi				

2. Based on the average of your five answers in problem 1, what has been the approximate rate of movement of the Pacific Plate in the area of the Hawaiian Islands over the last 5.1 million years?

3. Midway Island, to the northwest of Hawaiʻi, is also part of the Hawaiian chain and is believed to have been produced by the same hot spot. Midway is about 2430 kilometers (1510 miles) from the Kīlauea volcano on Hawaiʻi. Use the average rate of plate movement you calculated in problem 2 to estimate the age of volcanic rocks you would expect to find on Midway Island.

4. The actual age of the volcanic rock on Midway is about 27.7 million years. Suggest a reason why your answer for problem 3 differs noticeably from this.

EXERCISE **34**
Volcanoes

Objective:	To compare different kinds of volcanic mountains.
Materials:	Lens stereoscope.
Resources:	Internet access or mobile device QR (Quick Response) code reader app (optional).
Reference:	Hess, Darrel. *McKnight's Physical Geography*, 12th ed., pp. 416–422.

TYPES OF LAVA AND STYLES OF ERUPTION

There is a close relationship between the shape of a volcanic mountain, the style of a volcano's eruption, and the mineral composition of the **magma** associated with the volcano. In general, **lava** with relatively little **silica** (*mafic* lava), such as forms the volcanic rock **basalt**, tends to be quite fluid. (The viscosity of a lava, its mineral composition, and its temperature are all interrelated.) Basaltic lava usually flows from a volcano in effusive or "quiet," nonexplosive eruptions. The Hawaiian volcanoes erupt in this fashion. The term "quiet" is of course relative, and refers to the nonexplosive venting of fluid lava.

In contrast, lavas with greater amounts of silica, such as **andesite** (an "intermediate" lava whose plutonic equivalent is diorite), and high silica (*felsic*) lavas such as **rhyolite** (whose plutonic equivalent is granite), tend to be more viscous. Unlike the fluid basaltic lavas, gas bubbles can rise only slowly through these viscous lavas. As the magma moves toward the surface during an eruption, the reduction of pressure causes these gas bubbles to expand and escape in an explosive fashion. Explosive eruptions entail the ejection of **pyroclastics**—solid pieces of shattered volcanic rock of various sizes.

TYPES OF VOLCANIC MOUNTAINS

Shield Volcanoes: Effusive eruptions of fluid basaltic lava tend to produce wide, gently sloping mountains known as **shield volcanoes** (Figure 34-1b). The Hawaiian volcanoes are of this type. Shield volcanoes can be very high, but they are not steep-sided.

Mauna Loa volcano in Hawai'i Volcanoes National Park is an exceptionally large shield volcano. Map T-1 is a topographic map and Map T-21a is a satellite image showing the southern portion of the island of Hawai'i. Mauna Loa has been active many times over the last century, but since the 1950s, Kīlauea (seen on the eastern part of the map) has been the most active of the Hawaiian volcanoes.

Composite Volcanoes: Explosive eruptions are common with volcanoes that emit higher silica intermediate lavas, such as andesite. These volcanoes tend to develop into symmetrical, steep-sided mountains known as **composite volcanoes** (also called **stratovolcanoes**). Mount Fuji in Japan and Mount Rainier in Washington are examples of composite volcanoes.

These mountains develop steep sides by the buildup of alternating layers of ejected pyroclastic material (ash and cinders) from explosive eruptions, and lava flows from nonexplosive

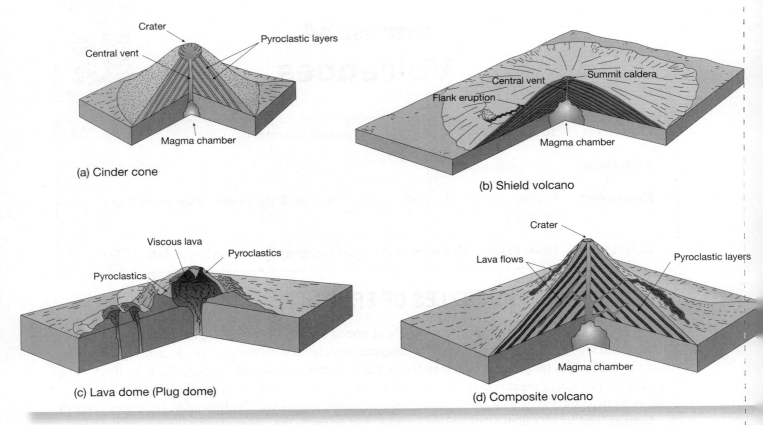

Figure 34-1: The four principal types of volcanoes. (From McKnight and Hess, *Physical Geography*, 9th ed.)

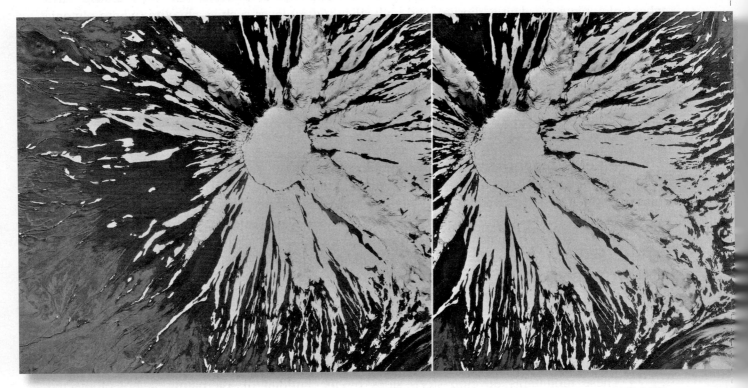

Figure 34-2: Stereogram of Mt. Vsevidof in the Aleutian Islands, Alaska (scale 1:60,000; USGS photographs, 1983; N↗).

eruptions. The explosively erupted ash and cinders tend to produce the steep slopes, while the lava flows tend to hold this loose material together (Figure 34-1d).

Mount Vsevidof in the Aleutian Islands is a typical moderate-sized composite volcano. Map T-2 is a topographic map and Map T-21b is a satellite image of Umnak Island, Alaska, showing Mount Vsevidof, along with other volcanic mountains. A stereogram of Mount Vsevidof is shown in Figure 34-2. Mount Vsevidof was most recently active in 1957 with an eruption of ash and in 1999 with a small steam plume.

Plug Domes: Lava domes, or **plug domes**, develop from masses of very viscous lava, such as rhyolite, that are too thick to flow very far. Instead, lava bulges up from the vent, and the dome grows largely by expansion from below and from within (Figure 34-1c). When viscous lava does flow from the vent of a plug dome, it tends to produce a short, steep-sided lava flow called a "coulee."

The Mono Craters are a chain of very young, mostly rhyolitic plug domes south of Mono Lake in California. Map T-23c is a topographic map showing the Mono Craters; Map T-23a is a satellite image showing the same area; Map T-23b is a stereogram of several of the plug domes in the Mono Craters chain. A flow of rhyolite and rhyolitic obsidian is shown on the map below the words "Mono Craters Tunnel."

Cinder Cones: Cinder cones are the smallest kind of volcanic mountain. They are cone-shaped peaks that build up from pyroclastics ejected into the air from a small volcanic vent (Figure 34-1a). The steepness of the slope of a cinder cone is generally related to the size of the particles being ejected. Volcanic ash (particles less than 2 mm in diameter) can produce slopes as steep as 35 degrees, while the larger cinders (particles between 2 mm and 64 mm in diameter) will produce slopes up to about 25 degrees. Cinder cones are generally less than 500 meters (1600 feet) high, and are often found in association with other volcanoes. Occasionally lava flows issue from the same vent that produces a cinder cone.

SP Mountain in Arizona is a young cinder cone in a volcanic field at the southern margin of the Colorado Plateau (Map T-22b is a topographic map and Map T-22a is a matching aerial photograph of SP Mountain). A lava flow of basaltic andesite (the dark region on the map labeled "lava") flowed north from the cinder cone vent. Note that the scale of this map is 1:24,000. Figure 34-3 is a stereogram of the same region.

GRADIENT

A simple way to compare the steepness of slopes is to use the **gradient**. If working with English units of measure, the gradient is usually stated in feet of elevation change per mile. For example, if a mountain increases elevation by 3200 feet over a distance of 2.5 miles, the gradient is as follows:

$$\text{Gradient} = \frac{\text{Elevation Change}}{\text{Number of Miles}} = \frac{3200 \text{ ft}}{2.5 \text{ miles}} = 1280 \text{ ft/mi}$$

When calculating slope gradients from topographic maps, it may be easiest to use the graphic map scale to measure out a set distance on the map (for example, 2 miles), and then determine the elevation change over that distance. For example, when determining gradients in the problems that follow (Exercise 34, Part I), it is recommended that you calculate the gradient in problem 1 over a distance of 10 miles; calculate the gradient in problem 2 over a distance of 2 miles; and calculate the gradient in problem 3 over a distance of 0.2 miles.

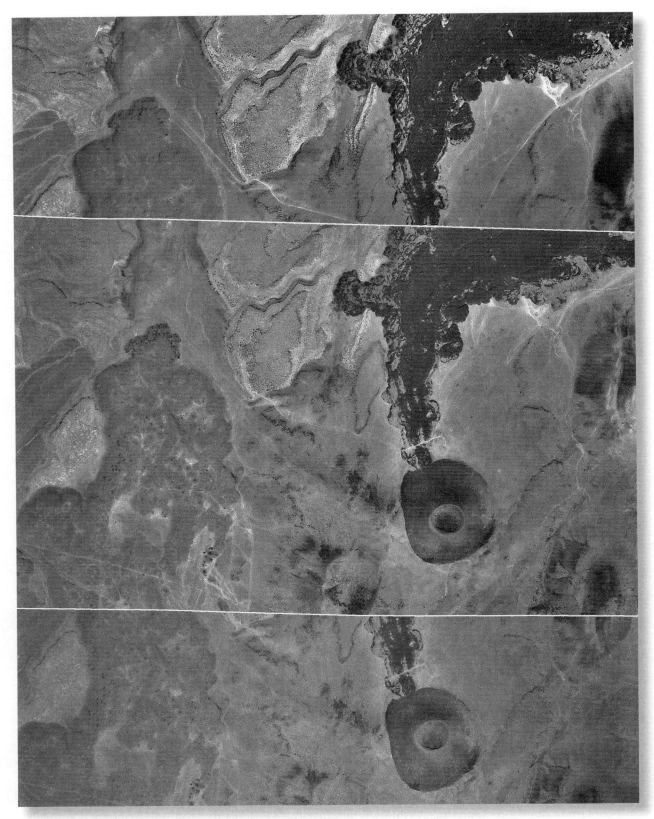

Figure 34-3: Stereogram of SP Mountain, Arizona, (scale 1:40,000; USGS photographs, 1992; ↑N).

Name _____   Section _____

EXERCISE 34 PROBLEMS—PART I

In the following problems you will compute the gradients of Mauna Loa in Hawai'i, Mount Vsevidof in Alaska, and SP Mountain in Arizona. (You may also determine the elevation changes and distances needed to calculate these gradients by using Google Earth™.) The recommended starting point and distance to measure are given for all three volcanoes.

1. Using Map T-1, the "Hawaii, Hawaii," topographic map (scale 1:250,000; contour interval 200 feet), calculate the gradient of Mauna Loa (19°28'17"N, 155°35'41"W) along line AB, from the 5000', contour (near Point A) toward the summit. On the edge of a piece of paper measure out a distance of 10 miles using the graphic map scale (inside front cover), and then determine the elevation change over that distance.

$$\frac{}{\text{(Elevation Change)}} \text{ feet} \div \frac{10}{\text{(Number of Miles)}} = \frac{}{\text{(Gradient)}} \text{ feet/mile}$$

2. Using Map T-2, the "Umnak, Alaska," topographic map (scale 1:250,000; contour interval 200 feet), calculate the gradient of Mount Vsevidof (53°07'35"N, 168°41'16"W) along line CD, from the 1000' contour (near Point C) toward the summit. On the edge of a piece of paper measure out a distance of 2 miles using the graphic map scale, and then determine the elevation change over that distance.

$$\frac{}{\text{(Elevation Change)}} \text{ feet} \div \frac{2}{\text{(Number of Miles)}} = \frac{}{\text{(Gradient)}} \text{ feet/mile}$$

3. Using Map T-22b, the "SP Mountain, Arizona," quadrangle (scale 1:24,000; contour interval 40 feet), calculate the gradient of SP Mountain (35°34'56"N, 111°37'55"W) from the north base of the cone (at the 6200' contour) toward the crater rim. On the edge of a piece of paper measure out a distance of 0.2 miles using the graphic map scale, and then determine the elevation change over that distance.

$$\frac{}{\text{(Elevation Change)}} \text{ feet} \div \frac{0.2}{\text{(Number of Miles)}} = \frac{}{\text{(Gradient)}} \text{ feet/mile}$$

4. (a)   Which of the three volcanoes has the steepest slope?   _____

    (b)   Why does this kind of volcano have the steepest slope?

    (c)   Which of the three volcanoes has the gentlest slope?   _____

    (d)   Why does this kind of volcano have the gentlest slope?

Name _____   Section _____

EXERCISE 34 PROBLEMS—PART II

The following problems are based on Map T-1 and Map T-2. Both of these maps have a scale of 1:250,000. Using the graph at the right, construct topographic profiles between the appropriate lettered points. Vertical exaggeration of topographic profiles is approximately 2× (an explanation of vertical exaggeration is found in Exercise 30).

1.   Construct a topographic profile of the Mauna Loa volcano in Hawai'i (Map T-1) from Point A to Point B. Except for the area around the summit, you only need to plot index contours.

2.   Construct a topographic profile of Mount Vsevidof (Map T-2) from Point C to Point D. Except for the area around the summit and around the base, you only need to plot index contours.

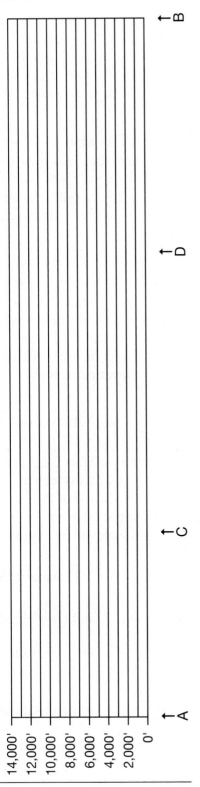

Name _____   Section _____

EXERCISE 34 PROBLEMS—PART III

Look at the following volcanoes (note that the maps are of several different scales):

- **Mount Vsevidof** (53°07'35"N, 168°41'16"W), a composite volcano in Alaska (shown on Map T-2, Map T-21b, and Figure 34-2)
- **Crater Mountain** (37°52'42"N, 119°00'25"W), the highest peak of the Mono Craters, a chain of rhyolitic plug dome volcanoes in California (shown on Maps T-23a and T-23c)
- **SP Mountain** (35°34'56"N, 111°37'55"W), a basaltic cinder cone in Arizona (shown on Maps T-22a and T-22b and Figure 34-3)
- **Mauna Loa** (19°28'17"N, 155°35'41"W), a large shield volcano on the Big Island of Hawai'i (shown on Maps T-1 and T-21a)

1. Why are the summits of Mount Vsevidof and Crater Mountain so different? (Hint: Contrast the formation of each.)

2. What indicates that the lava flows associated with Crater Mountain (and the other Mono Craters) were viscous?

3. What indicates that the lava that flowed north from SP Mountain was very fluid?

4. Some geologists think that the large lava flow to the north of SP Mountain occurred before the final formation of the SP Mountain cinder cone itself. What evidence *that you can see* supports this conclusion? (Hint: Think about the structure, and so the relative strength, of a cinder cone.)

5. (a)   Did all of the lava flows associated with Mauna Loa originate at the top of the volcano?

    (b)   How do you know?

6. What is the relationship of lava flows on Mauna Loa to the "Rift Zones" along the slopes of the volcano?

Name _____   Section _____

EXERCISE 34 PROBLEMS—PART IV—GOOGLE EARTH™

To answer the following questions, go to the Hess *Physical Geography Laboratory Manual*, 12th edition, website at **www.MasteringGeography.com**, then Exercise 34, and select "Exercise 34 Part IV Google Earth™" to open a KMZ file in Google Earth™, or scan the QR (Quick Response) code for this exercise to open the "Exercise 34 Part IV Google Earth™ video."

1. (a)   Fly to Point 1 at the base of a lava flow near the rhyolitic plug dome volcano of Crater Mountain in California. Compare the elevation of the bottom and top of the lava flow edge here to determine its thickness. Do the same at Point 2, a lava flow near the basaltic cinder cone volcano SP Mountain in Arizona.

Crater Mtn. flow: _____ ft. thick   SP Mtn. flow: _____ ft. thick

   (b)   What explains the difference in lava flow thicknesses?

2. (a)   Fly to Point 3, another cinder cone near SP Mountain, and then to Point 4, a third cinder cone in the area. Which of the cinder cones—SP Mountain, the cinder cone at Point 3, or the cinder cone at Point 4—is likely to be the oldest, and which is likely to be the youngest?

   (b)   How can you tell?

EXERCISE 34 PROBLEMS—PART V—INTERNET

The following questions are based on Figures 34-4, 34-5, 34-6, and 34-7. Go to the Hess *Physical Geography Laboratory Manual*, 12th edition, website at **www.MasteringGeography.com**, then Exercise 34, or scan the QR (Quick Response) code for this exercise, to view these photographs.

1. (a)   Which of the four photographs shows a composite volcano?   Figure 34-_____
   (b)   Describe the evidence you see in the photograph that supports your answer.

2. (a)   Which of the four photographs shows a shield volcano?   Figure 34-_____
   (b)   Describe the evidence you see in the photograph that supports your answer.

3. (a)   Which of the four photographs shows a plug dome volcano?   Figure 34-_____
   (b)   Describe the evidence you see in the photograph that supports your answer.

4. (a)   Which of the four photographs shows a cinder cone?   Figure 34-_____
   (b)   Describe the evidence you see in the photograph that supports your answer.

EXERCISE 35
Volcanic Calderas

Objective:	To study the features of volcanic calderas.
Materials:	Lens stereoscope.
Resources:	Internet access or mobile device QR (Quick Response) code reader app (optional).
Reference:	Hess, Darrel. *McKnight's Physical Geography*, 12th ed., pp. 422–423.

CALDERAS

While craters are found at the summit of many volcanoes, some composite and shield volcanoes have developed much larger depressions known as **calderas**. A caldera is a large, generally circular, basin-shaped depression that can be several kilometers across. Calderas form when the upper part of a volcano collapses, often catastrophically, during or following an eruption.

During a major eruption, the magma chamber below a volcano may be emptied, or nearly emptied. The volcano may then be unable to support its own weight, and so will collapse in on itself, leaving a wide, steep-walled caldera (Figure 35-1). After the formation of the caldera, volcanic activity may continue. It is common to see a series of small volcanic cones develop in and around a caldera.

The most famous caldera in North America is Crater Lake in Oregon. This caldera formed about 7700 years ago when one of the Cascade volcanoes (known as "Mount Mazama") collapsed. The caldera is now filled with water, forming a deep lake (Figure 35-2).

Although calderas may develop following major explosive eruptions of **composite volcanoes**, large **shield volcanoes** may develop calderas at their summits in a slightly different manner. Both Mauna Loa and Kīlauea in Hawai'i (Maps T-1 and T-21a) have well-developed "summit" calderas. Calderas such as these develop when fluid lava is vented from rift zones along the flanks of the volcano. As the magma chamber empties, the summit area collapses and a caldera is formed.

OKMOK CALDERA

Okmok Caldera, on Umnak Island in the Aleutian Islands (Map T-2 and Map T-21b), formed from the catastrophic collapse of a large volcano. Figure 35-3 is a digital shaded relief map and Figure 35-4 is a stereogram of the Okmok Caldera. The composition of Okmok varies, but it is mostly basaltic. Although the structure is that of a collapsed shield volcano, there are interbedded layers of pyroclastic deposits as well.

The geologic history of the Okmok Caldera is complex, but evidently involved two caldera-forming eruptions. The first event, about 8250 years ago, formed the outer rim visible on the map, and deposited pyroclastic debris widely throughout the region. The second caldera-forming eruption occurred about 2400 years ago and entailed the subsequent collapse of younger lava flows within the old caldera walls. The last significant activity in Okmok took place in 1997 when a lava flow was vented onto the caldera floor. An earthquake swarm was reported in 2001, and ash was ejected from the caldera in 2008.

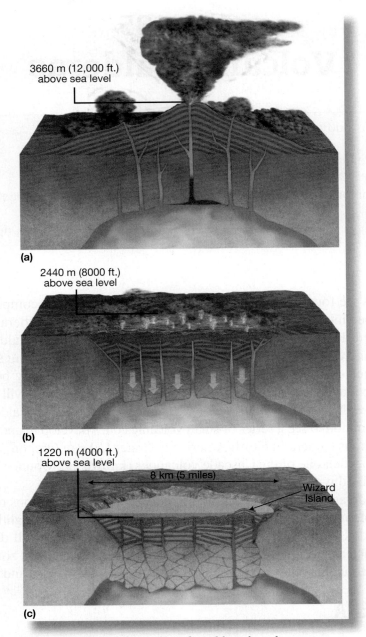

3660 m (12,000 ft.)
above sea level

(a)

2440 m (8000 ft.)
above sea level

(b)

1220 m (4000 ft.)
above sea level

8 km (5 miles)

Wizard
Island

(c)

Figure 35-1: The formation of a caldera, based on
Oregon's Crater Lake. (a) Mount Mazama about 7700 years
ago. (b) During cataclysmic eruption, an enormous volume
of pyroclastic material was erupted from the magma chamber and the volcano collapsed, forming a caldera. (c) The
caldera partially filled with water forming a lake; a new
fissure formed the volcano known as Wizard Island.
(From Hess, *McKnight's Physical Geography*, 12th ed.)

Figure 35-2: Digital shaded relief map of Crater Lake, Oregon. Wizard Island, on the western side of the lake, formed after the initial collapse of the caldera. (From U.S. Geological Survey; scale approximately 1:300,000; ↑N)

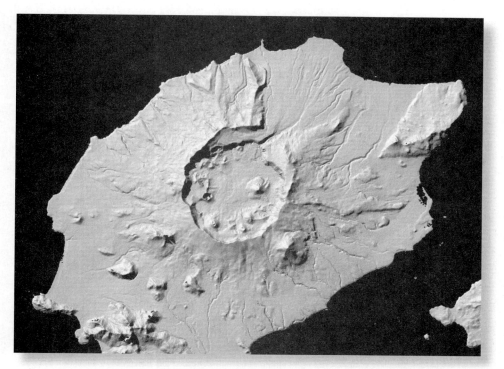

Figure 35-3: Digital shaded relief map of Okmok Caldera on Umnak Island, Alaska. (Image courtesy of Alaska Volcano Observatory/State of Alaska Division of Geological & Geophysical Surveys/Janet Schaefer; ↑N)

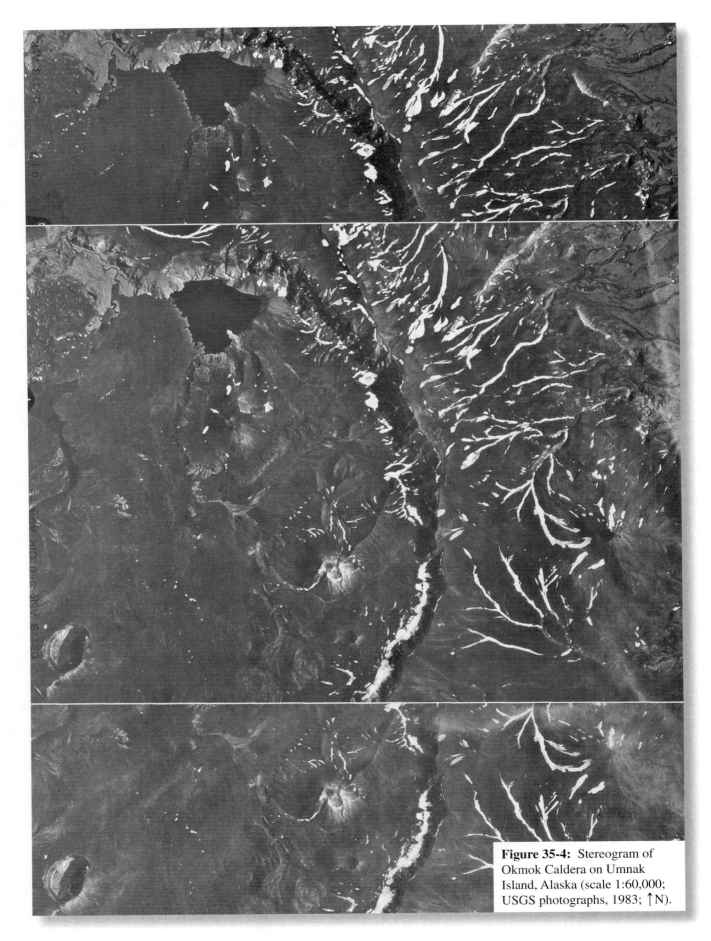

Figure 35-4: Stereogram of Okmok Caldera on Umnak Island, Alaska (scale 1:60,000; USGS photographs, 1983; ↑N).

Name _____   Section _____

EXERCISE 35 PROBLEMS—PART I

The following questions are based on Map T-2, the "Umnak, Alaska," topographic map (scale 1:250,000; contour interval 200 feet), Map T-21b, a satellite image; and Figures 35-3 and 35-4, a shaded relief map and a stereogram showing the Okmok Caldera (53°25′33″N, 168°07′38″W) on Umnak Island in the Aleutian Islands.

1.   What is the approximate diameter of the rim of the Okmok Caldera?   _____ miles

2.   What is the approximate depth of the caldera (from the rim down to the floor)?

_____ feet

3.   A caldera similar to this one filled with water to become Crater Lake in Oregon (42°56′42″N, 122°06′01″W; see Figure 35-2). Why isn't the Okmok Caldera presently filled with water?

4.   What evidence from the map and stereogram suggests that further volcanic activity took place inside the caldera after the final collapse of the volcano? Be as specific as you can.

EXERCISE 35 PROBLEMS—PART II—GOOGLE EARTH™

Go to the Hess *Physical Geography Laboratory Manual*, 12th edition, website at **www .MasteringGeography.com** and open the KMZ file for Exercise 35 Part II, or scan the QR (Quick Response) code for this exercise to view the "Exercise 35 Part II Google Earth™ video." You can also look at Map T-21a, a satellite image showing the summit caldera, lava flows, and rift zones of Mauna Loa.

1.   Fly to Point 1, overlooking Mokuʻāweoweo, the summit caldera of Mauna Loa on the Big Island of Hawaiʻi. Does it appear that many recent (dark black) lava flows have spilled out over the sides of the summit caldera?

2.   Fly to Point 2, the small depression of Lua Hou, and Point 3, the Southwest Rift Zone of the volcano. What is the apparent relationship between Lua Hou and the Southwest Rift Zone?

3.   From where do most of the recent lava flows on Mauna Loa appear to originate?

Name _____   Section _____

EXERCISE 35 PROBLEMS—PART III

The following questions are based on Map T-2, the "Umnak, Alaska," topographic map (scale 1:250,000; contour interval 200 feet). In this problem, you will estimate the height of the original Okmok volcano before the caldera was formed. (You may also determine the elevation changes and distances needed to make these calculations by using Google Earth™.)

Gradients can be used to estimate an elevation increase over a given distance. (See Exercise 34 for a review of gradients.) For example, if a mountain has a gradient of 1000 feet per mile, we can estimate that over a horizontal distance of five miles, the elevation will increase by 5000 feet.

$$5 \text{ miles} \times 1000 \text{ ft/mi} = 5000 \text{ foot elevation increase over five miles}$$

1. Compute the gradient of the outer base of the Okmok Caldera along line EF. Use the coastline closest to Point E as the starting point for measuring horizontal distance, and the northwest rim of the caldera as the stopping point. It may be easiest to use the graphic map scale to measure out a distance of four or five miles, and then determine the elevation change over that distance.

$$\frac{}{\text{Elevation Change}} \text{ feet} \div \frac{}{\text{Number of Miles}} \text{ miles} = \frac{}{\text{Gradient}} \text{ feet/mile}$$

2. Estimate the height of the original volcano. Assume that the present gradient of the Okmok Caldera (calculated in problem 1) is the same as that of the volcano before it collapsed. Given the complex history of the Okmok Caldera, this is a very simplistic assumption, but it can be used to provide a crude estimate of the volcano's previous height.

    (a) Following line EF, what is the horizontal distance from
         the coastline (closest to Point E) to the center of the caldera?   _____ miles

    (b) Multiply the horizontal distance (from problem 2a) by the gradient
         of the present caldera (problem 1). This is an estimate of the
         original elevation of the volcano before its collapse.   _____ feet

    (c) Approximately how much higher was the original
         volcano than the present caldera rim?   _____ feet

EXERCISE 35 PROBLEMS—PART IV—GOOGLE EARTH™

Go to the Hess *Physical Geography Laboratory Manual*, 12th edition, website at **www .MasteringGeography.com** and open the KMZ file for Exercise 35 Part IV, or scan the QR (Quick Response) code for this exercise to view the "Exercise 35 Part IV Google Earth™ video." Using the measurement and elevation functions in Google Earth™ and the same calculation procedures as in Part III, estimate the original height of Mount Mazama before it collapsed to form the Crater Lake caldera (this will provide only a crude estimate of the height of the original volcano).

1. Gradient of southern slope of Crater Lake caldera: _____ feet/mile

2. Approximately how much higher was Mount Mazama
    than the present caldera rim?   _____ feet

EXERCISE 36
Faulting

Objective:	To review the different kinds of faults and to study faulted landscapes.
Materials:	Lens stereoscope.
Resources:	Internet access or mobile device QR (Quick Response) code reader app (optional).
Reference:	Hess, Darrel. *McKnight's Physical Geography*, 12th ed., pp. 432–436.

FAULTS

Faulting occurs when stresses forcibly break apart and displace a rock structure. The displacement along a fault can be horizontal, vertical, or a combination of the two. This displacement takes place along a fracture zone in the rock called a *fault plane* (Figure 36-1). The intersection of the fault plane with the surface is called a *fault line*. With some large faults, there may not be a simple fault plane, instead, a zone of crushed rock many meters wide called a *fault zone* will be present.

If vertical movement has taken place along a fault, the side that has dropped down relative to the other side is called the *downthrown block,* and the side that has been relatively uplifted is known as the *upthrown block*. A steep *fault scarp* may be present where the fault plane on the *upthrown block* is exposed.

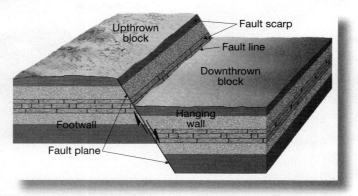

Figure 36-1: A simple fault structure. (From Hess, *McKnight's Physical Geography*, 12th ed.)

TYPES OF FAULTS

Although there are many different kinds of faults, they all can be placed into four general categories.

Normal Faults: As shown in Figure 36-2a, along a **normal fault** the movement is primarily vertical, exposing a steep fault plane. Normal faulting is the result of extension ("tension")—stresses working to stretch or pull apart the landscape (direction of stress shown with wide arrows).

Reverse and Thrust Faults: Movement along **reverse faults** (Figure 36-2b) is also mainly vertical, but in this case compressional stresses have produced the fault displacement. **Thrust** (or "overthrust") faults (Figure 36-2c) also result from compression, but the upthrown block overrides the downthrown block at a low angle.

Strike-Slip Faults: The movement along a **strike-slip fault** (Figure 36-2d) is primarily horizontal and is produced by shear stresses. Strike-slip faults are discussed in more detail in Exercise 37.

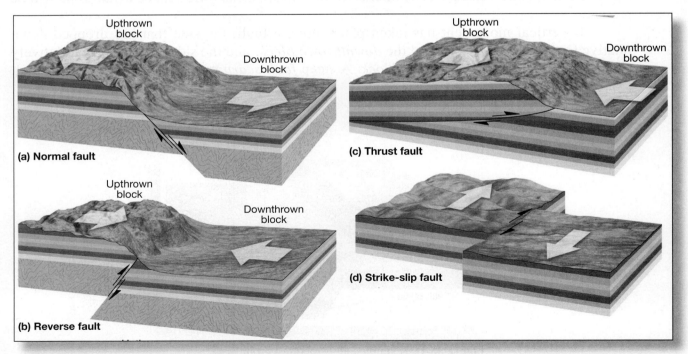

Figure 36-2: The principal types of faults. The large arrows show the direction of stress; the small arrows show the relative direction of displacement along the fault plane. (From Hess, *McKnight's Physical Geography*, 12th ed.)

LANDFORMS PRODUCED BY NORMAL FAULTING

There are many conspicuous landforms associated with normal faulting. **Tilted fault block mountains** such as the Sierra Nevada (Figure 36-3) are produced by fault displacement along one side. When a block of land is downdropped between two roughly parallel faults, a **graben** is formed (Figure 36-4). When a basin is tilted down along just one side, it is sometimes referred to as a *half graben*. A mountain block between two parallel down-dropped blocks is known as a **horst**.

Predominantly normal faulting throughout much of the Basin and Range province of the western United States has produced a series of fault-block mountains and down-dropped basins. Figure 36-5 and Map T-24 show a faulted landscape in the Basin and Range province in northeastern California.

Figure 36-3: The Sierra Nevada is a tilted fault block mountain range in California. (From Hess, *McKnight's Physical Geography*, 12th ed.)

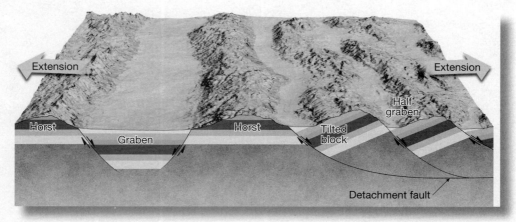

Figure 36-4: Horsts and grabens are bound by faults on both sides, whereas tilted blocks have been faulted along just one side, forming tilted fault-block mountains and *half grabens*. Extension may also result in *detachment faults*, where a steep normal fault ties in with a nearly horizontal fault below. (From Hess, *McKnight's Physical Geography*, 12th ed.)

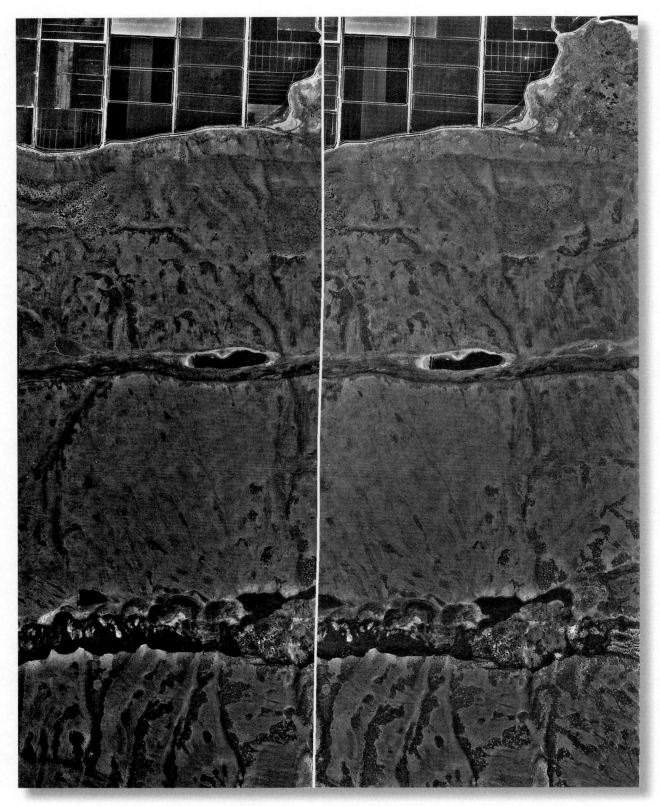

Figure 36-5: Stereogram of fault scarps near Mt. Dome, California. North is to left side of page (scale 1:40,000; USGS photographs, 1993; ← N).

Name _____   Section _____

EXERCISE 36 PROBLEMS—PART I

The following question is based on the "Mt. Dome, California" quadrangle (Map T-24; enlarged to scale 1:48,000; contour interval 40 feet). Three prominent fault scarps can be seen running north to south across the map, and will be referred to as the "western" fault, the "central" fault, and the "eastern" fault. These faults scarps are the result of normal faulting. Figure 36-5 is a stereogram of the same region (in Figure 36-5, north is to the left side of the page).

Using the graph below, construct a topographic profile from Point "A" to Point "B." Plot the index contours, as well as the crest and bottom of the fault scarps. The vertical exaggeration of the profile is approximately 6.7×.

> Hint: Because the contour lines are very close together, it may be difficult to discern the elevation of the top and bottom of a scarp. To determine these elevations, find an index contour in the gently sloping area between two scarps, and then count the number of contour lines to the top or bottom of a scarp.

EXERCISE 36 PROBLEMS—PART II

The following questions are based on the "Mt. Dome, California," quadrangle (Map T-24; enlarged to scale 1:48,000; contour interval 40 feet) and the stereogram of the same region (Figure 36-5; north is to the left side of the page). Three prominent fault scarps can be seen running north to south across the map, and will be referred to as the "western" fault, the "central" fault, and the "eastern" fault. These fault scarps are the result of normal faulting (41°49′39″N, 121°33′15″W).

1. Determine the approximate amount of visible vertical displacement along each of the three fault scarps at their intersection with line AB (to determine the amount of visible displacement, measure the height [relief] of each fault scarp; it may be easiest to count the number of contour lines shown on each scarp to determine the elevation change).

   Western Fault: _____ feet   Central Fault: _____ feet   Eastern Fault: _____ feet

2. (a) Describe the variation in the height of the "central fault" scarp going from north to south.   _____

   (b) What might explain this variation in scarp height?

Name _____   Section _____

EXERCISE 36 PROBLEMS—PART III—GOOGLE EARTH™

To answer the following questions, go to the Hess *Physical Geography Laboratory Manual*, 12th edition, website at **www.MasteringGeography.com**, then Exercise 36, and select "Exercise 36 Part III Google Earth™" to open a KMZ file in Google Earth™, or scan the QR (Quick Response) code for this exercise to view "Exercise 36 Part III Google Earth™ video."

1.   Fly to Point 1 at the base of a fault scarp near Mount Dome in northeastern California (also shown in Map T-24 and Figure 36-5). Compare the height of the fault scarp here to that at Point 2.

    (a)   What happens to the height of the fault scarp as you move north from Point 1 to Point 2?

    (b)   What might explain this variation in scarp height?

2.   Fly to Point 3, near the Gillem Lakes. Based on the surrounding topography, why have lakes formed here?

EXERCISE 36 PROBLEMS—PART IV—GOOGLE EARTH™

Go to the Hess *Physical Geography Laboratory Manual*, 12th edition, website at **www.MasteringGeography.com** and open the KMZ file for Exercise 36 Part IV, or scan the QR (Quick Response) code for this exercise to view the "Exercise 36 Part IV Google Earth™ video." Here you see the western side of the Wasatch Range near Salt Lake City, Utah. The triangular "facets" along the mountain front mark the location of the eroded fault plane along which the range was uplifted.

1.   Fly from Point 1 to Point 2, and then to Point 3.

    (a)   Does the fault along the foot of the range run in a straight line?

    (b)   If not, describe its course.

2.   At Point 3, notice that the valleys between the triangular facets of the mountain front have been cut down nearly to the level of the basin floor in the foreground. Fly back to Point 1.

    (a)   Have all the streams along the mountain front near Point 1 cut their valleys down to the basin floor between the triangular facets?

    (b)   Why might this be the case?

EXERCISE 37
The San Andreas Fault

Objective:	To study features produced by strike-slip faults and to examine the displacement of a stream by the San Andreas Fault.
Materials:	Lens stereoscope.
Resources:	Internet access or mobile device QR (Quick Response) code reader app (optional).
Reference:	Hess, Darrel. *McKnight's Physical Geography*, 12th ed., pp. 409-411 and 435-436.

STRIKE-SLIP FAULTS

In contrast to normal and reverse faults in which the dominant movement along the fault plane is vertical, displacement along **strike-slip faults** is predominantly lateral. Figure 37-1 is a block diagram showing the buildup of stress along a strike-slip fault and finally the rupture and displacement along the fault. The fault shown in the diagram below is known as a "right-lateral" strike-slip fault. This means that the displacement, looking across the fault, is to the right. In other words, relative to our position on one side of the fault, things on the other side of the fault appear to have been offset to the right.

LANDFORMS PRODUCED BY STRIKE-SLIP FAULTS

Strike-slip faults can produce a wide variety of landforms (Figure 37-2). The trace of a large strike-slip fault may be expressed by a narrow **linear fault trough** or valley that can extend for many kilometers. Repeated movement along the fault crushes the rock in the fault zone, and this rock is more easily eroded, leaving a linear trough. Small depressions caused by the settling of small blocks with the fault zone also often develop and become filled with water to form **sag ponds**.

Linear features can develop along the trace of a strike-slip fault in a number of ways. Slight compressional stresses can squeeze up small linear ridges, parallel to the trace of the fault.

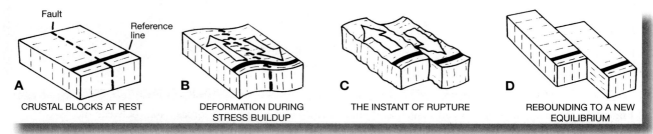

Figure 37-1: Movement along a strike-slip fault. (From U.S. Geological Survey Circular 1045)

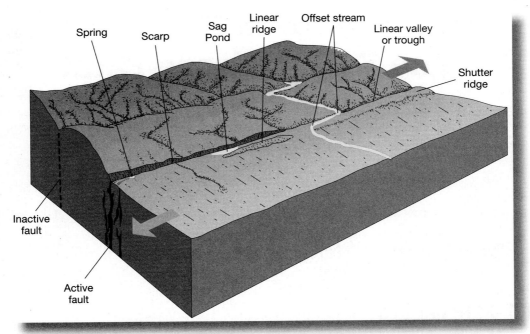

Figure 37-2: Common landforms produced by strike-slip faults. Diagram shows right-lateral offset (fault displacement shown with large arrows). (From McKnight and Hess, *Physical Geography*, 9th ed.; after U.S. Geological Survey Map I-575)

A **shutter ridge** is a hill that has been faulted laterally, closing off a valley. In addition to the dominant lateral movement, many strike-slip faults also exhibit a limited amount of vertical movement ("dip-slip") that can produce **scarps** along the trace of the fault.

One of the most conspicuous landforms produced by a strike-slip fault is an **offset stream** or offset drainage channel. Through repeated fault movement, streams flowing across a fault gradually have their courses displaced (Figure 37-3).

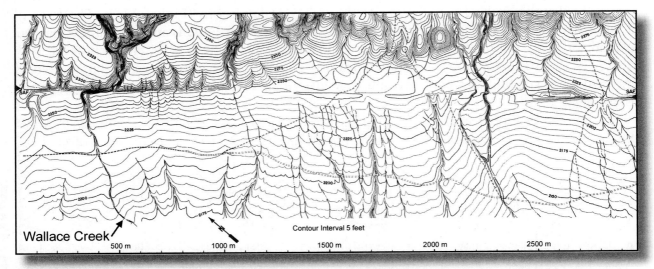

Figure 37-3: Scarp and offset streams (such as Wallace Creek) along the San Andreas Fault (between the "SAF" triangles on both sides of the map) in the Carrizo Plain of California. (Adapted from Sieh, K.E., and R.E. Wallace, 1987. "The San Andreas Fault at Wallace Creek, San Luis Obispo County, California" in *Geological Society of America, Cordilleran Section Centennial Field Guide*, p. 235)

THE SAN ANDREAS FAULT

The San Andreas Fault is perhaps the most prominent strike-slip fault in the world. This right-lateral strike-slip fault extends for approximately 1300 kilometers (800 miles) from the Gulf of California, north through the Coast Ranges of California, and then finally out to sea.

Although commonly called the "plate boundary" between the North American Plate and the Pacific Plate, this is somewhat simplistic. The San Andreas Fault system is the most important and most obvious component of this plate boundary, but it is likely that stresses associated with this boundary have resulted in more widespread faulting—possibly including faulting within the Basin and Range province of western North America (Figure 37-4). Perhaps only 60 percent of the approximately 5.6 centimeters (2.2 inches) per year of relative motion between the Pacific and North American Plates is released along the San Andreas Fault itself.

The San Andreas Fault is not a simple break. In places it has broken into strands, producing a fault zone several hundred meters wide. In addition, the fault system includes dozens of other mostly parallel right-lateral strike-slip faults, such as the Hayward and Calaveras Faults in the San Francisco Bay Area, and the San Jacinto and Elsinore Faults in southern California.

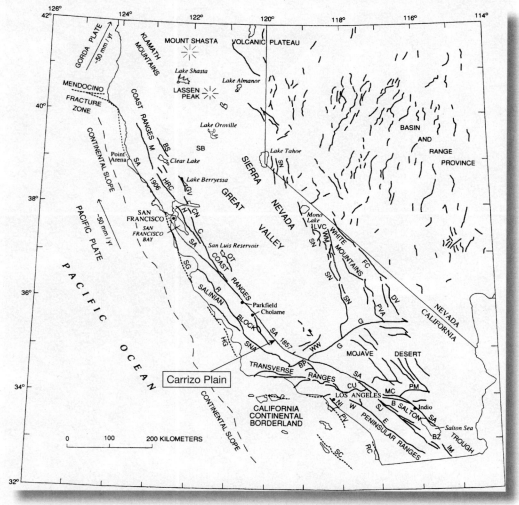

Figure 37-4: The San Andreas and other faults in California and Nevada. Labeled faults include C—Calaveras; E—Elsinore; H—Hayward; SA—San Andreas; SJ—San Jacinto. (From U.S. Geological Survey Professional Paper 1515)

THE SAN ANDREAS FAULT IN THE CARRIZO PLAIN

Figure 37-6 (on page 258) shows a portion of the "McKittrick Summit, California" quadrangle. A color map and aerial imagery of this area is also shown on Maps T-25a and T-25b, and in the stereogram in Figure 37-5. This quadrangle is located in the southern Coast Ranges of California, where the Temblor Range (northeast portion of map and stereogram) runs through the arid Carrizo Plain National Monument (southwest portion of map and stereogram). The San Andreas Fault runs diagonally through this area, from the upper left corner to the lower right corner (approximately between the two arrows on the map). Ephemeral streams are shown as dashed and dotted lines on the map; graded dirt roads are shown as double lines on the map.

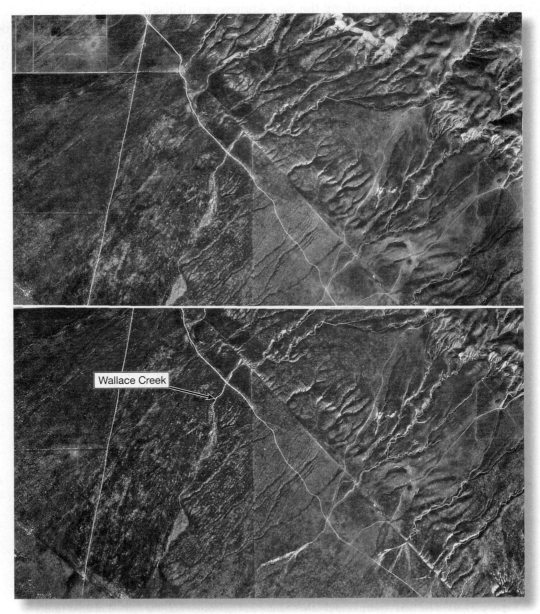

Wallace Creek

Figure 37-5: Stereogram showing San Andreas Fault in the Carrizo Plain, California (scale 1:40,000; USGS photographs, 1989; ↑N).

Name _____   Section _____

EXERCISE 37 PROBLEMS—PART I

The following questions are based on Maps T-25a and T-25b, Figure 37-3, a detailed topographic map of the San Andreas Fault, Figure 37-5, a stereogram of the San Andreas Fault in the Carrizo Plain National Monument, and Figure 37-6, a portion of the "McKittrick Summit, California," quadrangle reproduced on the next page. The stereogram shows the area in the southern half of the topographic map.

1. One offset stream (known as "Wallace Creek") has been labeled for you on the map (35°16′16″N, 119°49′38″W). On the map on the following page, use an arrow labeled "1" to identify another stream that has been offset by displacement along the fault. The contour pattern of dry gullies may also provide you with clues.

    Note: Several streams have had their courses deflected to the *left* rather than to the right. This can happen, for example, when a small shutter ridge is carried along one side of the fault, blocking a stream's course and causing it to flow along the trace of the fault (in this case, to the southeast) until it finds an outlet.

2. On the map on the following page, use an arrow labeled "2" to identify a small linear ridge running parallel to the trace of the fault. (Hint: Look for tiny closed contour lines along the fault trace.) The stereogram, detailed topographic map, and aerial imagery may also be helpful in identifying ridges.

3. On the map on the following page, use an arrow labeled "3" to identify a sag or sag pond. The stereogram may be helpful in identifying sags and dry sag ponds.

4. What stream features, seen in the stereogram, aerial imagery, and detailed topographic map but not the Figure 37-6 map, indicate the position of the fault? (Hint: Look for small stream gullies that flow down toward the fault.)

5. Based on the topographic clues you have identified above, as well as the aerial imagery in Map T-25b and stereogram of the same region, use a colored pencil to sketch in a line showing the position of the San Andreas Fault across the map on the following page.

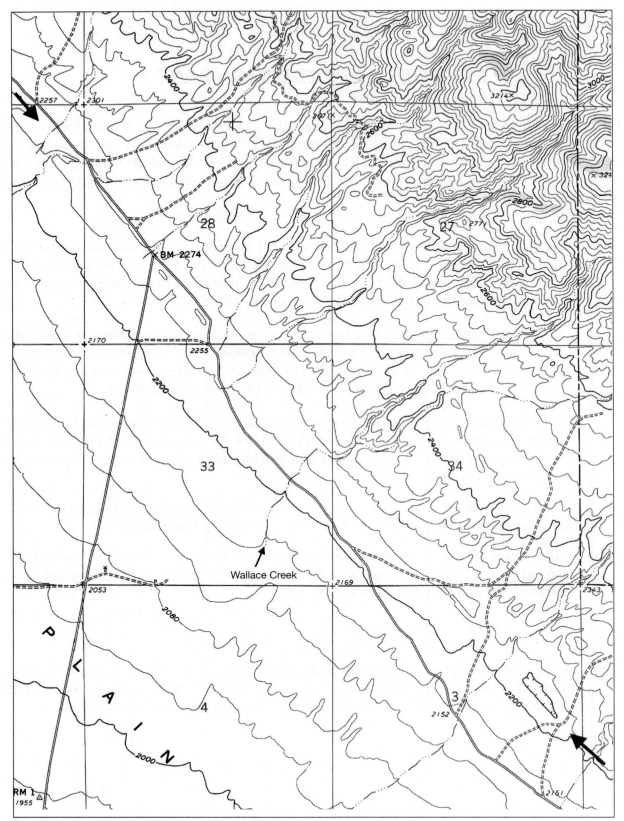

Figure 37-6: USGS "McKittrick Summit, California" quadrangle (scale 1:24,000; contour interval 40 feet; ↑N). The San Andreas Fault runs approximately between the two arrows. Map T-25a is a color map and Map T-25b is aerial imagery of this same area.

Name _____   Section _____

EXERCISE 37 PROBLEMS—PART II

The following questions are based on Map T-25a, Figure 37-3, a detailed topographic map of the San Andreas Fault, Figure 37-5, a stereogram of the San Andreas Fault in the Carrizo Plain National Monument, and Figure 37-6, a portion of the "McKittrick Summit, California," quadrangle. "Wallace Creek" is an offset stream that has been labeled for you on Figure 37-6.

Most of the movement between the Pacific Plate side and the North American Plate side of the San Andreas Fault is released abruptly after years of accumulating strain—at which time the fault ruptures with a displacement of several meters.

Further, it appears that different segments of the fault remain locked for different periods of time. In a general way, the size of an earthquake expected to occur along a segment of the fault system relates to the length of time the fault accumulates strain before rupturing. The longer the strain accumulates, the larger the earthquake when the fault finally ruptures.

1. Wallace Creek has been offset by 130.0 meters (430 feet) to the right along the fault. The age of stream deposits indicates that this offset began 3700 years ago. Based on these figures, calculate the average rate of movement along this segment of the fault in centimeters per year (or inches per year). Keep in mind that your answer is an estimate of the long-term average, *not* the expected movement *each* year.

   (a) Total offset in centimeters (or inches): _____ centimeters (or inches)

   (b) Average offset per year: _____ centimeters per year (or in./yr.)

2. The great Fort Tejon earthquake of January 9, 1857 (magnitude 7.9) was the last major earthquake in this region. It ruptured a 370 kilometer (220 mile) segment of the San Andreas Fault and produced 10.0 meters (33 feet) of offset in this area. Based on the average rate of fault movement calculated in problem 1b, estimate how many years of accumulated strain were released during that earthquake. (Note: This answer is based on a very simplistic assumption.)

   _____ years of accumulated strain

3. Assuming that this segment of the San Andreas Fault ruptures at fairly regular intervals, and paleoseismic studies suggest that this may be true, estimate the year when the next great earthquake might occur along this section of the fault. (Note: This answer is also based on a very simplistic assumption.)

   Approximate year of next great earthquake: _____

Name _____   Section _____

EXERCISE 37 PROBLEMS—PART III—GOOGLE EARTH™

To answer the following questions, go to the Hess *Physical Geography Laboratory Manual*, 12th edition, website at **www.MasteringGeography.com**, then Exercise 37 and select "Exercise 37 Google Earth™" to open a KMZ file in Google Earth™, or scan the QR (Quick Response) code for this exercise to view "Exercise 37 Part III Google Earth™ video."

1.   Fly to Point 1 where Wallace Creek flows out along the San Andreas Fault at the foot of the Temblor Range in California (also shown in Maps T-25a and T-25b, Figures 37-3, 37-5, and 37-6), and then to Point 2. Wallace Creek has been offset by about 130 meters (430) feet between Point 1 and Point 2; from Point 2 its channel continues to the southwest. Look for evidence of an older, now abandoned, offset channel of Wallace Creek farther to the northwest along the fault. Use the ruler function to measure the additional amount of offset from Point 2 to this abandon channel.

Additional offset from Point 2 to the abandoned channel: _____ meters (or feet)

2.   (a)   Fly to Point 3. What happens to the tiny gullies on the slope here when they reach the fault?

(b)   Why don't the channels continue on the other side of the fault?

3.   Fly to Point 4. This stream channel begins at the fault, sloping downhill to the southwest. It is known as a *beheaded stream* because it starts abruptly at the fault. How did it form?

Lab Exercise
38

https://goo.gl/jvrXyg

EXERCISE **38**
Mass Wasting

Objective:	To study landform features produced by mass wasting.
Materials:	Lens stereoscope.
Resources:	Internet access or mobile device QR (Quick Response) code reader app (optional).
Reference:	Hess, Darrel. *McKnight's Physical Geography*, 12th ed., pp. 453–462.

MASS WASTING

Mass wasting involves the relatively short-distance, downslope movement of weathered rock, primarily under the influence of gravity. There are many types of mass wasting, each produced by a different combination of factors—such as the steepness of the slope, the coherency of the bedrock, and the amount of water present—and each leaving a different kind of mark in the landscape.

Rockfall: With **rockfall**, loose, weathered rock falls to the foot of a cliff or steep slope. The angular, largely unsorted material that accumulates is known as **talus** or **scree**, and over time it may build up into a steeply sloping **talus cone** at the foot of the slope (Figure 38-1).

Landslide: **Landslides** involve the movement of masses of rock—occasionally coherent sections of a hillside—sliding downslope as a unit along a largely flat sliding plane. A landslide leaves a prominent scar at its point of origin, as well as an irregular jumble of rock where it comes to rest (Figure 38-2). Landslides are frequently triggered by heavy rainfall, which adds mass to a hillside and makes it more susceptible to sliding. However, slides do not involve the flow of material in a mixture of water.

Figure 38-1: Talus cones accumulating from rockfall. (From Hess, *McKnight's Physical Geography*, 12th ed.)

Figure 38-2: Landslide. (From Hess, *McKnight's Physical Geography*, 12th ed.)

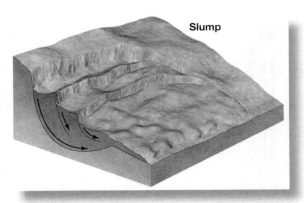

Figure 38-3: Slump. (From Hess, *McKnight's Physical Geography*, 12th ed.)

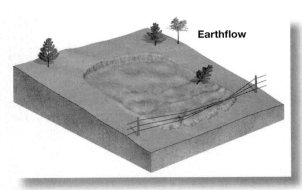

Figure 38-4: Earthflow. (From Hess, *McKnight's Physical Geography*, 12th ed.)

Slump: A **slump** is a "rotational" slide that involves movement along a curved slide plane (Figure 38-3).

Earthflow: When the quantity of water is great, a section of hillside consisting of loose or highly weathered material may begin to flow downhill as a waterlogged mass. Like a slide, an **earthflow** leaves a scar at its starting point, but in this case very wet material is flowing—rather than sliding—downslope (Figure 38-4).

Mudflow and Debris Flow: A **mudflow** is a very wet mixture of water and rock that amasses in a stream valley, then flows down-valley with the consistency of wet concrete. When mudflows contain large boulders, they are often called **debris flows** (although the terms "debris flow" and "mudflow" are sometimes used interchangeably). Mudflows and debris flows are common in arid and semi-arid mountainous regions where a heavy rain can trigger a flashflood. The flashflood quickly moves down a dry canyon picking up loose weathered material, eventually depositing the mud and rock at the mouth of the canyon in an **alluvial fan** (Figure 38-5; additional discussion of alluvial fans is found in Exercise 45). Note that whereas landslides, slumps, and earthflows take place on hillsides, mudflows and debris flows develop in stream valleys.

Creep: **Creep** or **soil creep** entails the very slow downslope movement of the surface layer of a hillside, often as the result of repeated freeze-thaw (or wetting and drying) cycles of the soil layer. Often the most visible signs of soil creep are fence posts or telephone poles that are leaning downhill (Figure 38-6).

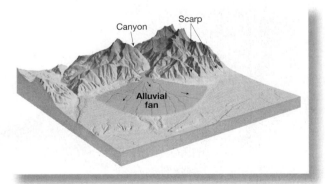

Figure 38-5: Alluvial fan developing from mudflow and debris flow deposits at the mouth of a canyon. (From Hess, *McKnight's Physical Geography*, 12th ed.)

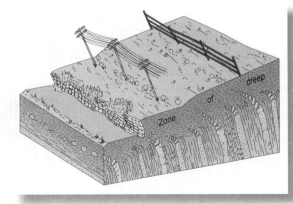

Figure 38-6: Soil creep. (From Hess, *McKnight's Physical Geography*, 12th ed.)

Figure 38-7: Photograph taken on March 27, 2014, showing the Oso landslide in Washington. (Adapted from USGS photograph taken by Jonathan Godt.)

OSO LANDSLIDE

On March 22, 2014, an enormous landslide occurred along the North Fork Stillaguamish River in Washington (Figure 38-7). Mud and debris buried homes in the community below, killing 43 people.

Landslides have occurred in this region in the past. A slide in 2006 shifted the river course more than 200 meters (700 feet) south, and material from that slide moved again in 2014 after a period of heavy rain, ultimately becoming a "runout slide" that moved a great distance from its origin point (Figure 38-8).

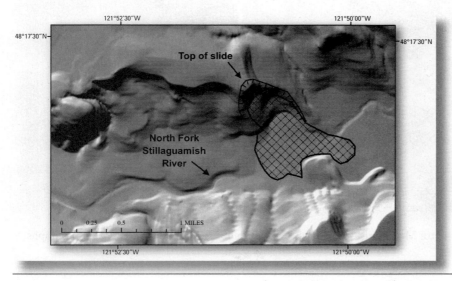

Figure 38-8: Shaded relief digital elevation model showing the extent of the 2014 Oso landslide in Washington. (Adapted from USGS Open-File Report 2014-1065.)

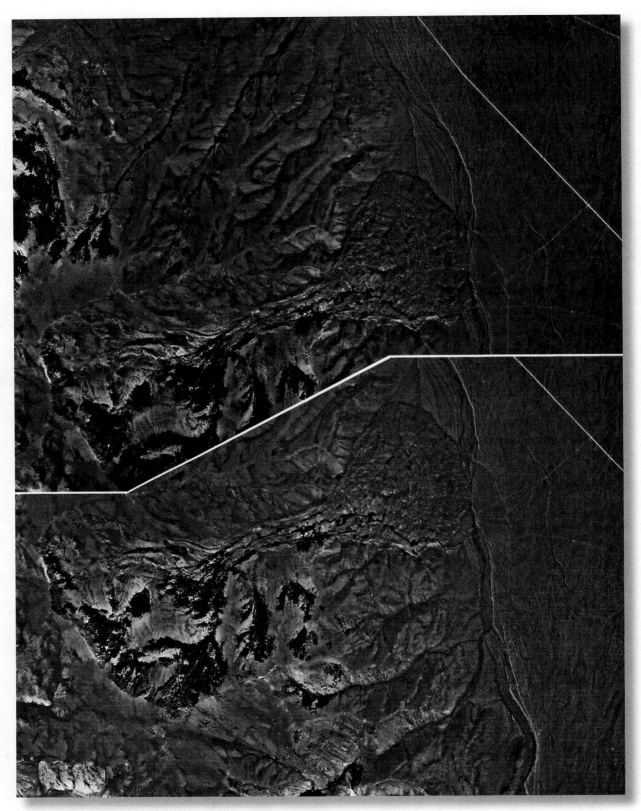

Figure 38-9: Stereogram of Pahsimeroi Mountains, near Spring Hill in Idaho (scale 1:40,000; USGS photographs, 1999; ↑N).

Name _____   Section _____

EXERCISE 38 PROBLEMS—PART I

The following questions are based on Map T-26a, a section of the USGS "Spring Hill, Idaho," quadrangle (scale 1:24,000; contour interval 20 feet), and Figure 38-9, a stereogram of the same region. The map and stereogram show a large mass wasting deposit at the foot of the Pahsimeroi Mountains in Idaho (44°17′39″N, 113°43′02″W).

1.   What evidence shown on the map and stereogram suggests that the material in the deposit does not consist of coherent bedrock? (Hint: Contrast the appearance of the deposit with that of the surrounding mountain front.)

2.   (a)   Which kind of mass wasting most likely left this deposit,
                a type of *flow* or a type of *slide*?   _____

       (b)   What evidence shown on the map and stereogram supports your answer? (Hint: Note the path taken by the material on its way downslope.)

EXERCISE 38 PROBLEMS—PART II

The following questions are based on Map T-26a, a section of the USGS "Spring Hill, Idaho," quadrangle (scale 1:24,000; contour interval 20 feet). You are going to calculate the approximate volume of the mass wasting deposit shown at the foot of the Pahsimeroi Mountains.

1.   How thick is the deposit at the end (the "toe") of the deposit?
       (Take the 6800′ contour to be the top of the deposit).   _____ feet

2.   Although the deposit is somewhat irregular, assume that it is square or rectangular in shape. Estimate the length of the two sides of the deposit that are in contact with the mountain front (the northwest [NW] side and the southwest side [SW]), then calculate its surface area.

$$\underset{\text{(NW side length)}}{\underline{\hspace{3cm}}} \text{feet} \times \underset{\text{(SW side length)}}{\underline{\hspace{3cm}}} \text{feet} = \underset{\text{(total surface area)}}{\underline{\hspace{3cm}}} \text{square feet}$$

3.   Assuming that your answer to problem 1 represents the uniform thickness of the deposit (a simplistic assumption), calculate the total volume of the deposit.

$$\underset{\text{(deposit thickness)}}{\underline{\hspace{3cm}}} \text{feet} \times \underset{\text{(total surface area)}}{\underline{\hspace{3cm}}} \text{square feet} = \underset{\text{(approx. total volume)}}{\underline{\hspace{3cm}}} \text{cubic feet}$$

Name _____   Section _____

EXERCISE 38 PROBLEMS—PART III—INTERNET

The following questions are based on Figures 38-10, 38-11, 38-12, and 38-13, photographs you can view on the Lab Manual website. Go to the Hess *Physical Geography Laboratory Manual*, 12th edition, website at **www.MasteringGeography.com**, then Exercise 38, or scan the QR (Quick Response) code for this exercise.

1.   (a)   Which of the four photographs shows the result of rockfall? Figure 38-_____

    (b)   Describe the evidence you see in the photograph that supports your answer.

2.   (a)   Which of the four photographs shows the result of landside or slump?

                                                     Figure 38-_____

    (b)   Describe the evidence you see in the photograph that supports your answer.

3.   (a)   Which of the four photographs shows the result of earthflow? Figure 38-_____

    (b)   Describe the evidence you see in the photograph that supports your answer.

4.   (a)   Which of the four photographs shows the result of soil creep? Figure 38-_____

    (b)   Describe the evidence you see in the photograph that supports your answer.

Name _____   Section _____

EXERCISE 38 PROBLEMS—PART IV

The following questions are based on Figure 38-8, a digital elevation model showing the 2014 Oso, Washington, landslide along the North Fork Stillaguamish River, and color Map T-26b in the back of the Lab Manual, showing a portion of the "Mount Higgins, Washington" quadrangle (scale 1:24,000; contour interval 40 feet); this same topographic map is reproduced in black-and-white as Figure 38-14.

1.   Using Figure 38-8 for reference, use a colored pencil to draw in the extent of the 2014 Oso landslide on Figure 38-14.

2.   Why did the debris separate into two lobes at the far end of the slide?

3.   Using the graphic map scale, estimate how far the slide moved across the valley floor of the North Fork Stillaguamish River from the bottom of the mountain side.

_____ mile(s)

4.   How much of Steelhead Drive (and the houses along it) were covered by the landslide?

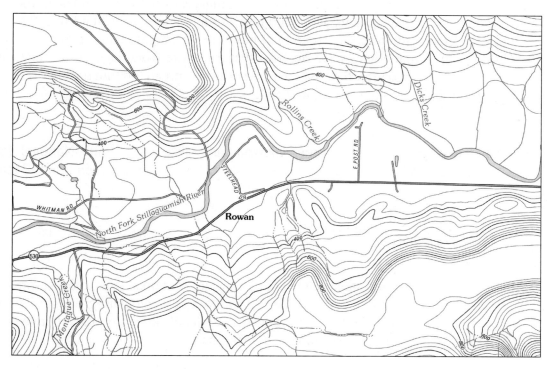

Figure 38-14: USGS "Mount Higgins, Washington" quadrangle (scale approximately 1:24,000; contour interval 40 feet; ↑N; this map is shown in color as Map T-26b in the back of the Lab Manual).

Name _____   Section _____

EXERCISE 38 PROBLEMS—PART V—GOOGLE EARTH™

To answer the following questions, go to the Hess *Physical Geography Laboratory Manual*, 12th edition, website at **www.MasteringGeography.com**, then Exercise 38, and select "Exercise 38 Part V Google Earth™" to open a KMZ file in Google Earth™, or scan the QR (Quick Response) code for this exercise and view "Exercise 38 Part V Google Earth™ video."

1. (a) Fly to Point 1, a large mass wasting deposit at the foot of the of the Pahsimeroi Mountains in Idaho (also shown in Map T-26a and Figure 38-9), and then to Point 2. At least some of this material originated up near the crest of the mountains near Point 2. Did this material more likely move down out of the mountain side as a slide of dry material, or as a flow of fairly wet material?

    (b) What evidence do you see that supports your answer? (Hint: Look at the likely source area of the material, as well as the path it took down to the mountain front.)

2. Fly to Point 3, showing the Blackhawk landslide in the Mojave Desert of California, and then to Point 4. This prehistoric slide—one of the largest known landslides on Earth—began as a rockfall off Blackhawk Mountain near Point 3. It then moved north off the mountain front, perhaps riding on a cushion of compressed air, before coming to rest near Point 4. What evidence that you can see suggests that the entire front edge of the slide came to an abrupt stop (perhaps after the cushion of air became spread too thin to support the mass) rather than stopping gradually or unevenly?

3. (a) Fly to Point 5 in Lee Vining Canyon, just east of Yosemite National Park in California. What kind of mass wasting deposit is shown between Points 5 and 6?

    (b) How does such a landform develop?

    (c) Fly to Point 6 and then determine the **angle of repose** (the steepest angle loose material can achieve before it begins to move) by calculating the gradient of the deposit. Use the ruler function to determine the distance (in miles) between Points 5 and 6, and compare the elevation difference.

    _____ ÷ _____ = _____ feet/mile gradient
    (Elevation change)   (Distance in miles)

EXERCISE **39**
Drainage Basins

Objective:	To study stream drainage basins, stream order, and stream gradients.
Materials:	Lens stereoscope.
Resources:	Internet access or mobile device QR (Quick Response) code reader app (optional).
Reference:	Hess, Darrel. *McKnight's Physical Geography*, 12th ed., pp. 468–471.

STREAM SYSTEMS

Running water is the most important agent of erosion on Earth's land surface. Most of this fluvial erosion is accomplished by the action of **streams**. One important consideration in the study of fluvial geomorphology is the way in which streams come together as a system.

Several different characteristics of stream systems can be recognized. Tiny streams flow together to form larger streams, and these streams in turn join to become still larger streams. We see that within any river system there is a hierarchy of streams, and within this hierarchy we can see differences in the gradient, the length, the area of land being drained, and the amount of water being carried by a stream.

DRAINAGE BASINS

A **drainage basin**, or **watershed**, is an area within which all water flows toward a single stream. Figure 39-1 is a diagram showing the drainage basins of three adjacent streams. The dashed line represents the **drainage divide** that delimits the drainage basin of the middle stream from the drainage basins of the streams on either side. Drainage divides are typically the high ground that separates streams flowing into one drainage basin from streams flowing into another.

As shown in Figure 39-1, a drainage divide may also include an area of **interfluve**—the part of a landscape where water moves downslope as unchanneled **overland flow** rather than as the channeled **streamflow** found in **valleys**. Figure 39-3 is a stereogram of the Eds Creek drainage basin near Deer Peak, Montana. Map T-3 is a color topographic map of the same region.

STREAM ORDERS

One way of analyzing patterns of tributaries within a stream system is with the concept of **stream order**. A "first-order" stream is the smallest stream in a stream system and is defined as a stream without tributaries. Where two first-order streams join, a second-order stream is formed. Where two second-order streams join, a third-order stream is formed (Figure 39-2).

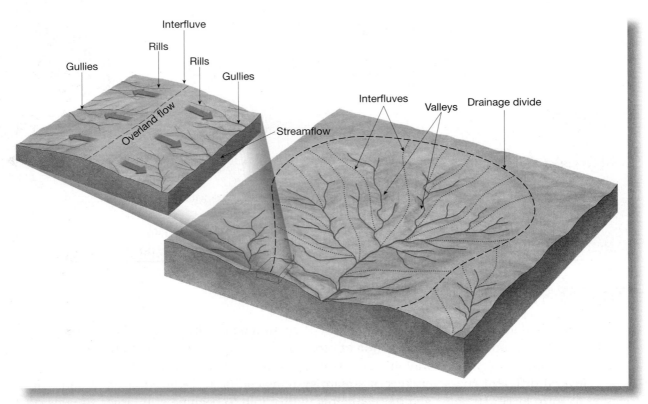

Figure 39-1: Drainage basins, valleys, and interfluves. Streams and valleys within a drainage basin are enclosed by a surrounding drainage divide. (From Hess, *McKnight's Physical Geography*, 12th ed.)

Notice that when a first-order and second-order stream meet, a third-order stream is not formed. Two second-order streams are required to form a third-order stream, two third-order streams are required to form a fourth-order stream, and so on.

In most well-established stream systems, there will be more first-order streams than all other orders combined, and each successively higher order will contain fewer and fewer streams. Also, as stream order increases, stream length and stream drainage areas tend to increase, while the gradient of streams tends to decrease.

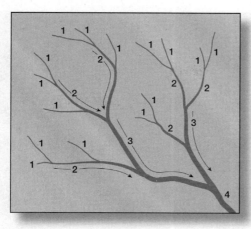

Figure 39-2: Stream orders. (From McKnight and Hess, *Physical Geography*, 9th ed.)

STREAM GRADIENT

The most common way to express the slope of a stream is by its **gradient**. If English measurement units are being used (as on the topographic maps in this exercise), the gradient of a stream is usually stated in feet of elevation change per mile. For example, if a stream drops 678 feet over a distance of 1.5 miles, the gradient is as follows:

$$\text{Gradient} = \frac{\text{Elevation Change}}{\text{Number of Miles}} = \frac{678 \text{ feet}}{1.5 \text{ miles}} = 452 \text{ ft/mi}$$

Figure 39-3: Stereogram of Eds Creek drainage basin near Deer Peak, Montana. North is to left side of page (scale 1:40,000; USGS photographs, 1995; ← N).

Name _____   Section _____

EXERCISE 39 PROBLEMS—PART I

The following questions are based on Map T-3, the "Deer Peak, Montana" quadrangle (scale 1:24,000; contour interval 40 feet), and Figure 39-3, a stereogram of the same area showing the drainage basin of Eds Creek (46°54′26″N, 114°31′02″W).

1. The unimproved dirt road (the double-gray and white line) along "Ed's Ridge" roughly follows which natural feature associated with the Eds Creek drainage basin?

    _____

2. The stream pattern within the drainage basin of Eds Creek on Map T-3 is reproduced at right at a smaller scale. On this small map, trace the length of all first-order streams with blue lines, the length of all second-order streams with red lines, and the length of Eds Creek (as a third-order stream) with a green line. If you don't have colored pencils, number each segment as 1, 2, or 3.

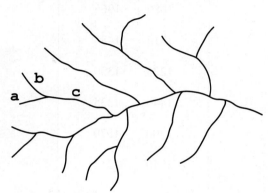

3. (a) How many first-order streams are shown?   _____

    (b) How many second-order streams are shown?   _____

4. Using Map T-3, determine the gradients of the two first-order streams labeled "a" and "b" on the map above, the second-order stream labeled "c" on the map above, as well as Eds Creek after it has become a third-order stream. Determine stream length to the nearest 0.1 mile (take your measurements from Map T-3, *not* the small map above) using the graphic scale provided.

Stream	Order	Elevation Drop (feet)	Length (miles) (to nearest 0.1 mile)	Gradient (feet/mile)
a	1st			
b	1st			
c	2nd			
Eds Creek	3rd			

Name _____   Section _____

EXERCISE 39 PROBLEMS—PART II

Answer the following questions after completing the problems in Part I.

5.   The table below gives the gradients of 12 more first-order streams and 4 more second-order streams in the Eds Creek drainage basin. Fill in the gradients of the streams calculated in Part I, problem 4 (streams "a" and "b" under first-order; stream "c" under second-order; Eds Creek under third-order).

First-Order Stream Gradients (ft/mi)		Second-Order Stream Gradients (ft/mi)	Third-Order Stream Gradient (ft/mi)
1680	1965	800	
1715	1485	680	
1225	1300	665	
1440	1335	800	
1200	1370		
1210	1355		
"a" _____ ft/mi		"c" _____ ft/mi	_____ ft/mi
"b" _____ ft/mi			(Eds Creek)

6.   Using the data from the table above, compute the following:

   (a)   Average gradient of all first-order streams: _____ ft/mi

   (b)   Average gradient of all second-order streams: _____ ft/mi

7.   What generally happens to the gradients of streams as the stream order increases?   _____

8.   Describe the general width and shape (cross section) of the valley floors of first-order streams in the Eds Creek drainage basin.

9.   How is the valley floor of Eds Creek different from the valley floors of the first-order streams? (Hint: Look at the difference in valley floor width.)

Name _____   Section _____

EXERCISE 39 PROBLEMS—PART III

The following questions are based on Figure 39-4, a section of USGS "Dane Canyon, Arizona" quadrangle below, showing the southern edge of the Mogollon Mesa (34°24′33″N, 111°10′52″W), formed by a nearly flat-lying layer of resistant rock (you can view this map in color by going to the Lab Manual website or by scanning the QR code for this exercise).

1. Compute the gradient of any first-order stream north of the mesa edge and the gradient of any first-order stream south of the mesa edge. On the map label the northern stream "a" and the southern stream "b."

Stream	Elevation Drop (ft)	Length (miles)	Gradient (ft/mi)
(a) North			
(b) South			

2. If gradient were the only factor controlling the erosive power of these streams, what should happen to the position of the mesa edge with time?

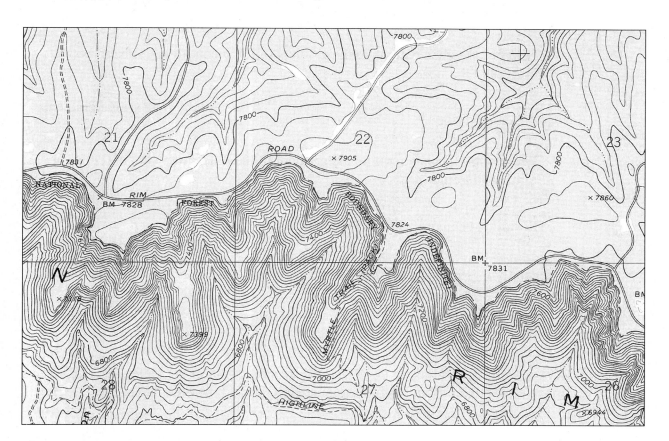

Figure 39-4: USGS "Dane Canyon, Arizona" quadrangle (scale 1:24,000; contour interval 40 feet; ↑N).

Name _____   Section _____

EXERCISE 39 PROBLEMS—PART IV—GOOGLE EARTH™

To answer the following questions, go to the Hess *Physical Geography Laboratory Manual*, 12th edition, website at **www.MasteringGeography.com**, then Exercise 39 and select "Exercise 39 Part IV Google Earth™" to open a KMZ file in Google Earth, or scan the QR (Quick Response) code for this exercise and view "Exercise 39 Part IV Google Earth™ video." In this exercise you'll compare the characteristics of streams in the Eds Creek drainage basin in Montana (also shown in Map T-3 and Figure 39-3).

1. Determine the gradients of three segments of the Eds Creek drainage basin: a first-order stream between Points 1 and 2; a second-order stream between Points 2 and 3; and a third-order stream (Eds Creek) between Points 3 and 4. Use the ruler function to determine distances (to the nearest 0.01 miles) and the elevation change between points.

    (a) Fly to Point 2, then determine the gradient of the first-order stream (between Point 1 and Point 2).

    $$\frac{\text{_____}}{\text{(Elevation change)}} \div \frac{\text{_____}}{\text{(Distance in miles)}} = \text{_____ feet/mile gradient}$$

    (b) Fly to Point 3, then determine the gradient of the second-order stream (between Point 2 and Point 3).

    $$\frac{\text{_____}}{\text{(Elevation change)}} \div \frac{\text{_____}}{\text{(Distance in miles)}} = \text{_____ feet/mile gradient}$$

    (c) Fly to Point 4, then determine the gradient of the third-order stream (between Point 3 and Point 4).

    $$\frac{\text{_____}}{\text{(Elevation change)}} \div \frac{\text{_____}}{\text{(Distance in miles)}} = \text{_____ feet/mile gradient}$$

2. What generally happens to stream gradient as you move downstream from the first-order stream?

3. Fly back to Point 2, and then to Point 4. How does the width of the valley floor change as you move downstream from the first-order stream to the third-order stream?

EXERCISE **40**
Floodplains

Objective:	To study the formation and characteristics of floodplain landforms.
Materials:	Lens stereoscope.
Resources:	Internet access or mobile device QR (Quick Response) code reader app (optional).
Reference:	Hess, Darrel. *McKnight's Physical Geography*, 12th ed., pp. 486–488.

MEANDERING STREAMS

In the upper reaches of a typical river system, or in other places where the gradient of a stream is steep, erosion (often downcutting) is the most prominent fluvial process. In contrast, in the lower reaches of a typical river system, or in other places where a stream is flowing down a gentle slope, a stream will usually begin to meander and depositional features become much more common.

Over time, a stream will meander back and forth across its flat alluvial valley floor known as a **floodplain**. Figure 40-1 illustrates how a **meandering stream** shifts its course through the process of lateral erosion.

Erosion is concentrated on the outside bank of a meander because the water is moving fastest here as it flows into the turn. At the same time, deposition takes place on the inside bank, where the water is moving most slowly. Through this process of lateral erosion on the outside bank and deposition on the inside bank, the position of the stream channel gradually shifts back and forth across the floodplain, generally in the direction of the outside bank of each meander.

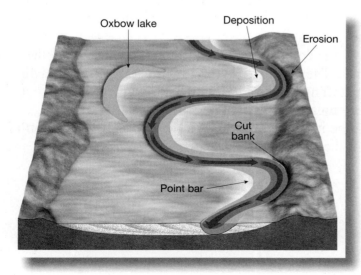

Figure 40-1: Lateral erosion and deposition of a meandering stream. Erosion occurs on the outside of the meander bend where water flow is fastest. (From Hess, *McKnight's Physical Geography*, 12th ed.)

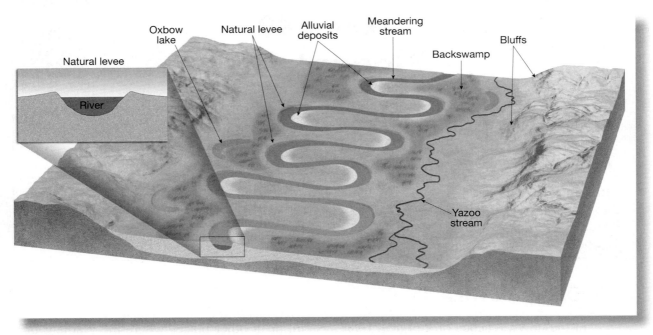

Figure 40-2: Common floodplain landforms. (From Hess, *McKnight's Physical Geography*, 12th ed.)

FLOODPLAIN LANDFORMS

The width of a major floodplain is often due, at least in part, to valley widening through lateral erosion, and floodplains are typically marked by low **bluffs** on both sides. However, most landforms found on a floodplain are largely the result of deposition. The floodplain itself is periodically inundated with flood waters. Along the course of the river, **natural levees** build up from the **alluvium** deposited along its banks during these floods (Figure 40-2).

Because of the nearly level terrain and natural levees that prevent water from draining back into the main river, poorly drained areas and swamps are typical of floodplains. Tributaries known as **yazoo streams** may run parallel to a main river for many kilometers. Yazoo streams are unable to enter because of the main river's natural levees.

When a stream is meandering tightly, it is common for one meander to cut into another, allowing the stream to take a shorter course. The resulting **cutoff meander** may be initially isolated as an **oxbow lake**. Eventually these cutoff meanders will fill up with sediment and dry out, first becoming a swamp, and then finally a dry **meander scar** (Figure 40-3).

Figure 40-4 is a stereogram showing the meandering course of the Souris River in North Dakota, and Map T-4 is a topographic map of the same region.

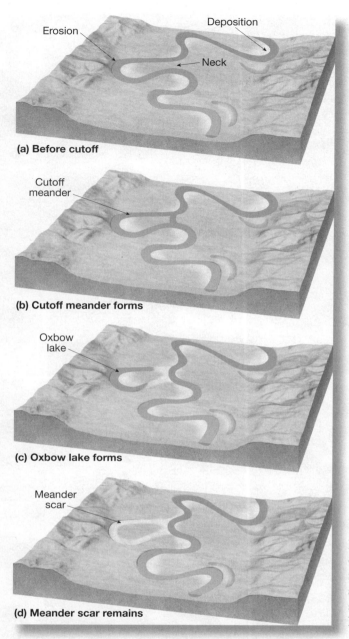

Figure 40-3: The formation of cutoff meanders on a floodplain. As the river cuts across the narrow neck of a meander, the cutoff river bend becomes an oxbow lake, which over time becomes an oxbow swamp, which in turn becomes a meander scar. (From Hess, *McKnight's Physical Geography,* 12th ed.)

MEANDERING RIVERS AS POLITICAL BOUNDARIES

Because meandering rivers often form political boundaries between states or counties, there is a need to fix these boundaries to avoid conflicts when the river shifts its course. For example, by the early 1900s, political boundaries had been permanently established along the Mississippi River. These boundaries remain fixed even if the river changes course. Today, boundaries often follow an old abandoned river channel rather than the present channel.

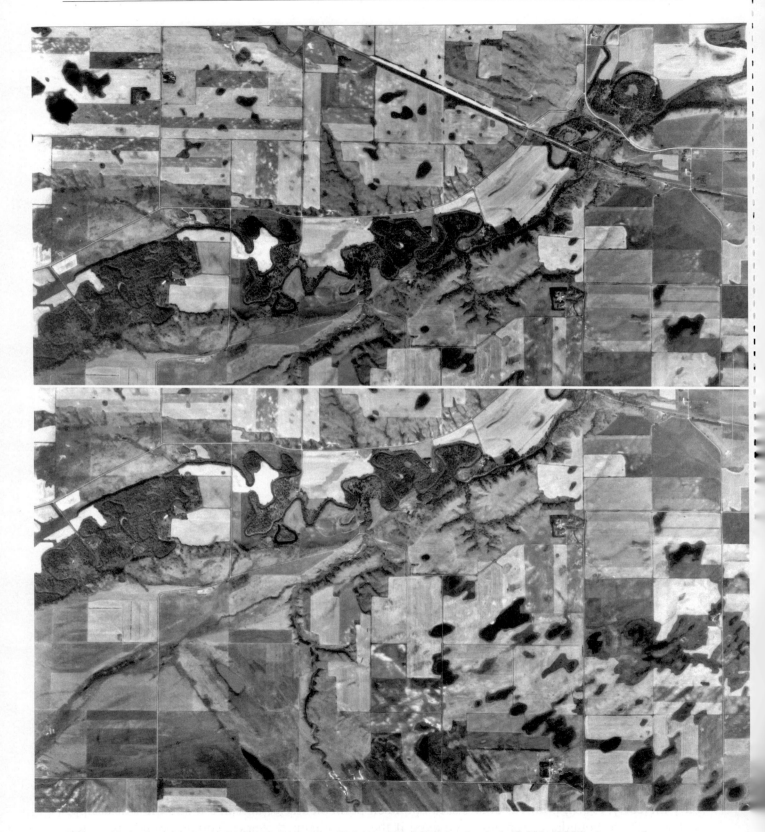

Figure 40-4: Stereogram of the Souris River near Voltaire, North Dakota (scale 1:40,000; USGS photographs, 1997; ↑N).

Name _____   Section _____

EXERCISE 40 PROBLEMS—PART I

The following questions are based on Map T-4, a portion of the "Voltaire, North Dakota" quadrangle (scale 1:24,000; contour interval 5 feet) and Figure 40-4, a stereogram of the same region showing the Souris River (48°06′23″N, 100°48′05″W).

1. (a)   In which direction is the Souris River flowing? From _____ to _____.

    (b)   How can you tell?

2. (a)   What is the approximate width of the Souris River channel?   _____ feet

    (b)   What is the approximate width of the Souris River floodplain?   _____ feet

    (c)   What evidence suggests that the river is widening its valley through lateral erosion?

3. Compare the present length of the meandering Souris River course shown on the map with the length of a river flowing down this valley if it were *not* meandering. (You can measure the distance with a piece of string, or your instructor can show you a more precise method.)

    (a)   Length of present meandering course: _____ miles

    (b)   Length of a straight course down the valley: _____ miles

4. Natural levees can be seen along the Souris River in the center portion of the map (north of Westgaard Cemetery). Approximately how high are the levees?

    _____ feet

Name _____   Section _____

EXERCISE 40 PROBLEMS—PART II

The following questions are based on Map T-4, a portion of the "Voltaire, North Dakota" quadrangle (scale 1:24,000; contour interval 5 feet) and Figure 40-4, a stereogram of the same region showing the Souris River (48°06′23″N, 100°48′05″W).

1. (a) Explain the formation of the narrow, triangle-shaped lake just to the east of Westgaard Cemetery (in the NE ¼ of Section 3). You may use a sketch to illustrate your answer.

   (b) Describe the location of two more lakes, swamps, or topographic depressions along the Souris River that formed in a similar way. (You may refer to Public Land Survey township quarter sections to simplify your location description.)

2. (a) Describe a location where the formation of a new cutoff meander in the Souris River appears imminent.

   (b) Sketch the current river course.

   (c) Sketch the new river course after the cutoff.

Name _____   Section _____

EXERCISE 40 PROBLEMS—PART III

The following questions are based on Map T-5, a portion of the "Jackson, Mississippi-Louisiana" topographic map (scale 1:250,000; contour interval 50 feet; dashed lines are supplementary contours at 25-foot intervals). This map shows the Mississippi River near the town of Vicksburg, Mississippi (32°49′56″N, 91°11′07″W). A color satellite image of this same region is shown in Figure 40-5 (scan the QR code for this exercise or go to the Lab Manual website and Exercise 40). The eastern bluffs of the Mississippi River floodplain can be seen in the southeast corner of the map. The dashed black line along the Mississippi River shows the state boundary between Mississippi and Louisiana. Notice that the "boundary course" of the Mississippi River is quite different from the present course of the river. Some of the changes were natural, others were artificial. There is an extensive set of artificial levees along the course of the river (shown with closely spaced brown tick marks).

1. (a) Find "Willow Cut-off" and "Albemarle Lake" (an oxbow lake) in the center of the map. What natural features shown on the map suggest that a channel of the river once followed a course *between* the boundary course and the present course?

    (b) Name two other oxbow lakes shown on the map (either natural or artificially produced).

2. Compare the length (in miles) of the "boundary course" of the river with the main channel of the present course. (You can measure the distance with a piece of string, or your instructor may show you a more precise method.)

    (a) Length of present course: _____ miles

    (b) Length of "boundary course":   _____ miles

    (c) Amount of shortening: _____ miles

    (d) The present course length represents what percentage
        of the boundary course length? ("present" ÷ "boundary" × 100) _____ %

3. In general, how is the gradient of the present course
    different from the gradient of the boundary course?
    (In other words, is the present course steeper or less
    steep than the boundary course?)   _____

4. Compared to the "boundary course," why would the present course of the Mississippi River be an advantage to river traffic such as barges?

Name _____   Section _____

EXERCISE 40 PROBLEMS—PART IV

The following questions are based on Map T-5, a portion of the "Jackson, Mississippi-Louisiana" topographic map (scale 1:250,000; contour interval 50 feet; dashed lines are supplementary contours at 25-foot intervals). A color satellite image of this same region is shown in Figure 40-5 (scan the QR code for this exercise or go to the Lab Manual website and Exercise 40). For reference, the dashed contour line on "Paw Paw Island" in the southern part of the map shows an elevation of 75 feet.

1.　(a)　To the west of Hollybrook (south of Lake Providence along Highway 65), find the streams named "Otter Bayou" and "Swan Lake." What kind of floodplain landform explains the curved paths taken by these streams? (Hint: The landform is too low to be shown with this map's contour interval.)

　　(b)　Which human-built feature nearby appears to have
　　　　been influenced by the same kind of landform?　_____

2.　Other than the levees, does there appear to be any land more than
　　25 feet higher than the river between Alsatia (west of the Mississippi
　　River along Highway 65) and the Mississippi River itself?　_____

3.　The Yazoo River (from which the geographic term "yazoo stream" was taken) flows into the Mississippi River just south of this portion of the map. Name another yazoo stream (or "creek," "bayou," etc.) shown on the map that parallels the Mississippi for at least 10 miles.

EXERCISE 40 PROBLEMS—PART V—GOOGLE EARTH™

To answer the following questions, go to the Hess *Physical Geography Laboratory Manual*, 12th edition, website at **www.MasteringGeography.com**, then Exercise 40 and select "Exercise 40 Part V Google Earth™" to open a KMZ file in Google Earth, or scan the QR (Quick Response) code for this exercise and view the "Exercise 40 Part V Google Earth™ video."

1.　Fly to Point 1 and then to Point 2 along the Mississippi River near Vicksburg, Mississippi (also shown in Map T-5). What explains the curved striations in the fields here?

2.　Fly to Point 3. Explain the likely origin of this curved patch of vegetation on the floodplain floor:

3.　Fly to Point 4 along the Souris River in North Dakota (also shown in Map T-4 and Figure 40-4). Compare the appearance of this abandoned segment of river channel in 2005, 2009, and 2010.

　　(a)　What could explain why water is filling this old channel in 2009, but not in 2005 and 2010?

　　(b)　How can such intermittent flooding of old channels help explain why oxbow lakes eventually disappear?

EXERCISE 41
Stream Drainage Patterns

Objective:	To study different patterns of stream drainage.
Resources:	Internet access or mobile device QR (Quick Response) code reader app (optional).
Reference:	Hess, Darrel. *McKnight's Physical Geography*, 12th ed., pp. 477–480.

DRAINAGE PATTERNS

The overall drainage pattern of a stream system is often strongly influenced by the underlying geologic structure of a region. The most common stream drainage patterns are shown in Figures 41-1, 41-2, and 41-3.

Dendritic Drainage: The common branching **dendritic** pattern indicates that the underlying structure is not exerting much influence on the courses of streams, and that all of the rock is more or less equally resistant to erosion (Figure 41-1).

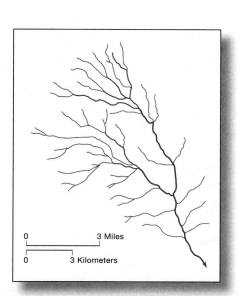

Figure 41-1: Dendritic drainage pattern, from the Pat O'Hara Mountain, Wyoming, topographic quadrangle. (From Hess, *McKnight's Physical Geography*, 12th ed.)

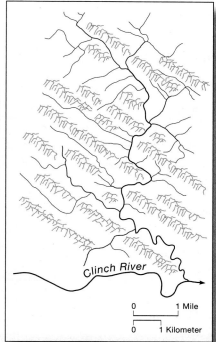

Figure 41-2: Trellis drainage pattern, from the Norris, Tennessee, topographic quadrangle. (From Hess, *McKnight's Physical Geography*, 12th ed.)

Trellis Drainage: In contrast to dendritic drainage, with **trellis** drainage, the courses of streams are largely controlled by a pattern of alternating parallel ridges and valleys (Figure 41-2). Tributaries usually flow parallel to each other until one can cut across a low spot in the separating ridge and join another stream.

Radial and Centripetal Drainage: **Radial** patterns develop where streams flow down and away from a central high area (Figure 41-3b). **Centripetal** drainage is the opposite of radial, and develops where streams converge as they flow down into a central basin (Figure 41-3c).

Annular Drainage: An **annular** pattern may develop around an eroded dome or basin in which alternating bands of soft and resistant rock have been exposed (Figure 41-d).

Rectangular and Parallel Drainage: **Rectangular** patterns often indicate that streams are following a pronounced set of joints or faults (Figure 41-3a). **Parallel** patterns can develop in a region with a uniform but usually gentle slope (Figure 41-3e).

Deranged Drainage: Sometimes the term *deranged drainage* is used to describe an apparently chaotic drainage system without any obvious order. Such a pattern can develop when a preexisting drainage system is disrupted by a process such as glaciation (see Exercise 47), and there hasn't been enough time for a more organized drainage pattern to develop.

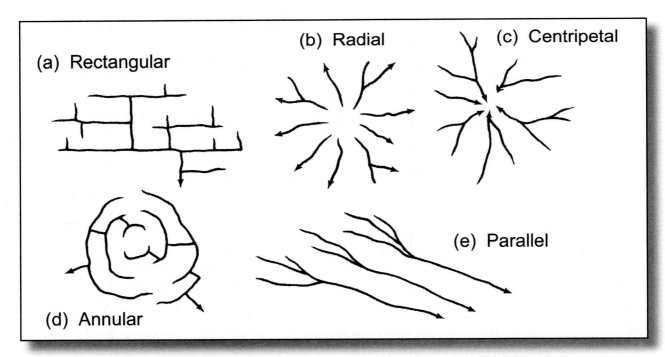

Figure 41-3: Common stream drainage patterns: (a) rectangular; (b) radial; (c) centripetal; (d) annular; and (e) parallel. (Adapted from McKnight, *Physical Geography*, 2nd ed.)

Name _____   Section _____

EXERCISE 41 PROBLEMS—PART I

The following questions are based on Map T-6, the "Johnson City, Tennessee-Virginia-Kentucky-North Carolina" topographic map (scale 1:250,000; contour interval 100 feet). This map shows a portion of the Appalachian Mountains at the Tennessee, Kentucky, and Virginia border (36°39′30″N, 83°01′05″W). Figure 41-4 is a satellite image of this same region (to view this image, scan the QR code for this exercise or go to the Hess Laboratory Manual website and Exercise 41).

1. (a)   Describe the general topography of the region between Clinch Mountain and Powell Mountain.

    (b)   What kind of stream drainage pattern has developed here? _____

    (c)   How has the road system been influenced by the topography?

2. (a)   Describe the general topography in the region north of Poor Valley Ridge.

    (b)   What kind of stream drainage pattern has developed here? _____

    (c)   How has the road system been influenced by the topography?

3. (a)   The Powell River has a meandering pattern.
           Is it flowing across a flat floodplain?   _____

    (b)   How do you know?

The following questions are based on Map T-2, the "Umnak, Alaska" topographic map (scale 1:250,000; contour interval 200 feet).

4.   What kind of drainage pattern has developed around the outside of the Okmok Caldera?   _____

5.   What kind of drainage pattern has developed inside the Okmok Caldera?   _____

Name _____   Section _____

EXERCISE 41 PROBLEMS—PART II—GOOGLE EARTH™

To answer the following questions, go to the Hess *Physical Geography Laboratory Manual*, 12th edition, website at **www.MasteringGeography.com**, then Exercise 41 and select "Exercise 41 Part II Google Earth™" to open a KMZ file in Google Earth, or scan the QR code for this exercise to view "Exercise 41 Part II Google Earth™ video."

1. (a) Fly to Point 1. The area shown is in the Appalachian Mountains near Pennington Gap, Virginia (also shown on Map T-6). Describe the general topography of this region.

   (b) What kind of stream drainage pattern has developed here?

   (c) How has the road system been influenced by the topography?

2. (a) Fly to Point 2 a short distance away from Point 1. Describe the general topography here.

   (b) What kind of stream drainage pattern has developed here?

   (c) How has the road system been influenced by the topography?

3. (a) Fly to Point 3 showing a volcano on Herbert Island in the Aleutian Islands of Alaska. Describe the general topography here.

   (b) What kind of stream drainage pattern has developed around the volcano?

EXERCISE 42
Stream Rejuvenation

Objective:	To study the consequences of stream rejuvenation.
Materials:	Lens stereoscope.
Resources:	Internet access or mobile device QR (Quick Response) code reader app (optional).
Reference:	Hess, Darrel. *McKnight's Physical Geography*, 12th ed., pp. 490–493.

STREAM REJUVENATION

The tectonic history of a region can significantly influence the fluvial landscape. Rejuvenation occurs when a stream undergoes renewed downcutting. For example, the tectonic uplift of a region can result in **stream rejuvenation**. As the gradient is increased through uplift, the downcutting ability of a stream will also increase. There are several kinds of landforms that may result from rejuvenation.

Entrenched Meanders: **Entrenched meanders** (Figure 42-1) may develop when a region is uplifted. In this case, a stream downcuts a gorge into its old floodplain while maintaining its old meandering pattern. The resulting "meandering gorge" is quite striking because we normally expect to see a meandering river only on top of a flat floodplain.

Figure 42-1: (a) The deeply entrenched meanders of the Green River in southeastern Utah. (b) Floodplain meanders before uplift and rejuvenation. (c) If the stream maintains its meandering pattern during uplift, renewed downcutting can produce entrenched meanders. (From Hess, *McKnight's Physical Geography,* 12th ed.)

289

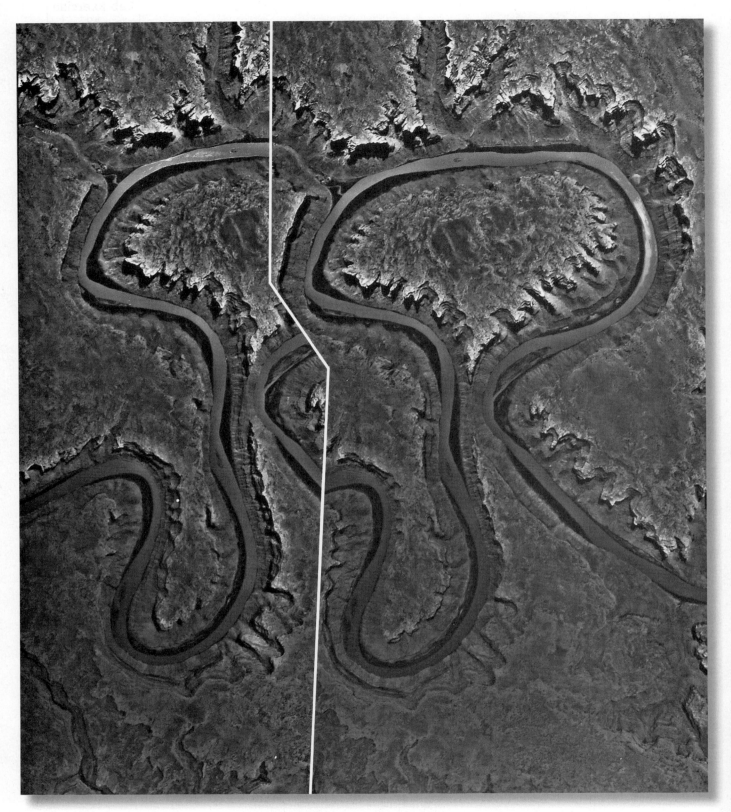

Figure 42-2: Stereogram of the entrenched meanders of the Green River in Utah. North is to the left side of the page (scale 1:40,000; USGS photographs, 1993; ← N).

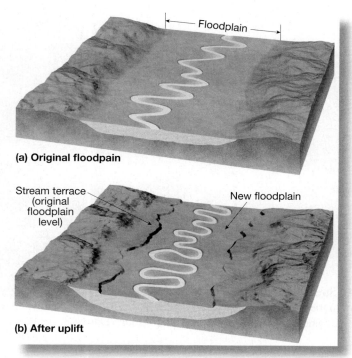

Figure 42-3: Formation of stream terraces. (a) Before uplift, a stream meanders across an alluvial floodplain. (b) After uplift, the stream downcuts and then widens a new floodplain, leaving the remnants of the original floodplain as a pair of stream terraces. (From Hess, *McKnight's Physical Geography,* 12th ed.)

Figure 42-2 and Figure 32-3 (on page 220) are stereograms showing the deeply entrenched meanders of the Green River in Utah. Map T-7 is a portion of the "Canyonlands National Park" topographic map showing this same region. Figure 42-1a is an oblique aerial photograph of this region.

Stream Terraces: Stream terraces along a river represent old valley floors. During rejuvenation, a stream downcuts into its old floodplain floor, leaving portions of its old floodplain as terraces above the present level of the stream (Figure 42-3).

Figure 42-4 shows stream terraces in the Transverse Ranges of southern California. Here, tectonic uplift caused a stream to downcut into its old alluvial floodplain.

Figure 42-4: Stream terraces along Cañada de Los Alamos in Lower Hungry Valley near Gorman in the Transverse Ranges of southern California formed when uplift caused this stream to downcut into its original floodplain floor. (Darrel Hess photo)

Name _____   Section _____

EXERCISE 42 PROBLEMS—PART I

The following questions are based on Map T-7, the "Canyonlands National Park, Utah" topographic map (scale 1:62,500; contour interval 80 feet), Figures 32-3 and 42-2, stereograms of this region, and Figure 42-1a, an oblique aerial photograph of this area. The map, stereograms, and photograph show the deeply entrenched meanders of the Green River north of Canyonlands National Park, Utah (38°33′40″N, 110°03′16″W).

1.   How deep is the gorge of the Green River? Use the top of the land inside "Bowknot Bend" as your upper reference point (determine the relief, not the elevation).   _____ feet

2.   (a)   Describe where a cutoff meander developed in the past (describe the location relative to labeled places on the map).

      (b)   Did the cutoff occur before or after the entrenchment began? _____

      (c)   How do you know?

3.   (a)   Describe where another cutoff meander might develop in the future (describe the location relative to labeled places on the map).

      (b)   How high is the land between the two loops of the river at this point? (Determine the relief, not the elevation.)   _____ feet

4.   The gorge wall on the outside of some meander turns is noticeably steeper than the gorge walls on the inside of the meander turn. For example, see "Cottonwood Bottom" and the meander to the west of "Spring Canyon Point." What might explain this? (Hint: Consider the processes of erosion associated with both meandering and entrenchment; before entrenchment, was the stream course directly above its current location?)

Name _____   Section _____

EXERCISE 42 PROBLEMS—PART II

The following questions are based on Map T-4, the "Voltaire, North Dakota" topographic map (scale 1:24,000; contour interval 5 feet). You may also want to look at Figure 40-4 on page 280, a stereogram of this same area. The Souris River has downcut into the glacial outwash deposits along its course leaving stream terraces (48°06′23″N, 100°48′05″W).

1.   Approximately how high are the stream terraces north of the
     Souris River in the SW¼ of Section 33?   _____ feet

2.   (a)   Is the stream terrace north of the Souris River of uniform width? _____

     (b)   If not, describe the pattern.

EXERCISE 42 PROBLEMS—PART III—GOOGLE EARTH™

To answer the following questions, go to the Hess *Physical Geography Laboratory Manual*, 12th edition, website at **www.MasteringGeography.com**, then Exercise 42 and select "Exercise 42 Part III Google Earth™" to open a KMZ file in Google Earth, or scan the QR code for this exercise to view "Exercise 42 Part III Google Earth™ video."

1.   (a)   Fly to Point 1 along the San Juan River in Utah. What kind of landforms are shown along this portion of the San Juan River?

     (b)   Briefly explain how they formed.

2.   Fly to Point 2. Explain the likely origin of this dry gorge along the San Juan River:

3.   Fly to Point 3 along Lower Hungry Valley in California's Transverse Ranges (the stream terraces here are also shown in Figure 42-4); then fly to Point 4. What might explain why the stream terraces here disappear farther up the valley near Point 4?

Flood Recurrence Intervals

Objective:	To study flood recurrence intervals.
Reference:	Hess, Darrel. *McKnight's Physical Geography*, 12th ed., pp. 473–474.

STREAMFLOW RECORDS

Although natural disasters such as floods seem to occur randomly, it is possible to calculate the probabilities of floods of various sizes occurring each year by analyzing long-term streamflow records.

The amount of water flowing through a stream is called its **discharge**, usually described in terms of cubic meters of water flow per second (cms). All streams vary in their discharge during a year, depending on such factors as seasonal precipitation and surface runoff conditions. In this exercise, we are interested in the annual peak discharge of a stream—the highest "flood flow" of a stream each year.

As stream discharge increases, the volume, speed, and height of a stream increase. With an increase in height during a flood, streams typically spill out of their channels and flood adjacent areas. For this reason the **gage height** at a monitoring station—which describes the height of a stream above its local level or **datum**—is one of the most common measures of flood size.

FLOOD RECURRENCE INTERVALS

In this exercise, we will study Cache Creek, a stream near Clear Lake in northern California, from 1980 to 2015 (a table with complete streamflow data is shown in Figure 43-2). The stretch of Cache Creek chosen for this exercise is in the upper reaches of the stream where its flow is not artificially regulated. These data represent only a 36-year period, so we must be careful about extrapolating beyond the data range and about drawing conclusions about the actual probabilities of flooding along Cache Creek. A much longer period of study is required to make meaningful estimates of actual flood height and recurrence.

The first step in our analysis is to relate the discharge of Cache Creek to its gage height. This is shown with a **rating curve** (Figure 43-1). For each of the 36 years, the gage height in meters was plotted against its discharge in cms; a smooth line was added through the plot to more clearly show the relationship. (Note: Because of the type of stream gage used to gather this data, when streamflow is less than about 30 cubic meters per second, gage height readings may not be accurate.) Notice that the "Peak Discharge" scale is logarithmic and so the rating curve line appears as a gentle curve. Using the rating curve, you can estimate the height of Cache Creek for any given discharge.

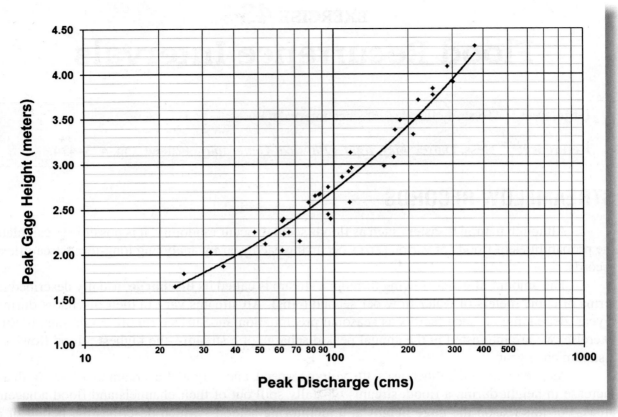

Figure 43-1: Rating curve for Cache Creek, California, 1980–2015. (U.S. Geological Survey streamflow data)

The next step is to calculate the **recurrence interval** or **return period** of floods along Cache Creek. The recurrence interval is an estimate of the number of years between times a peak stream discharge is reached or exceeded. For example, a "5-year" flood refers to the highest gage height likely to be reached once every 5 years, and a "100-year" flood is the highest gage height likely to be reached once every 100 years. However, because these estimates are based on long-term averages, it is possible to have more than one 10-year flood in a decade, and for several decades to pass without such a flood. Further, because accurate long-term streamflow data are lacking for many streams, estimates of 100-year floods and 500-year floods may be somewhat misleading.

To use streamflow data to calculate the recurrence interval (RI) of a flood, we first rank the peak flood discharge of the stream for each year. For example, in the data set shown in Figure 43-2, 1997 had the highest peak flow with 370 cubic meters per second, while 1986 was second with 302 cms. The rank of each year's peak discharge has been determined for you. We then use the following formula to calculate the recurrence interval:

$$RI = \frac{n + 1}{m}$$

In the formula "n" is the number of years in the data record (36 years in this case), and "m" is the rank of a given flood on the list of annual floods. For example, the recurrence interval for the 10th ranked flood (which took place in the year 2004) is (36 + 1) ÷ 10 = 3.7 years. This means that a flood with a discharge of about 176 cms occurs on average once every 3.7 years along Cache Creek. The recurrence intervals for most of the years have been calculated for you. You need to calculate the recurrence intervals for the years 1990, 1997, 2003, and 2013.

Once you have calculated the remaining recurrence intervals, plot the recurrence intervals against peak discharge on the chart in the exercise problem pages. Once your data points are plotted, draw a straight line through the data points, trying to get about half of the data points above this line, and half below—this is your **flood frequency curve**.

Year	Peak Gage Height in Meters	Peak Discharge in Cubic Meters per Second (cms)	Rank	Recurrence Interval (n + 1/m)
1980	2.98	159	12	3.1
1981	2.05	61	30	1.2
1982	2.58	116	15	2.5
1983	3.08	174	11	3.4
1984	2.40	97	18	2.1
1985	2.15	73	25	1.5
1986	3.91	302	2	18.5
1987	2.23	63	28	1.3
1988	1.87	36	33	1.1
1989	2.25	66	26	1.4
1990	1.65	23	36	
1991	2.45	95	20	1.9
1992	2.12	53	31	1.2
1993	3.52	222	6	6.2
1994	1.79	25	35	1.1
1995	3.77	250	5	7.4
1996	2.68	88	21	1.8
1997	4.31	370	1	
1998	3.71	219	7	5.3
1999	3.13	117	14	2.6
2000	2.65	84	23	1.6
2001	2.92	115	16	2.3
2002	2.86	108	17	2.2
2003	3.84	251	4	
2004	3.38	176	10	3.7
2005	2.38	62	29	1.3
2006	4.08	286	3	12.3
2007	2.40	64	27	1.4
2008	2.96	118	13	2.8
2009	2.25	48	32	1.2
2010	2.67	87	22	1.7
2011	2.75	96	19	1.9
2012	2.58	79	24	1.5
2013	3.49	185	9	
2014	2.03	32	34	1.1
2015	3.33	209	8	4.6

Figure 43-2: Streamflow data for Cache Creek at Hough Springs, California, 1980–2015. (U.S. Geological Survey streamflow data)

Name _____   Section _____

EXERCISE 43 PROBLEMS

The following questions are based on the rating curve and streamflow data for Cache Creek (Figures 43-1 and 43-2):

1. Using the data in Figure 43-2, calculate the recurrence interval for Cache Creek's stream-flow for the years 1990, 1997, 2003, and 2013. Once you have calculated the recurrence intervals, fill in your answers below and in Figure 43-2 (round your answers to 1 decimal place).

    1990: _____ years   1997: _____ years   2003: _____ years   2013: _____ years

2. Determine the Flood Frequency Curve for Cache Creek by plotting the "Recurrence Inter-val" against "Peak Discharge" on the chart below (note that the Recurrence Interval scale is logarithmic). (Hint: Your plot of data points should form a nearly straight line, so first plot the three largest floods, and then the three smallest—the rest of your points will line up along this trend.) Using a ruler, draw a straight line through the center of your plot, try-ing to have as many data points below the line as above.

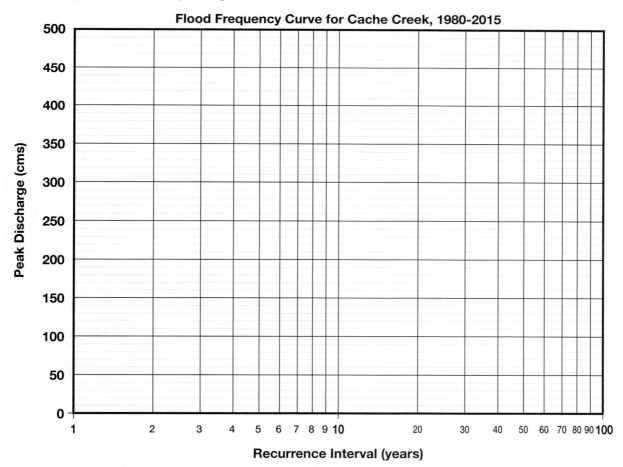

Flood Frequency Curve for Cache Creek, 1980-2015

Peak Discharge (cms)

Recurrence Interval (years)

3. Using the Rating Curve in Figure 43-1, estimate the gage height (to 1 decimal place) when Cache Creek has a discharge of

   (a) 40 cms: _____ meters   (b) 250 cms: _____ meters

4. Identify the approximate discharge of Cache Creek when it has a gage height of

   (a) 2.0 meters: _____ cms   (b) 3.5 meters: _____ cms

5. Using your Flood Frequency Curve, estimate the recurrence interval of a flood with a discharge of

   (a) 100 cms: _____ years   (b) 300 cms: _____ years

6. The discharge of the following floods will be

   (a) 5-year flood: _____ cms   (b) 10-year flood: _____ cms

7. (a) Extrapolate from your data and estimate the discharge of the
       100-year flood: _____ cms

   (b) Why might this be a poor estimate of the size of the 100-year flood?

8. In order to be above the water level of the 20-year flood, what is the minimum height above the stream level datum you should build a home? (Hint: Estimate the height in meters, not the discharge in cms.)

   _____ meters

9. On the topographic map in Figure 43-3, determine the extent of inundation for the 20-year flood along Cache Creek that you calculated in problem 8:

   (a) Convert your flood level height in problem 8 from meters to feet (the measurement units used on the map):

       Water height of 20-year flood:

       _____ feet

   (b) Using a blue pencil, shade in the extent of the 20-flood on the map. Note that elevation of the stream gage is 1534 feet and that dashed lines are 20-foot contours.

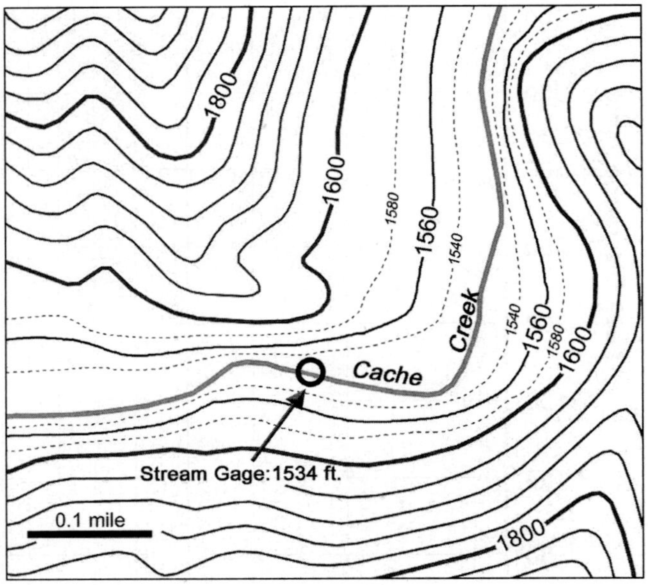

Figure 43-3: Portion of USGS "Hough Springs, California" quadrangle (enlarged to approximately scale 1:8000; contour interval 40 feet; dashed lines are 20-foot contours; N↑).

EXERCISE 44
Karst Topography

Objective:	To study the landforms that develop in limestone regions.
Materials:	Lens stereoscope.
Resources:	Internet access or mobile device QR (Quick Response) code reader app (optional).
Reference:	Hess, Darrel. *McKnight's Physical Geography*, 12th ed., pp. 516–506.

UNDERGROUND WATER

The landscapes of most regions are strongly influenced by the work of running water on the surface. However, the action of water below the ground can also influence the surface topography. This is especially true in humid regions underlain by soluble rock such as limestone. Water beneath the surface moves slowly through the limestone structure. Through chemical action, this water can dissolve rock, transport this material, and eventually deposit it in new forms.

Limestone and other rocks composed primarily of **calcium carbonate** ($CaCO_3$) are easily altered by water. Rainwater becomes mildly acidic when it combines with carbon dioxide to form **carbonic acid** (H_2CO_3). Carbonic acid reacts with calcium carbonate to produce calcium bicarbonate, which is easily dissolved and transported in solution.

The chemical action of underground water is aided by the joints and bedding planes that are common in limestone. These openings allow greater movement of water through the underground structure. Eventually, underground cave networks, commonly called **caverns**, may develop (Figure 44-1). Within these caves the precipitation of dissolved calcium carbonate can produce striking features such as **stalactites** (columns hanging from the ceiling) and **stalagmites** (columns rising from the floor).

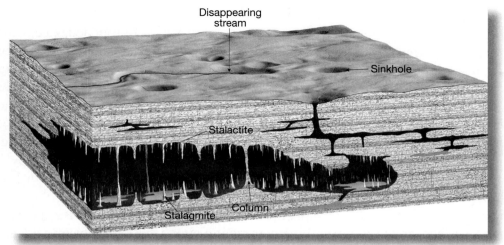

Figure 44-1: Caverns form by the solution action of underground water as it moves along bedding planes and joints. (From Hess, *McKnight's Physical Geography*, 12th ed.)

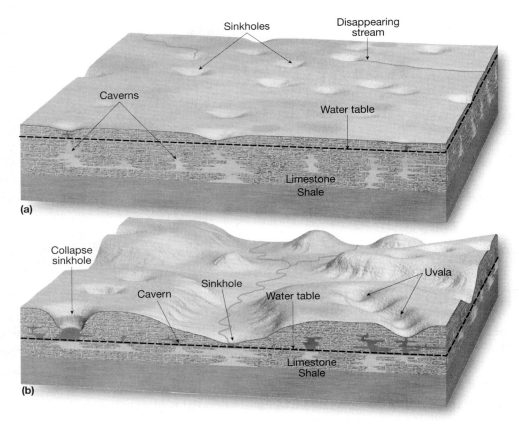

Figure 44-2: Karst landscapes dominated by (a) sinkholes and (b) collapse sinkholes (collapse dolines). (From Hess, *McKnight's Physical Geography*, 12th ed.)

KARST TOPOGRAPHY

Although most of the activity in limestone areas is taking place underground, there are also characteristic features that develop on the surface. This collection of landforms is known as **karst** topography (Figure 44-2). Karst regions in general are characterized by great local irregularities in topography with many small hills and depressions.

Often the most common surface feature in areas of karst is a rounded, steep-sided depression known as a **sinkhole** (or **doline**). Sinkholes develop when carbonate rocks are dissolved at the surface, often at a joint intersection. Sinkholes range in size from a few meters to several hundred meters across. A depression that results from the collapse of a subsurface cave is called a **collapse sinkhole** or **collapse doline**. Where a series of sinkholes or collapse dolines coalesce, a karst valley known as an **uvala** develops.

One of the most distinctive characteristics of karst regions is the lack of a well-established surface drainage system. The surface drainage system has been largely replaced by a complex underground drainage system. Much of the water from precipitation simply seeps down below the surface through the joints and bedding planes.

Where surface streams do develop, they often do not flow very far. **Disappearing streams** result when a surface stream flows into the bottom of a sink, where it seeps into a crack or flows into a **swallow hole**—a direct opening into an underground channel.

Map T-9 shows an area of sinkhole karst topography near Park City, Kentucky. The boundary of Mammoth Cave National Park is just beyond the northern margin of the map. Figure 44-3 is a stereogram of the same area. The surface shown in the southern two-thirds of the map is predominantly limestone, whereas in much of the northern third, a layer of sandstone caps the limestone.

Figure 44-3: Stereogram of area near Park City, Kentucky (scale 1:40,000; USGS photographs, 1998; ↑N).

Name _____   Section _____

EXERCISE 44 PROBLEMS—PART I

The following questions are based on Map T-8, the "Putnam Hall, Florida" quadrangle (scale 1:24,000; contour interval 10 feet). This map shows a portion of northeastern Florida underlain by limestone (29°39′10″N, 81°57′15″W). Here the water table (the upper level of groundwater) is close to the surface.

1. Describe the overall topography of this region.

2. What is the approximate depth (from top to bottom) of the sinkhole
    containing a lake to the southwest of the Melrose Landing Airstrip?   _____ feet

3. The elevation of the surface of a lake is a good indication of the level of the water table at that location. Assuming that the elevation of a lake is half a contour interval lower than the contour line closest to the shore, determine the elevations of the following lakes.

    (a)   Slipper Lake   _____ feet

    (b)   Lake Green Sills   _____ feet

    (c)   Whirlwind Lake   _____ feet

    (d)   Trotting Pond   _____ feet

    (e)   Lake Lucy   _____ feet

    (f)   Trout Lake   _____ feet

4. Based on the lake levels in problem 3, and noting the levels of other lakes and swamps, does it appear that the water table in this region is level or sloping? If sloping, in which direction?

5. Why isn't a lake found in the bottom of the sinkhole 0.4 mile (0.6 km) south of the eastern Whirlwind Lake?

Name _____   Section _____

EXERCISE 44 PROBLEMS—PART II

The following questions are based on Map T-9, the "Park City, Kentucky" quadrangle (scale 1:24,000; contour interval 10 feet) and Figure 44-3, a stereogram of the same area (37°05′03″N, 86°05′27″W). The boundary of Mammoth Cave National Park is just beyond the northern margin of the map. The surface shown in the southern two-thirds of the map is predominantly limestone, whereas in much of the northern third, a layer of sandstone caps the limestone.

1. (a) Briefly contrast the general topography in the northern third of the map, with the topography in the southern two-thirds of the map.

    (b) What might explain these differences?

2. Propose a reason why the hills in the northern part of the map are covered with vegetation, while those in the southern part have a much sparser cover.

3. In which direction does Gardner Creek flow?   From _____ to _____

4. What happens to all of the surface streams shown on the map?

EXERCISE 44 PROBLEMS—PART III—GOOGLE EARTH™

To answer the following questions, go to the Hess *Physical Geography Laboratory Manual*, 12th edition, website at **www.MasteringGeography.com**, then Exercise 44 and select "Exercise 44 Part III Google Earth™" to open a KMZ file in Google Earth, or scan the QR code for this exercise and view "Exercise 44 Part III Google Earth™ video."

1. Fly to Point 1 near Park City, Kentucky (this area is shown in Map T-9 and Figure 44-3). What are the shallow, dark depressions you see in this area?

2. (a) As you fly from Point 1 to Point 2, do you see any evidence of surface streams?

    (b) What might explain this?

Fly to Point 3 near Putnam Hall, Florida (this area is shown in Map T-8):

3. What happened to the size of lakes in this area between 1993 (the date of the map) and the year of the latest Google Earth™ imagery?

4. What kinds of natural or human-influenced factors could have produced this change in lake size?

EXERCISE **45**
Desert Landforms

Lab Exercise 45

https://goo.gl/7Gs7FC

Objective:	To study desert landforms of the Basin and Range province in the United States.
Materials:	Lens stereoscope.
Resources:	Internet access or mobile device QR (Quick Response) code reader app (optional).
Reference:	Hess, Darrel. *McKnight's Physical Geography*, 12th ed., pp. 516–522 and 530–533.

SPECIAL CONDITIONS IN DESERTS

The landscapes of deserts tend to be quite different from the landscapes found in humid areas. There are a number of special conditions in arid areas that strongly influence landform development.

Perhaps surprisingly, running water is the most important process of erosion and deposition in deserts. There are a number of factors in deserts that compensate for the lack of regular precipitation and allow fluvial processes to be significant. First, deserts typically have a sparse cover of vegetation and soil. Without this mantle of protection, the bedrock of deserts is often directly exposed to the processes of weathering and erosion. Second, impermeable surface layers are common in deserts. These hard surface layers effectively increase the erosional potential of a rainstorm. Because the rainwater doesn't soak in easily, the runoff into stream channels is increased. Finally, although rainfall is infrequent in deserts, it often comes from an intense thunderstorm. Such short-lived downpours cause the dry streambeds to fill with water quickly and flow rapidly. These **flash floods** can result in a great deal of erosion in a short period of time.

Other conditions also influence the development of landforms in deserts. For example, **basins of interior drainage** are common. This means that ephemeral streams typically flow toward the bottom of a basin, where alluvium is deposited and the water evaporates. Although the work of wind may be more prominent in deserts than in humid areas, it mostly moves around loose sand. Desert sand dunes are discussed in Exercise 46.

COMMON DESERT LANDFORMS

Although the landscapes of deserts vary greatly, the landforms in the Basin and Range province of the United States (covering all of Nevada and portions of surrounding states) illustrate the results of the special conditions and processes operating in arid regions. Figure 45-1 is a diagram showing some of the common desert landforms found in the Basin and Range province.

In the Basin and Range province, the mountains and basins have been produced by block faulting. The mountain fronts are typically steep and rocky and are dissected by V-shaped canyons. The basin floors are nearly flat, because they are slowly filling with alluvium washed down from the mountains.

Most of these basins have interior drainage, so at the lowest part of the basin, a dry lake bed known as a **playa** is found. A playa is an almost perfectly level expanse of dried mud, often

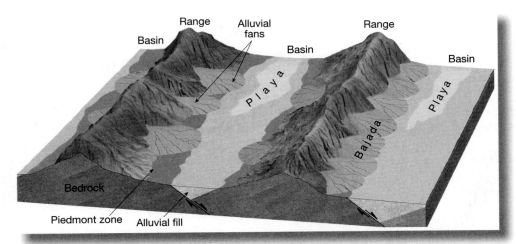

Figure 45-1: Typical Basin and Range landscape. (From Hess, *McKnight's Physical Geography*, 12th ed.)

covered with a thin crust of salt. At the bottom of a few basins, salts have accumulated to great thicknesses.

At the foot of the mountains, in what is known as the **piedmont zone**, several prominent landforms may be found. In some piedmont zones a gently sloping bedrock platform called a **pediment** is visible. Some pediments appear to be the remnants of the weathering and erosion of the mountain front, whereas others may form as a result of low-angle detachment faulting (Figure 45-2). In most desert regions, the pediment is covered with a mantle of alluvium. Figure 45-3 is a portion of the "Antelope Peak, Arizona" quadrangle (scale 1:62,500; contour interval 25 feet). In the region shown on the map, the pediment has been mostly covered with a veneer of alluvium.

Probably the most prominent landforms in the piedmont zone are **alluvial fans**. As the ephemeral streams flow out of their steep canyons and reach the gentle slope of the basin floor, the water velocity decreases and alluvium is deposited. With time, the alluvium accumulates into a fan-shaped landform consisting of the mud, sand, gravel, and even boulders, brought down by flash floods and **debris flows** from the mountains (also see Figure 38-5 in Exercise 38).

In some cases, the alluvial fan from one canyon coalesces with the alluvial fan from an adjacent canyon. When alluvial fans join to form a continuous alluvial apron along a mountain front, the resulting landform is called a **bajada**. The stereogram, topographic map, and aerial

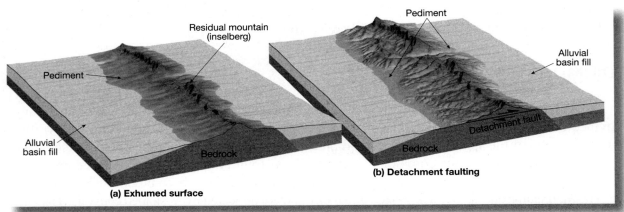

Figure 45-2: Desert pediments. (a) Some pediments may represent the top of a now exhumed weathered bedrock surface. (b) Other pediments may develop as a consequence of detachment faulting. The residual desert hills are sometimes called inselbergs. (From Hess, *McKnight's Physical Geography*, 12th ed.)

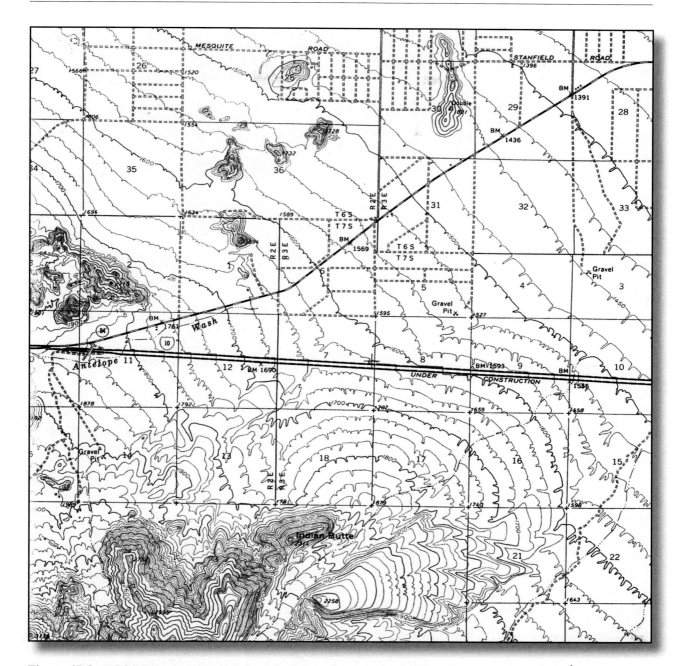

Figure 45-3: USGS "Antelope Peak, Arizona" quadrangle (scale 1:62,500; contour interval 25 feet; ↑N).

imagery of the Stovepipe Wells region of Death Valley, California (Figures 45-4 and 45-5, and Map T-27a), show several alluvial fans that are coalescing to form a bajada.

In many desert areas, isolated steep-sided landforms known as **inselbergs** are found. Inselbergs are the eroded remnants of mountain tops and now stand out as hills or ridges rising out of the surrounding plains.

A striking feature in many deserts is a dark brown, shiny coating on exposed rock surfaces known as **desert varnish**. This coating of iron and manganese oxides seems to be a consequence of biochemical processes involving bacteria and wind-delivered clay. Desert varnish is a relative dating tool for geomorphologists because the longer a rock surface has been exposed and undisturbed by fluvial erosion or deposition, the darker the coating of varnish.

309

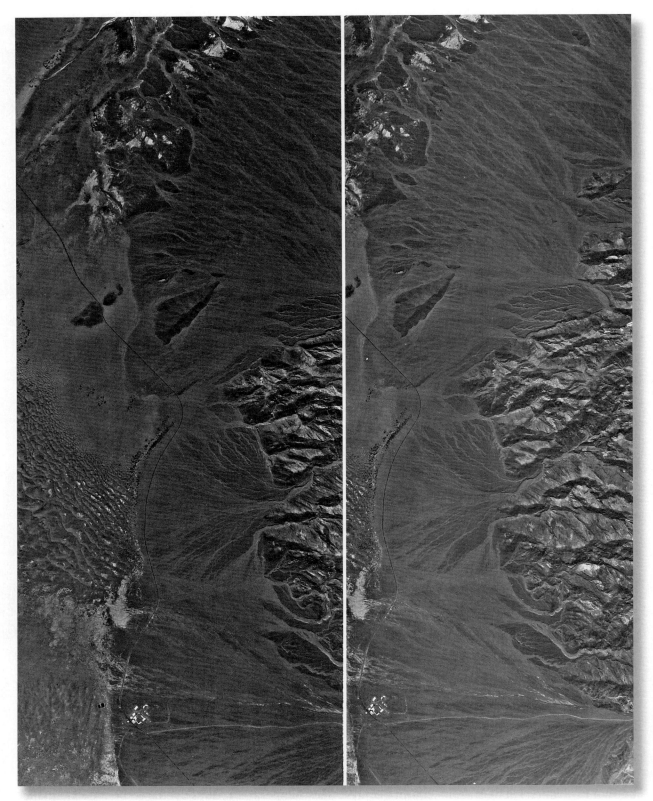

Figure 45-4: Stereogram of the Stovepipe Wells region of Death Valley, California. North is to the left side of the page (scale 1:40,000; USGS photographs, 1993; ← N). This area is also shown in color in Map T-27a.

Name _____   Section _____

EXERCISE 45 PROBLEMS—PART I

The questions on the following page are based on Figure 45-4, a stereogram of the Stovepipe Wells region of Death Valley, California, Figure 45-5, a portion of the "Stovepipe Wells, California" quadrangle (scale 1:62,500; contour interval 80 feet; north is to the left side of the stereogram and topographic map), and Map T-27a, color aerial imagery of this same region of Death Valley. Several large alluvial fans that come out of the Tucki Mountains are coalescing to form a bajada (36°35′17″N, 117°06′35″W).

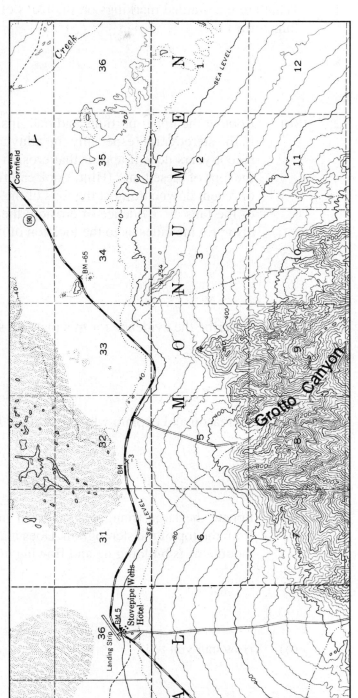

Figure 45-5: USGS "Stovepipe Wells, California" quadrangle. North is to the left side of the page (scale 1:62,500; contour interval 80 feet, dotted lines represent half-interval contours; ← N).

The following questions are based on Figures 45-4 and 45-5, and Map T-27a, a stereogram, topographic map and color aerial imagery of the Stovepipe Wells region of Death Valley, California. North is to the left side of the stereogram and map.

1. What are the braided markings on the faces of the fans? (Hint: Think about how alluvial fans form.)

2. (a) Look at the upper portions and margins of the alluvial fans shown on the map and in the stereogram (especially the alluvial fan at the mouth of "Grotto Canyon"). In recent years does it appear that erosion or deposition has been more prominent at the tops of these fans? (Hint: Look carefully for evidence of either recent deposition or recent downcutting at the tops of the fans. For example, gullies cut into the tops of the fans are evidence of erosion; the presence or absence of desert varnish can also provide clues as to the location of recent activity on the fans.)

    (b) Describe the evidence for this that you see in the photographs or map.

3. Suggest a reason and explain why the alluvial fans here vary so greatly in size.

4. (a) Locate the large triangle-shaped hill in the bajada, about 3 inches (8 cm) down from the top of the stereogram. Does it appear that alluvium brought down from the mountains is building up and flowing around this hill?

    (b) How can you tell?

Name _____   Section _____

EXERCISE 45 PROBLEMS—PART II

The following questions are based on Figure 45-3, a portion of the "Antelope Peak, Arizona" quadrangle (scale 1:62,500; contour interval 25 feet). In the northern portion of the map, a gently sloping pediment is visible (32°47′52″N, 112°05′21″W). You may view this map in color by going to the Lab Manual website or by scanning the QR code for this exercise.

1.   What evidence from the map suggests that the pediment is covered with a mantle of alluvium?

2.   (a)   What kind of landform is illustrated by the cluster of small hills in the northern portion of the map?

       (b)   Explain their development.

3.   (a)   Describe the location of one clearly defined alluvial fan on the map.

       (b)   Which landform—the pediment or the alluvial fan—has a steeper slope?

Name _____   Section _____

EXERCISE 45 PROBLEMS—PART III

The following questions are based on Map T-10, the "Furnace Creek, California" quadrangle (scale 1:62,500; contour interval 80 feet; dotted lines represent 20-foot contours). This portion of the Furnace Creek quadrangle shows the Panamint Range and the western side of Death Valley in Death Valley National Park (36°19′08″N, 116°53′25″W). Several large alluvial fans can be seen along the eastern front of the Panamint Range. The basin floor here is called the Death Valley "salt pan," because salts have accumulated here in great thicknesses.

1. (a) What is the lowest elevation shown on the basin floor?   _____ feet

    (b) What is the highest elevation shown in this
         part of the Panamint Range?   _____ feet

    (c) What is the local relief shown on the map?   _____ feet

2. (a) What is the maximum change in elevation (either up or down)
         you would experience along the basin floor traveling due
         south, from Section 9 to the bottom of the map?
         (Supplementary 20-foot contours are shown with dotted lines.)   _____ feet

    (b) Why is the basin floor so flat?

3. Suggest and explain a reason why the Trail Canyon alluvial fan and the alluvial fan just to the north are so different in size.

4. Find "Trail Canyon" coming out of the Panamint Range in the center of the map (a dirt road is shown coming down the fan of Trail Canyon). Why are the contours showing the Trail Canyon alluvial fan so irregular (many tiny zigzags)? (Hint: What surface features of the fan are these zigzags showing?)

5. How deep is the gully cut by the ephemeral stream (shown with a dashed blue line) that flows down the Trail Canyon alluvial fan where it crosses the "sea level" contour at the bottom of the fan? *(You can estimate the depth of a valley or gully on a slope by comparing the size of the "V" in the contour line to the overall spacing of contour lines on that slope. For example, if the V extends up to the position of the next higher contour line on the slope, the valley is about one contour interval deep; if it extends halfway to the next higher contour line, the valley is about one-half contour interval deep, and so on.)*

                                                    _____ feet deep

Name _____   Section _____

EXERCISE 45 PROBLEMS—PART IV

The following questions are based on Map T-10, the "Furnace Creek, California" quadrangle (scale 1:62,500; contour interval 80 feet; dotted lines represent 20-foot contours).

1. Using the graph at right, construct a topographic profile from Point A in the Panamint Range to Point B on the basin floor. Plot the index contours and any other contours needed to show key features along the profile.

2. On the topographic profile, label (a) the basin floor, (b) the alluvial fan, and (c) the mountain front.

Topographic profile for line AB on the "Furnace Creek, California" quadrangle. Vertical exaggeration of the profile approximately 2.6×.

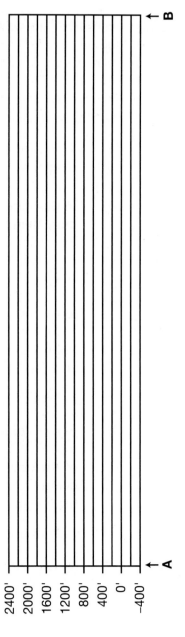

315

Name _____   Section _____

EXERCISE 45 PROBLEMS—PART V—GOOGLE EARTH™

To answer the following questions, go to the Hess *Physical Geography Laboratory Manual*, 12th edition, website at **www.MasteringGeography.com**, then Exercise 45 and select "Exercise 45 Part V Google Earth™" to open a KMZ file in Google Earth, or scan the QR (Quick Response) code for this exercise and view "Exercise 45 Part V Google Earth™ video." The opening view is looking south down Death Valley in Death Valley National Park, California. The Black Mountains are the lower range to the left (east), and the Panamint Range is the higher range to the right (west). A portion of this area is shown in Map T-10.

1. (a) Which mountain range—the Black Mountains or the Panamint Range—has the steeper mountain front?

   (b) In this part of Death Valley, one side of the basin floor is tilting down along faults more than the other side. Which side of the basin—the Black Mountain side or the Panamint Range side—appears to be dropping down more prominently?

2. (a) Fly to Point 1 along the Black Mountains, and then to Point 2. Point 1 marks the top of a small alluvial fan. Point 2 marks the top of the drainage basin that provides sediment to this fan. Using the ruler function, determine the length of the fan from Point 1 to its edge along the basin floor. Then determine the length of the drainage area from Point 1 to Point 2.

   Length of alluvial fan: _____ miles   Length of drainage area: _____ miles

   (b) Fly to Point 3 along the Panamint Range, and then to Point 4. Point 3 marks the top of a large alluvial fan within the bajada of Death Valley. Point 4 marks the top of the drainage basin that provides sediment to this fan. Using the ruler function, determine the length of the fan from Point 3 to its edge along the basin floor, and the length of the drainage area from Point 3 to Point 4.

   Length of alluvial fan: _____ miles   Length of drainage area: _____ miles

   (c) Why are the alluvial fans along the Panamint Range generally much larger than the fans along the Black Mountains?

3. Fly to Point 5 on the west side of the Panamint Range in Panamint Valley, and then to Point 6. Point 5 is near the lowest part of the Panamint Valley playa. Move the cursor around in this area to find the elevation of the lowest part of the playa. Then use the ruler to measure the distance to Point 6 at the southern edge of the playa. Determine the gradient of the playa.

   $$\frac{\text{_____}}{\text{(Elevation change)}} \div \frac{\text{_____}}{\text{(Distance in miles)}} = \text{_____ feet/mile gradient}$$

EXERCISE 46
Sand Dunes

Objective:	To study common types of desert sand dunes.
Materials:	Lens stereoscope.
Resources:	Internet access or mobile device QR (Quick Response) code reader app (optional).
Reference:	Hess, Darrel. *McKnight's Physical Geography*, 12th ed., pp. 522–527.

DESERT SAND DUNES

Sand dunes are the most conspicuous aeolian (wind-produced) landform in many deserts (although few extensive desert regions are completely covered with sand dunes). Sand dunes can take on a variety of forms. The shape of a sand dune depends on the amount of sand available, the persistence and direction of the wind, and the presence of vegetation. Sand dunes with little vegetation cover may move in response to the local winds. Grains of sand are blown up the gentle windward side of a dune (Figure 46-1) and then fall down the steep **slip face**, resting at the **angle of repose** for dry sand—about 32–34°. Common kinds of desert sand dunes are shown in Figure 46-2.

Barchan: The best known kind of sand dune is the **barchan**. These are isolated, crescent-shaped dunes. The "horns" or "cusps" of the barchan point downwind. Barchans form in places where the supply of sand is limited and the winds consistently blow in one direction. A barchan dune migrates downwind over a mostly nonsandy surface.

Transverse: Transverse dunes are related to the barchans. Transverse dunes form in places where the supply of sand is greater. They maintain the generally crescent-shaped form of a barchan, but are much less uniform and tend to develop into interconnected ridges of sand. Figure 46-3 is a stereogram of transverse sand dunes in southeastern California. Figure 46-7 on page 322 is a topographic map of this same region.

Seif: Seif or longitudinal dunes are long parallel ridges of sand. It is thought that seif dunes develop in areas where there is a significant shift in wind direction for part of the year. Figure 46-4 is an aerial photograph of seif dunes in central Australia.

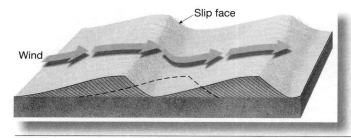

Figure 46-1: Profile of desert sand dunes, with arrows showing wind direction. Sand dunes migrate downwind as sand grains move up the gentle windward slope and are deposited on the steep slip face. (From Hess, *McKnight's Physical Geography*, 12th ed.)

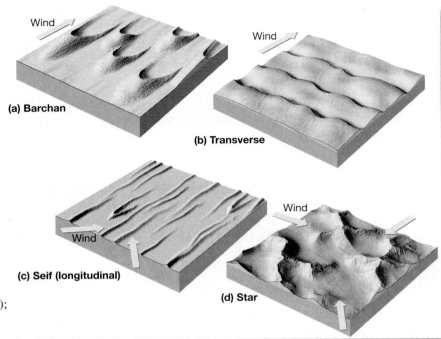

Figure 46-2: Common types of desert sand dunes: (a) barchan; (b) transverse; (c) seif (longitudinal); (d) star. (From Hess, *McKnight's Physical Geography*, 12th ed.)

Star Dunes: **Star dunes** are large pyramid-shaped dunes with arms radiating out in three or more directions. Star dunes develop in areas where the wind frequently varies in direction (Figure 46-5).

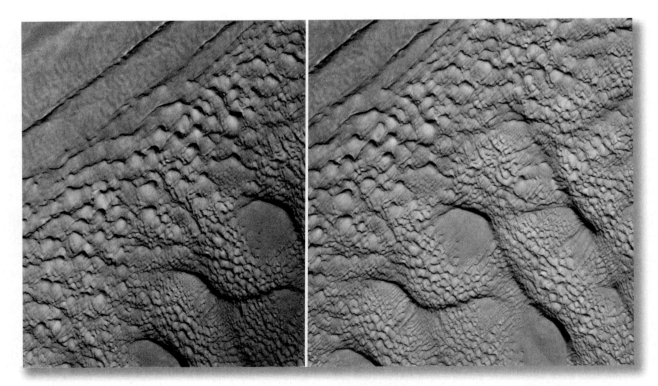

Figure 46-3: Stereogram of transverse sand dunes in southeastern California (scale 1:40,000; USGS photographs, 2002; ← N).

Figure 46-4: The parallel linearity of seif dunes is characteristic of some desert areas. This scene is from the Simpson Desert of central Australia. (Tom L. McKnight photo from Hess, *McKnight's Physical Geography*, 12th ed.)

Figure 46-5: Aerial photograph of star dunes in Death Valley National Park, California (Darrel Hess photo).

Figure 46-6 is a portion of the "Kane Spring NW, California" quadrangle, showing an active sand dune field near the Salton Sea in southern California.

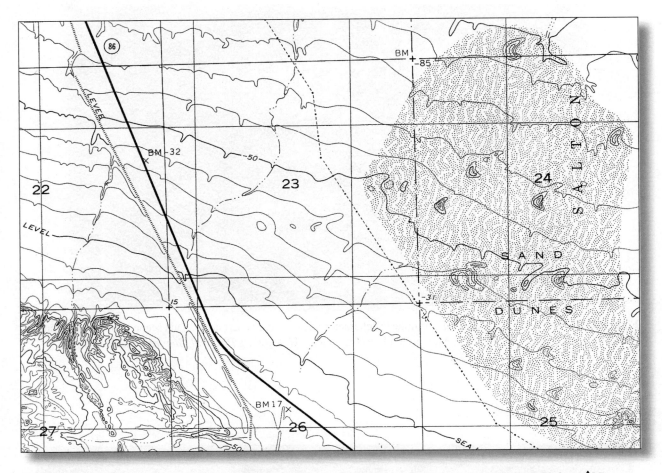

Figure 46-6: USGS "Kane Spring NW, California" quadrangle (scale 1:24,000; contour interval 10 feet; ↑N).

Name _____   Section _____

EXERCISE 46 PROBLEMS—PART I

The following questions are based on the portion of the "Kane Spring NW, California" quadrangle (scale 1:24,000; contour interval 10 feet) shown in Figure 46-6. These isolated dunes in the desert near the Salton Sea (33°11′02″N, 115°52′20″W) are moving over a surface that is only thinly covered with sand. To view this map in color, go to the Lab Manual website or scan the QR code for this exercise.

1.   (a)   What kind of sand dune appears to be most common in this dune field?

          (b)   How do you know?

2.   What is the height (the relief) of the tallest dune?   _____ feet

3.   What is the maximum width (or length) of the largest dune?   _____ feet

4.   (a)   From which direction does the prevailing wind blow? (The top of the map is north.)

                                                   From the _____

          (b)   How can you tell?

          (c)   If the prevailing wind direction remains the same, with time, in which direction will these dunes move?

                               From _____ to _____

5.   Find the large dune just south of "24" in Section 24. Assume that this dune is moving at an average rate of 90 feet (27 meters) per year. Calculate how long will it take for this dune to migrate out of the area shown on this map (a graphic map scale is found inside the front cover of the Lab Manual).

          (a)   Distance to edge of map: _____ feet (or meters)

          (b)   Time to move to edge of map: _____ years

Name _____   Section _____

EXERCISE 46 PROBLEMS—PART II

The following questions are based on the "Glamis SE, California" quadrangle shown below (Figure 46-7); a stereogram of the same area is shown in Figure 46-3. The map shows a region of mostly transverse sand dunes in southeastern California (32°50′46″N, 115°00′21″W). Note that many depression contours are used on this map, but they are difficult to read because of the sand stippling. To view this map in color, go to the Lab Manual website or scan the QR code for this exercise.

1.   With red lines, mark the location of at least five crescent-shaped dune crests.

2.   What is the maximum relief of the dune field? (Compare the height
      of dune crests with the bottom of the depression on the lee side.)   _____ feet

3.   What is the prevailing wind direction in this area?   From _____ to _____

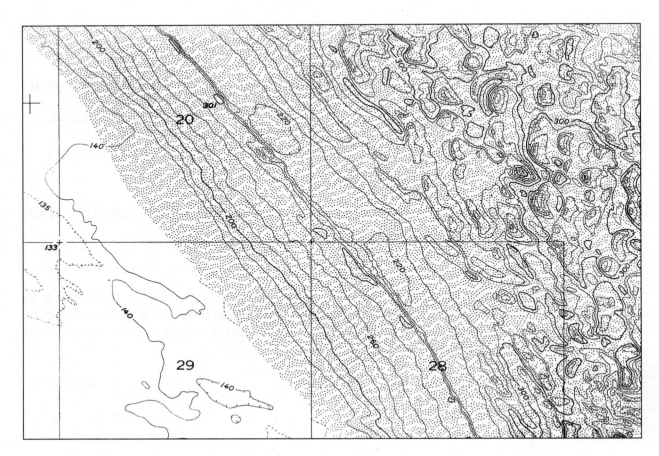

Figure 46-7: USGS "Glamis SE, California" quadrangle (scale 1:24,000; contour interval 20 feet; dashed lines represent 5-foot contours; ↑N).

EXERCISE **47**
Continental Glaciation

Objective:	To study landforms produced by continental glaciation.
Materials:	Lens stereoscope.
Resources:	Internet access or mobile device QR (Quick Response) code reader app (optional).
Reference:	Hess, Darrel. *McKnight's Physical Geography*, 12th ed., pp. 540–546 and 549–559.

PLEISTOCENE GLACIATIONS

By the end of the **Pleistocene Epoch**, about 11,700 years ago, Earth had undergone a series of extensive glaciations over a period of at least 2.6 million years. Most of this glaciation took place in the high latitudes of the Northern Hemisphere and in the high mountain areas of the world. During the Pleistocene there was a series of glacial advances, each followed by an *interglacial* period during which the glaciers melted and retreated (see Figure 24-5 in Exercise 24 for a chart showing temperature fluctuations during the later parts of the Pleistocene).

The continental and mountain (alpine) glaciers dramatically altered the preexisting landscape. Because the Pleistocene ended so recently (in terms of geologic time), the mark of these glaciations on the landscape is still very fresh in many parts of the world.

GLACIAL PROCESSES

The reshaping of the landscape by glaciers primarily takes place through three sets of processes.

Glacial Erosion: The direct force of the ice on the ground is enhanced by the abrasive action of the rocks being carried along the bottom of a glacier. **Glacial abrasion** often leaves a smooth or striated surface.

Probably more important overall than abrasion in removing rock is **glacial plucking**. Meltwater trickles down into the cracks in the rocks below a glacier and refreezes, fracturing the rock through **frost wedging**. These "fingers" of ice extending down and around the rocks below the glacier help pull out material as the glacier moves on. Glacial plucking tends to leave a blocky or irregular surface.

Glacial Transportation and Deposition: When ice is under pressure, as inside a large glacier, it can flow slowly, conforming to the topography. Debris can be transported on top of a glacier, and also within the flowing ice of a glacier.

Drift is a general term for any material deposited by glacial action. Glacial drift that has been deposited directly by the ice is known as **till**. Ridges of till can be deposited along the front and sides of a glacier, but till is also deposited in an unorganized fashion over the ground as a glacier retreats.

Glaciofluvial Processes: The meltwater from glaciers can also influence the landscape. Glaciofluvial deposits are often found many kilometers from the margin of the ice.

LANDFORMS PRODUCED BY CONTINENTAL GLACIATION

In the regions of the continents that were glaciated during the Pleistocene, ice sheets flowed out in all directions from their regions of accumulation, covering nearly the entire preexisting landscape. Continental ice sheets tended to flatten hills and reduce steep slopes. However, the ice sheets by no means produced a flat landscape. Areas that have experienced continental glaciation tend to be very irregular, but of low relief (see Exercise 48 for a description of alpine glaciation).

Typical landforms associated with the margin of a **continental ice sheet** are illustrated in Figure 47-1. The diagram shows the landscape as an ice sheet retreats.

Depositional Landforms: Landforms produced by deposition are often the most prominent ones in areas of continental glaciation. Some of the most conspicuous depositional features are accumulations of till known as **moraines**. The **terminal moraine** (sometimes called the **end moraine**) marks the maximum advance of a glacier, whereas a series of **recessional moraines** often develop behind the terminal moraine as the glacier retreats.

Drumlins are elongated mounds of till that evidently have been reworked by a subsequent advance of the ice. The long axis of the drumlin is oriented parallel to the flow of the ice, with the blunt, steep end of the mound facing the direction from which the ice came. Map T-12 is a topographic map of a drumlin field in New York; Figure 47-3 is a stereogram of the same area.

Glaciofluvial Landforms: Eskers are glaciofluvial features that formed when streams flowing under an ice sheet became clogged with glacial meltwater debris. Today, eskers are seen as sinuous ridges of glacial sand and gravel that can run for many kilometers.

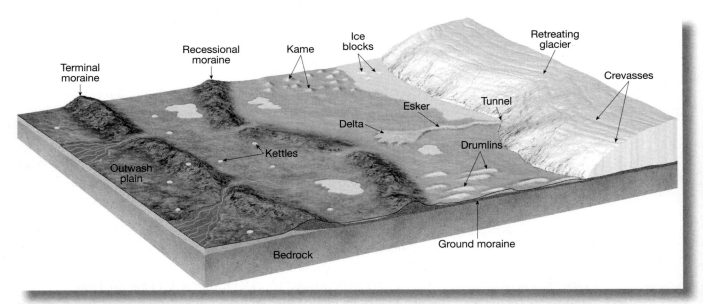

Figure 47-1: Glacial and glaciofluvial features left by a retreating continental ice sheet. (From Hess, *McKnight's Physical Geography*, 12th ed.)

Beyond the margin of the continental ice sheets, an **outwash plain** would develop. Outwash plains are often relatively flat areas covered with deposits of glacial sand and gravel. These outwash plains may extend well beyond the maximum advance of the ice. A **valley train** developed when glacial drift was washed down into a valley by meltwater. Today, these glaciofluvial deposits are seen along stream valleys, often great distances beyond the margin of the outwash plain.

Ice-Contact Deposits: Among the most varied features left by glaciers are those known as **ice-contact deposits** (Figure 47-2). These are the landforms that develop along the margin of the glacier, and include well-sorted sand and gravel, along with unorganized till. "Kame and kettle" topography often develops from ice-contact deposits.

A **kame** is a hill of debris that originally filled a hole in the ice. A **kettle** forms in the opposite manner, when a large piece of ice leaves a depression in the glacial drift. Today, many kettles are filled with water as small lakes.

Figure 47-2 shows landform development along the stagnant margin of an ice sheet. As a buried wedge of ice melts, the irregular topography of a **collapsed zone** develops in the outwash deposits.

Post-Glacial Drainage Patterns: Areas that have experienced continental glaciation are generally characterized by poor drainage. There are many lakes and swamps in the depressions left by the glaciation. The previous stream system was often totally disrupted, leaving a rather chaotic or **deranged drainage pattern**. The Pleistocene glaciations took place so recently, in geologic terms, that there has not been enough time for a new drainage pattern to become well established.

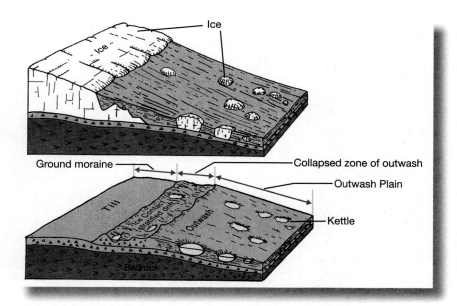

Figure 47-2: The relationship of buried ice to a collapsed zone in an outwash plain. (Adapted from USGS, "Geologic History of Cape Cod, Massachusetts," 1992)

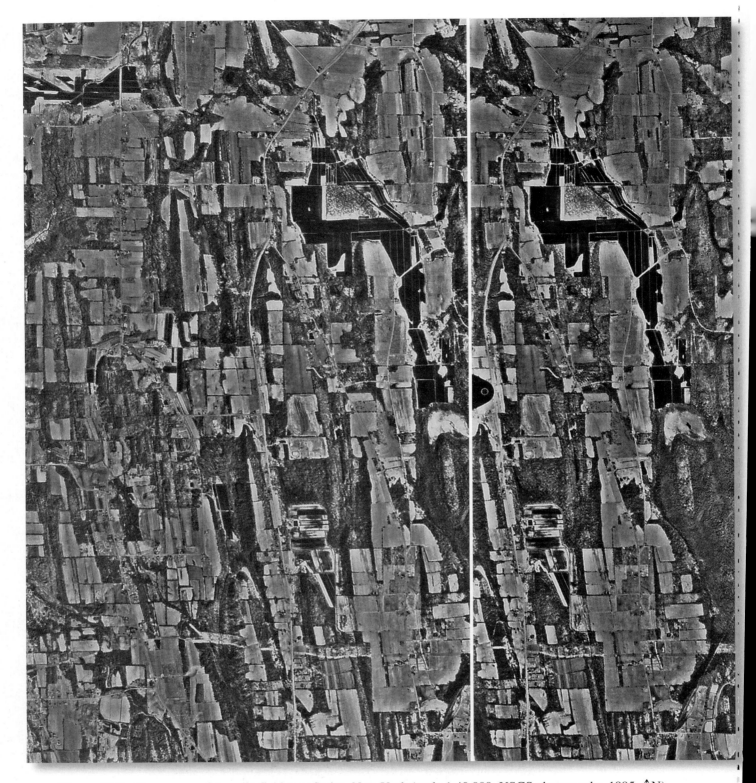

Figure 47-3: Stereogram of drumlin field near Sodus, New York (scale 1:40,000; USGS photographs, 1995; ↑N).

Name _____   Section _____

EXERCISE 47 PROBLEMS—PART I

The following questions are based on Map T-11, the "Whitewater, Wisconsin" quadrangle (scale 1:62,500; contour interval 20 feet; dotted lines represent 10-foot contours). The irregular high ground of "Kettle Moraine State Forest," running diagonally from the lower left to the right center of the map, is a terminal moraine left by a Pleistocene ice sheet (42°51′34″N, 88°35′59″W). The region northwest of the terminal moraine was once covered by the ice. The region in the southeast corner of the map is the edge of the glacial outwash plain.

1. Visualize a general topographic profile across this landscape, from the upper left corner to the lower right corner of the map, by determining the elevations of the following locations.

    (a) McLery Cemetery: _____ feet

    (b) Pumping Stations near Scuppernong: _____ feet

    (c) Blue Spring Lake: _____ feet

    (d) Crest of the terminal moraine about 0.5 mile
         (0.8 km) south of Blue Spring Lake: _____ feet

    (e) Round Prairie Cemetery: _____ feet

2. Estimate the minimum thickness of the ice at the terminal moraine in the area of Blue Spring Lake. Assume that the ice extended up at least to the highest point on the moraine. (Determine the thickness of the ice, not its elevation.)

                                                                    _____ feet

3. Suggest one reason why the elevation of the outwash plain in the southeast is higher than the ground that was below the ice, northwest of the terminal moraine.

4. Briefly describe the current extent of drainage system development around the terminal moraine and in the glacial outwash plain to the southeast.

Name _____   Section _____

EXERCISE 47 PROBLEMS—PART II

The following questions are based on Map T-11, the "Whitewater, Wisconsin" quadrangle (scale 1:62,500; contour interval 20 feet; dotted lines represent 10-foot contours), and Map T-27b, a shaded-relief map showing a portion of this area in greater detail. The irregular high ground of "Kettle Moraine State Forest," running diagonally from the lower left to the right center of the map, is a terminal moraine left by a Pleistocene ice sheet (42°51′34″N, 88°35′59″W). The region northwest of the terminal moraine was once covered by the ice. The region in the southeast corner of the map is the edge of the glacial outwash plain.

1. (a) What are the landforms shown as a series of small hills near the upper left corner of the map? (Map T-27b will be especially helpful here.)

    (b) How did they form?

2. (a) From which direction did the ice sheet advance toward what is now the terminal moraine? Be specific.

         From _____ to _____

    (b) What evidence on the map led you to your answer?

3. Provide one explanation of the formation of the cluster of small lakes about 1 mile (1.6 km) northeast of La Grange (near the southern margin of the map).

4. (a) The straight blue lines are artificial channels being used to drain swampy areas. Suggest one explanation of the formation of these swamps.

    (b) What is the evidence that these drainage channels are having their intended result?

Name _____   Section _____

EXERCISE 47 PROBLEMS—PART III

The following questions are based on Map T-12, the "Sodus, New York" quadrangle (scale 1:24,000; contour interval 10 feet) and Figure 47-3, a stereogram of this same area (43°09′55″N, 77°04′51″W). The map and stereogram show a number of drumlins, each with its long axis oriented approximately north-south.

1. Compare the dimensions of three of the drumlins on the map, identified below by their peak elevation marked with an x. These drumlins are found from bottom center to upper right on the map. "Drumlin Height" refers to the relief of the drumlin, not its elevation. Estimate the "length" of the drumlins from north to south (a graphic map scale is found inside the front cover of the Lab Manual). Finally, note whether the north-facing or south-facing side of the drumlin has the steepest slope.

Drumlin Peak Elevation	Drumlin Height	Length (in feet)	Steepest Slope (North or South)
615 feet			
680 feet			
547 feet			

2. Most geomorphologists suggest that the long axis of a drumlin reflects the direction of ice flow, with the steepest end facing the direction from which the ice came. Based on this assumption, from which direction did the ice flow over this region?

From _____ to _____

3. (a) Does it appear that these drumlins have been significantly altered by stream erosion?

   (b) Explain the reasons for your answer.

4. Briefly describe the current extent of drainage system development in this area.

5. To what extent have human features been influenced by the local topography?

Name _____   Section _____

EXERCISE 47 PROBLEMS—PART IV—GOOGLE EARTH™

To answer the following questions, go to the Hess *Physical Geography Laboratory Manual*, 12th edition, website at **www.MasteringGeography.com**, then Exercise 47 and select "Exercise 47 Part IV Google Earth™" to open a KMZ file in Google Earth, or scan the QR code for this exercise and view "Exercise 47 Part IV Google Earth™ video." The opening view is looking north toward the Finger Lakes region of New York. During the Pleistocene, this region was extensively glaciated by ice sheets.

1.   Fly to Point 1 in the area of the Finger Lakes. What is the general orientation of this series of lakes?

From _____ to _____

2.   Fly to Point 2, a drumlin in the region north of the Finger Lakes, noting the general orientation of this feature (a portion of this area is shown in Map T-12). Then fly to Point 3 and Point 4, again noting the general orientation of the drumlins in this area.

(a)   Most geomorphologists suggest that the long axis of a drumlin reflects the direction of ice flow, with the steepest end facing the direction from which the ice came. Based on this assumption, from which direction did the ice flow over this region?

From _____ to _____

(b)   Does the general orientation of the Finger Lakes match this? If not, describe the difference in orientation.

3.   To what extent have roads and other human-built features been influenced by the pattern of drumlins?

4.   Fly to Point 5 near Jackson, Michigan. This area was extensively glaciated during the Pleistocene. Fly to Point 6 along Blue Ridge, and then to Point 7, and finally to Point 8 along the same ridge.

(a)   What kind of glaciofluvial landform is Blue Ridge? _____

(b)   How does such a feature form beneath the ice of a glacier?

(c)   What kind of useful construction material has likely been excavated from Point 8 and the other quarries along Blue Ridge?

EXERCISE 48
Alpine Glaciation

Objective:	To study the landforms produced by alpine glaciation.
Materials:	Lens stereoscope.
Resources:	Internet access or mobile device QR (Quick Response) code reader app (optional).
Reference:	Hess, Darrel. *McKnight's Physical Geography*, 12th ed., pp. 546–549 and 556–563.

MOUNTAIN GLACIATION

During the Pleistocene, most of the high mountain areas of the world were glaciated. Many of these regions are still undergoing glaciation, although on a greatly reduced scale.

Alpine glaciers are individual glaciers that develop high on a mountain and then flow downvalley some distance, whereas other mountain glaciers develop and flow out of **highland ice fields** that can be many kilometers across. The emphasis of this exercise will be primarily alpine glaciation—see Exercise 47 for additional information about the Pleistocene and glacial processes. Alpine glaciers don't necessarily develop at the top of the highest mountain peaks. They generally need protected valleys, often below the summits of peaks, in order to accumulate enough snow to form glacier ice.

MOVEMENT OF GLACIERS

As with continental ice sheets, the size of an alpine glacier is a consequence of the balance between the **accumulation** of snow and ice and the annual wasting away of the ice through melting, evaporation, and sublimation—collectively known as **ablation**.

As shown in Figure 48-1, in the upper portion of a glacier there is greater annual accumulation of ice than ablation, whereas in the lower portion of the glacier there is greater ablation than accumulation. The boundary between these two zones of a glacier is called the **equilibrium line** or **annual snowline** (or **firn limit**). Firn (or névé) is partially compacted snow that has survived from a previous winter. The location of the equilibrium line can vary from year to year, depending on the amount of precipitation and temperature levels.

As a glacier flows down its valley beyond the equilibrium line, it increasingly ablates until it reaches its **terminus**, the point beyond which it can no longer advance. The terminus of a glacier is the location where the ice is melting as quickly as it is flowing in. It is important to note that when the terminus of a glacier is stationary (or even when it is retreating up a valley), the ice within a glacier continues to flow forward.

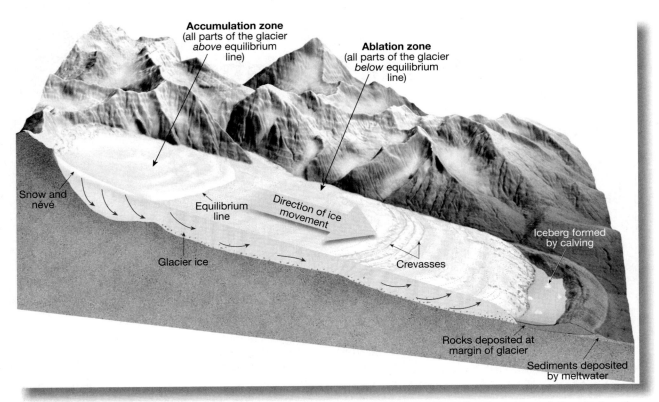

Accumulation zone
(all parts of the glacier *above* equilibrium line)

Ablation zone
(all parts of the glacier *below* equilibrium line)

Snow and névé

Equilibrium line

Direction of ice movement

Glacier ice

Crevasses

Iceberg formed by calving

Rocks deposited at margin of glacier

Sediments deposited by meltwater

Figure 48-1: Cross section through an alpine glacier. The upper portion is an area of net ice accumulation. Below the equilibrium line there is more ablation than accumulation. (From: Hess, *McKnight's Physical Geography*, 12th ed.)

EROSIONAL LANDFORMS FROM ALPINE GLACIATION

While depositional features tend to dominate the landscape once covered by continental ice sheets, erosional landforms are usually the most conspicuous features left by alpine glaciers.

The most basic of all alpine glacial landforms is the bowl or amphitheater-shaped valley head called a **cirque** (Figure 48-2). The head and side walls of a cirque may be nearly vertical, while the floor has often been carved into a shallow basin. Cirques grow through **glacial plucking** and **frost wedging** along the **headwall**. Especially in summer, a crevice at the head of the glacier known as the **bergschrund** opens and exposes the headwall to the freeze–thaw cycle. When water seeps down into the cracks and joints of the rock and freezes, the expansion of the ice further pries apart the opening.

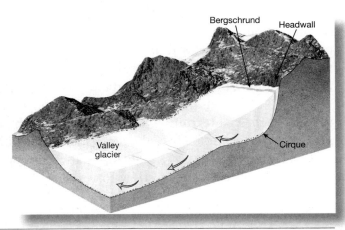

Bergschrund Headwall

Valley glacier

Cirque

Figure 48-2: The development of a cirque at the head of a valley glacier. (From: Hess, *McKnight's Physical Geography*, 12th ed.)

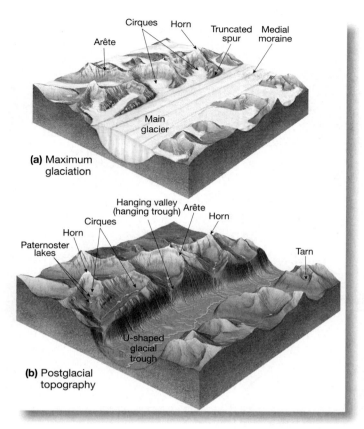

Figure 48-3: The development of landforms by alpine glaciation. (a) Landscape during glaciation. (b) Landscape after glaciation. (From Hess, *McKnight's Physical Geography*, 12th ed.)

When cirques are being worn back into a ridge from opposite sides through glacial erosion and frost wedging, a jagged ridge crest called an **arête** is formed (Figure 48-3). If erosion from both sides continues enough to remove part of the arête itself, a sharp-crested pass known as a **col** is formed.

Horns are another prominent alpine glacial landform. A horn is a pyramid-shaped mountain peak that develops as cirques are worn into the peak from three or four sides.

Alpine glaciers alter the shape and profile of former stream valleys. Glaciers tend to deepen and steepen valleys somewhat, often changing them from a V-shaped valley into a **glacial trough** with a U-shaped cross section (see Figure 48-3b). In addition, glaciers may partially straighten the sinuous pattern typical of stream valleys.

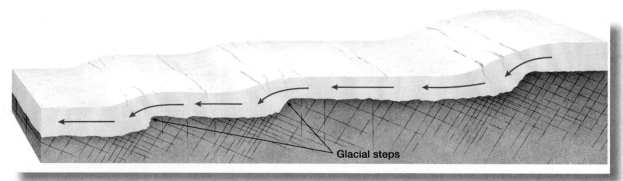

Figure 48-4: A longitudinal profile of a glaciated valley in mountainous terrain showing a sequence of glacial steps. (From Hess, *McKnight's Physical Geography*, 12th ed.)

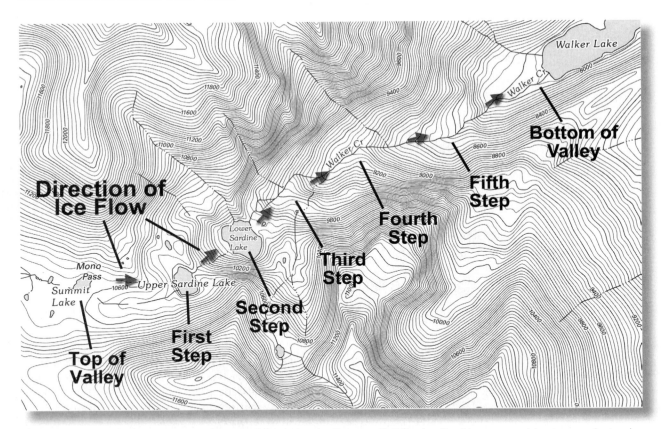

Figure 48-5: A series of glacial steps is found in Bloody Canyon, California. On this topographic map, each step is shown with abrupt change in slope (where the contour lines are close together). USGS "Koip Peak, California," *US Topo* quadrangle (scale approx. 1:30,000; contour interval 40 feet; ↑N).

In contrast, the downvalley profiles of glacial troughs are often somewhat irregular. Many glacial valleys have pronounced **glacial steps** (Figure 48-4) that evidently result from differences in the resistance of the valley floor rocks. Figure 48-5 is a topographic map showing a series of glacial steps in Bloody Canyon on the east side of the Sierra Nevada in California. Map T-13, the "Mono Craters, California," quadrangle, shows this same region of the Sierra.

The location of the upper edge of the ice, or "trimline," of a glacier is sometimes visible on the valley walls. Repeated cycles of frost wedging above the glacier tend to leave a rather jagged landscape, whereas glacial erosion below the trimline tends to leave a more smoothly sculpted landscape.

Hanging valleys (or "hanging troughs") are left when tributary glaciers enter the main valley glacier. The smaller tributary glaciers cannot deepen their valleys as much as the main "trunk" glacier, so after the ice is gone, these tributary valleys are left entering high above the main valley floor (Figures 48-3a and 3b).

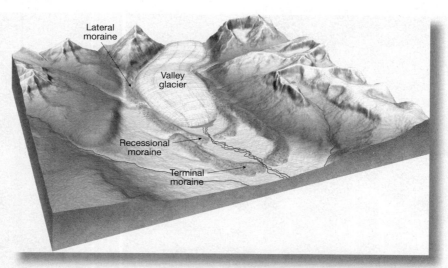

Figure 48-6: Moraines formed by an alpine glacier. (From Hess, *McKnight's Physical Geography*, 12th ed.)

DEPOSITIONAL LANDFORMS FROM ALPINE GLACIATION

Although erosional landforms are usually the most prominent features left by alpine glaciers, a number of important depositional landforms may be seen as well.

The maximum advance of an alpine glacier may be marked by a **terminal moraine** (Figure 48-6), and pauses in the retreat of a glacier are marked by **recessional moraines**. Often, the most distinctive depositional features left by alpine glaciers are **lateral moraines**. These ridges—made of unsorted, angular rock debris known as **till**—accumulate along the sides of a valley glacier. Where two glaciers come together, their adjacent lateral moraines may join and continue down the middle of a glacier, forming a **medial moraine**.

Figure 48-7 is a stereogram showing Bloody Canyon and the eastern crest of the Sierra Nevada in California. Walker Lake lies between two sets of large lateral moraines (north is to the left side of the stereogram). Map T-13, the "Mono Craters, California," quadrangle (scale 1:62,500; contour interval 80 feet) and Map T-23a, a satellite image, show this same region of the Sierra.

LAKES FROM ALPINE GLACIATION

Lakes are common in regions that have experienced alpine glaciation. A **tarn** is a lake that forms in the depression of a cirque. A chain of lakes may develop in a valley with glacial steps. Often each step will have a lake, connected to the lake on the step below by a stream (see Figure 48-5). These chains of lakes are known as **paternoster lakes**, because they resemble the beads on a rosary. Finally, terminal or recessional moraines can act as dams to form lakes at the mouths of glacial valleys.

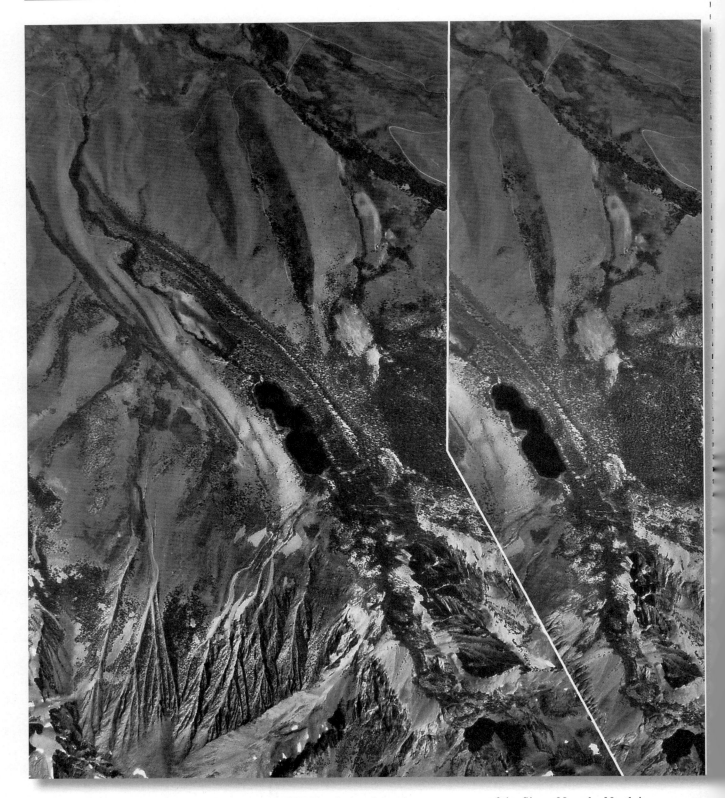

Figure 48-7: Stereogram of Bloody Canyon and Walker Lake along the eastern crest of the Sierra Nevada. North is to the left side of the page (scale 1:40,000; USGS photographs, 1993; ← N).

Name _____   Section _____

EXERCISE 48 PROBLEMS—PART I

The following questions are based on Map T-13, the "Mono Craters, California" quadrangle (scale 1:62,500; contour interval 80 feet), and Figures 48-5 and 48-7, a detailed map and stereogram, and Map T-23a, a satellite image, of the same region. The heavily glaciated eastern crest of the Sierra Nevada is seen along the western sides of the map and stereograms (37°52′37″N, 119°12′40″W). During the Pleistocene, glaciers in this region flowed down the valleys on the eastern slope of the Sierra toward the Mono Lake basin to the northeast. Large lateral moraines are found at the mouths of each canyon.

1. (a) A small glacier is located just north of the Dana Plateau (in the northwest section of the map). In what kind of glacial landform is this glacier found?

    (b) What evidence suggests this feature was eroded by a glacier that was larger than the present one?

2. (a) What is the name for the kind of glacial landform illustrated by Mt. Gibbs? (On the topographic map, the dashed black line showing the boundary between Mono and Tuolumne Counties runs through the summit of Mt. Gibbs.)

    (b) How does this kind of glacial landform develop?

Large lateral moraines can be see at the mouth of Bloody Canyon (Walker Lake and Walker Creek are between these moraines; 37°52′28″N, 119°09′53″W). The glacier that left these moraines flowed down Bloody Canyon from near Mono Pass. Sawmill Canyon (just south of Bloody Canyon) consists of two large lateral moraines—these moraines were left by a glacier that at one time also flowed down Bloody Canyon.

3. How deep (thick) was the ice in the glacier that formed the lateral moraines at the bottom of Bloody Canyon? You may assume that the ice reached the top of the lateral moraines. Estimate the height of the lateral moraine just south of the word "Lake" in "Walker Lake."

    _____ feet

4. Based on the evidence you see in the map and in the stereogram, explain the formation of Walker Lake.

5. (a) Which set of moraines formed first: the Bloody Canyon moraines or the Sawmill Canyon moraines?   _____

    (b) How do you know?

Name _____   Section _____

EXERCISE 48 PROBLEMS—PART II

The following questions are based on Map T-13, the "Mono Craters, California" quadrangle (scale 1:62,500; contour interval 80 feet), and Figures 48-5 and 48-7, a detailed map and stereogram, and Map T-23a, a satellite image, of the same region. The heavily glaciated eastern crest of the Sierra Nevada is seen along the western sides of the map and stereograms (37°52′37″N, 119°12′40″W). During the Pleistocene, glaciers in this region flowed down the valleys of the eastern slope of the Sierra toward the Mono Lake basin to the northeast. Large lateral moraines are found at the mouths of each canyon.

1.   There are five small glaciers shown on this map (glaciers are shown as white patches with blue contour lines). Why can glaciers survive today in these locations but not in others?

2.   What evidence suggests that the edges of the Dana Plateau were extensively glaciated in the past?

3.   Find and describe the location of a hanging valley.

4.   (a)   Name a valley (or "canyon") other than
               Bloody Canyon that has glacial steps: _____

       (b)   How many steps are there? _____

       (c)   How many lakes are found in the series of steps? _____

5.   (a)   What is the name for the kind of glacial landform marked by the black dashed line halfway between Mt. Dana and Mt. Gibbs?

       (b)   How did it form?

6.   How many tributary glaciers clearly flowed into Bloody Canyon? (Hint: Look for U-shaped valleys that enter Bloody Canyon; these valleys are now reoccupied by streams.)

Name _____   Section _____

EXERCISE 48 PROBLEMS—PART III

The following problems are based on the section of the "Mt. Tom, California" quadrangle reproduced below (Figure 48-8). The map shows a pair of large lateral moraines at the mouth of Pine Creek coming down the eastern slope of the Sierra Nevada (37°24'50"N, 118°36'11"W). A road (shown as a double black line) follows the course of Pine Creek out of the valley between the lateral moraines. You can view this map in color by going to the Lab Manual website or by scanning the QR code for this exercise.

1.   With a red pencil, mark the crests of each lateral moraine. Also mark the crests of any branches of these moraines.

2.   From the pattern of moraines, describe the evidence of more than one glacial advance.

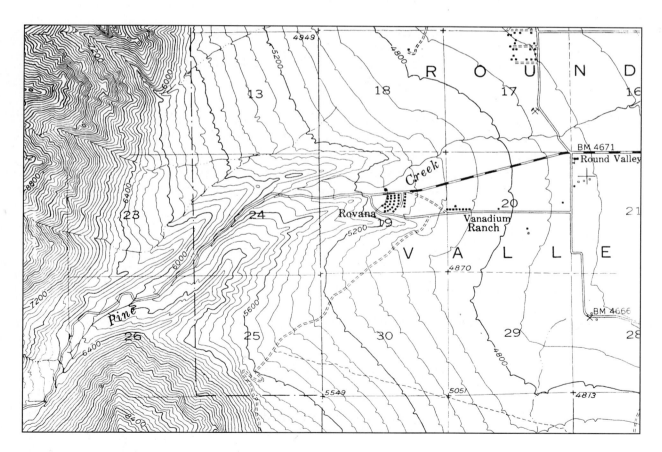

Figure 48-8: USGS "Mt. Tom, California" quadrangle (enlarged to scale 1:48,000; contour interval 80 feet; ↑N).

339

Name _____   Section _____

EXERCISE 48 PROBLEMS—PART IV

The following questions are based on Map T-14, the "Sumdum (D-4), Alaska" quadrangle (scale 1:63,360; contour interval 100 feet). The map shows two glaciers entering the "Tracy Arm." The Tracy Arm is a fjord just south of Juneau, Alaska (see Exercise 49 for a brief discussion of fjords). The "Sawyer Glacier" enters the Tracy Arm from the north and the "South Sawyer Glacier" enters the Tracy Arm from the southeast (57°52′27″N, 133°10′33″W).

1. What topographic evidence (other than the glaciers entering the water) suggests that the Tracy Arm is a fjord?

2. Find the tributary glacier that enters the South Sawyer Glacier from the south (the tributary enters just east of the word "Wilderness").

    (a) What are the brown stripes along the sides of the glacier (shown with brown stippling)? _____

    (b) What are the brown stripes down the middle of the glacier? _____

    (c) Explain how these stripes came to be in the middle of the glacier.

3. Find the word "South" in the label for the "South Sawyer Glacier" (in the east central part of the map). Assume that the rate of movement of this glacier is 9 inches (22 cm) per day (a hypothetical rate). How long would it take the ice to move from the location marked by the "S" in the word "South" to the water of the Tracy Arm? It may be easiest if you determine the distance in inches or centimeters, rather than in feet, miles, or kilometers.

    (a) Distance (from "S" to the water): $\dfrac{\rule{3cm}{0.4pt}}{\text{feet (or meters)}} = \dfrac{\rule{4cm}{0.4pt}}{\text{inches (or cm)}}$

    (b) Time for ice to move to the Tracy Arm: _____ years

4. Describe a location where a horn has developed at the top of a mountain peak.

5. Describe the location where an arête is being formed.

6. Describe the location of a glacier that is currently confined to a cirque.

340

Name _____   Section _____

EXERCISE 48 PROBLEMS—PART V—GOOGLE EARTH™

To answer the following questions, go to the Hess *Physical Geography Laboratory Manual*, 12th edition, website at **www.MasteringGeography.com,** and select "Exercise 48 Part V Google Earth™" to open a KMZ file in Google Earth, or scan the QR code for this exercise and view "Exercise 48 Part V Google Earth™ video." The opening view is looking north over Glacier National Park in northern Montana. The mountains here were extensively glaciated by alpine glaciers during the Pleistocene.

1.   Fly to Point 1. In what kind of glacial landform is this lake found?   _____

2.   (a)   Fly to Point 2. What kind of glacial landform is shown here?   _____

       (b)   Fly to Point 3. What might explain why this valley enters the valley of Lake Sherburne closer to the present level of the lake than the valley marked by Point 2?

3.   (a)   Fly to Point 4. What kind of glacial landform is shown here?   _____

       (b)   How does this kind of glacial feature develop?

4.   Fly to Point 5 along the eastern side of the Sierra Nevada Mountains in California. This view shows large lateral moraines at the bottom of Lee Vining Canyon. During the Pleistocene, glaciers flowed to the east through Lee Vining Canyon from the crest of the Sierra down into the Mono Lake basin. Notice that there is a pair of moraines on each side of the canyon (this area is also shown in Maps T-13 and T-23a). Point 6 marks the crest of one of the inner moraines formed by a different glacial advance during the Pleistocene than those that left the moraines marked by Point 5.

       (a)   Which moraine formed first, the one marked Point 5 or Point 6?   _____

       (b)   How do you know?

       (c)   How deep (thick) was the ice in the glacier that formed the lateral moraines at Point 5? You may assume that the ice reached the top of this lateral moraine.

       (d)   How much deeper was the glacier that left the moraines marked by Point 5 than those marked by Point 6? (Compare the moraine heights at the same place in the valley.)

5.   Fly to Point 7 in the Tracy Arm fjord in Alaska. Compare the location of the South Sawyer glacier's terminus shown in Map T-14 (mapped in 1960) to that shown in the latest Google Earth imagery. Using the "ruler" function in Google Earth™ or the graphic map scale in the front of the Lab Manual, estimate how far the glacier has retreated.

Name _____   Section _____

EXERCISE 48 PROBLEMS—PART VI—INTERNET

The following questions are based on Figures 48-9, 48-10, 48-11, 48-12, 48-13, and 48-14, photographs you can view from the Hess *Physical Geography Laboratory Manual*, 12th edition, website at **www.MasteringGeography.com**, then select Exercise 48, or scan the QR code for this exercise.

1.　(a)　Which photograph—Figure 48-9 or Figure 48-10—shows a deposit of glacial till?

Figure 48- _____

　　(b)　Describe the evidence you see in the photographs that supports your answer.

2.　(a)　Which photograph—Figure 48-11 or Figure 48-12—shows a landscape predominantly shaped by glacial erosion?

Figure 48- _____

　　(b)　Describe the evidence you see in the photographs that supports your answer.

3.　Look at Figure 48-13, a photograph showing a glaciated landscape on the eastern side of the Sierra Nevada mountains in California.

　　(a)　What erosional feature is marked by the letter "A"? _____

　　(b)　What erosional feature is marked by the letter "B"? _____

4.　Look at Figure 48-14, a photograph showing the terminus of the Emmons Glacier on Mount Rainier in Washington. The end of the Emmons Glacier is heavily mantled with rock debris.

　　(a)　What kind of depositional feature is forming in the area marked with the letter "C"? _____

　　(b)　What kind of depositional feature is marked by the letter "D"? _____

　　(c)　Was the feature marked "D" formed by the present-day Emmons Glacier? _____

　　(d)　How do you know?

EXERCISE 49
Coastal Landforms

Objective:	To study the special processes and landforms found along coastlines.
Materials:	Lens stereoscope.
Resources:	Internet access or mobile device QR (Quick Response) code reader app (optional).
Reference:	Hess, Darrel. *McKnight's Physical Geography*, 12th ed., pp. 570–588.

SPECIAL COASTAL PROCESSES

Waves: Waves are by far the most important erosional force shaping coastlines. Wave erosion along a coastline is influenced by a phenomenon known as **wave refraction**. As illustrated in Figure 49-1, when a wave approaches shore, the segment of the wave to reach shallow water first begins to slow, while the segment of the wave in deeper water continues to move quickly. Because of wave refraction, waves typically strike parallel to shore and tend to concentrate their erosive power on headlands.

Coastal Sediment Transport: Sediment is transported along a shoreline in two main ways (Figure 49-2). **Longshore currents** are commonly the key transportation mechanism of

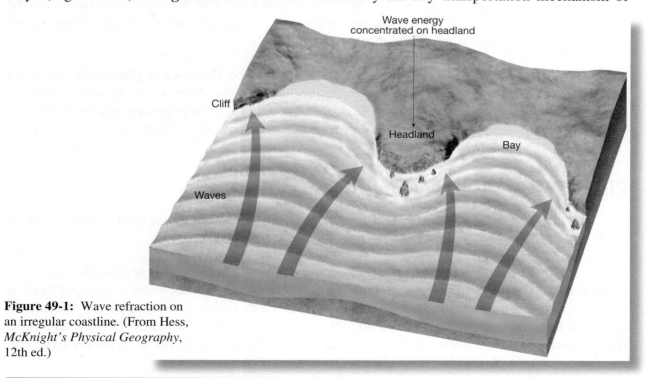

Figure 49-1: Wave refraction on an irregular coastline. (From Hess, *McKnight's Physical Geography*, 12th ed.)

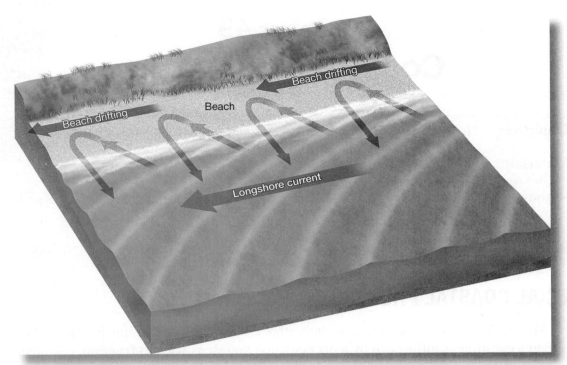

Figure 49-2: Coastal sediment transport. Beach drifting involves the zigzag movement of sand along a beach. Sand is brought obliquely onto the beach by the wave, and then is returned seaward by the backwash. Longshore currents develop just offshore and move sediment parallel to shore. Both beach drifting and longshore currents are set up by the action of waves striking a shoreline at a slight angle. (From Hess, *McKnight's Physical Geography*, 12th ed.)

sediment along a shoreline. Longshore currents are set up by the action of the waves, and because most waves are generated by the wind, these currents generally flow downwind and parallel to the shore. Sediment also moves through **beach drifting** in a "zigzag" fashion through wave **swash** onto the beach, and **backwash** off the beach.

Changes in Sea Level: During the height of the Pleistocene glaciations, so much water was locked up on the continents as glacier ice that the level of the ocean dropped as much as 130 meters (430 feet). At the end of the Pleistocene, roughly 10,000 years ago, sea level rose and flooded many low-lying coastal areas. Evidence of this recent rise in sea level is visible along many coastlines around the world.

COMMON COASTAL LANDFORMS

While coastlines exhibit a great variety of landforms, there are several common coastal landform assemblages.

Shorelines of Submergence: Because of the recent rise in sea level at the end of the Pleistocene, many shorelines show evidence of being submerged. When stream-cut topography is submerged, it forms what is called a **ria coastline**. In ria coastlines, what were once stream valleys have now become inlets and bays, and what were once hilltops have now become islands. A **fjorded coast** develops when glacial valleys are flooded by the sea. Fjords are typically long, narrow inlets with steep valley walls leading up from the water.

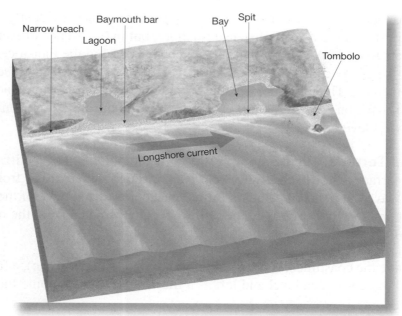

Figure 49-3: Common depositional landforms along a coastline. (From Hess, *McKnight's Physical Geography*, 12th ed.)

Depositional Landforms: There are many different kinds of depositional features found along coastlines. In general, deposition takes place where the power of currents (especially longshore currents) or waves is diminished.

One of the most common depositional features is the **spit** (Figure 49-3). A spit typically forms where a longshore current (shown with an arrow) moves over deeper water, such as at the mouth of a bay. As the current moves out over deeper water, it loses some of its power, and deposition takes place. For this reason, a spit generally points down current. Some spits develop a "hook" when currents carry some sediment back toward shore. **Baymouth bars** develop where a spit extends across an inlet, closing off a **lagoon**. A **tombolo** is formed where deposition connects an offshore island with the mainland.

Barrier Islands: **Barrier islands** (also called **barrier bars** or **offshore bars**) are found on gently sloping, shallow coastal shelves, such as along much of the Gulf and Atlantic coast areas of the United States. Barrier islands are low-lying and generally run parallel to the mainland, although they are often found several kilometers offshore. Lagoons are commonly formed behind barrier islands (Figure 49-4). These lagoons will generally fill up with sediment over time.

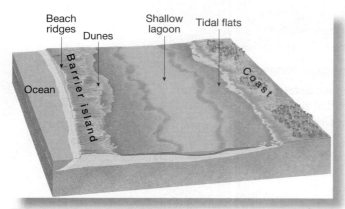

Figure 49-4: A typical relationship among the ocean, barrier island, and lagoon. (From Hess, *McKnight's Physical Geography*, 12th ed.)

The formation of barrier islands is not completely understood. They evidently develop where waves break and deposit sediment some distance offshore, but the existence of some of the larger barrier islands appears to be associated with sea level changes during the Pleistocene.

Map T-16 is a portion of the "Corpus Christi, Texas" topographic map (scale 1:250,000; contour interval 10 meters). The map shows Padre Island, a prominent barrier island just south of the city of Corpus Christi along the Gulf coast of Texas. South of the area shown on the map, Padre Island has been heavily developed.

Cliff/Marine Terrace: Figure 49-5 illustrates the combination of a cliff, wave-cut platform, and a marine terrace. The cliff tends to wear back through undercutting from wave action. At the same time, a wave-cut platform is being formed just below the surface of the water. As erosion proceeds, it is common for portions of the cliff to become isolated from the mainland as **sea stacks** (Figure 49-6).

In places where the coastline has been rising (a shoreline of emergence), a former wave-cut platform can be uplifted above sea level and left as a **marine terrace**. In some locations, a series

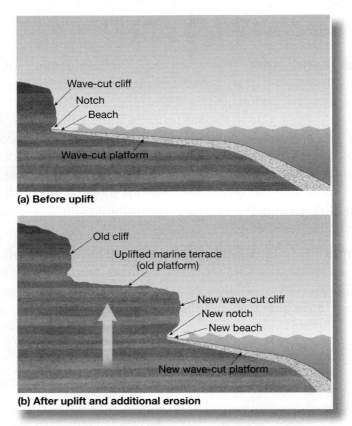

Figure 49-5: A marine terrace develops when a wave-cut platform is tectonically uplifted above sea level. (From Hess, *McKnight's Physical Geography*, 12th ed.)

of marine terraces can be found. The development of marine terraces in some areas is apparently associated with both a tectonically uplifted coastline and the fluctuations of sea level that took place during the Pleistocene.

A series of marine terraces is shown in Figure 49-7, a stereogram showing the southwestern side of San Clemente Island in southern California. Figure 49-8 is a topographic map of the same region. Each gently sloping marine terrace level is separated from the next higher terrace level by a steep slope. The well-preserved lower terrace levels are the easiest ones to recognize in the stereogram.

Figure 49-6: Sea stacks and coastal cliffs in Port Campbell National Park, Victoria, Australia. (Tom L. McKnight photo from McKnight's and Hess, *Physical Geography*, 9th ed.)

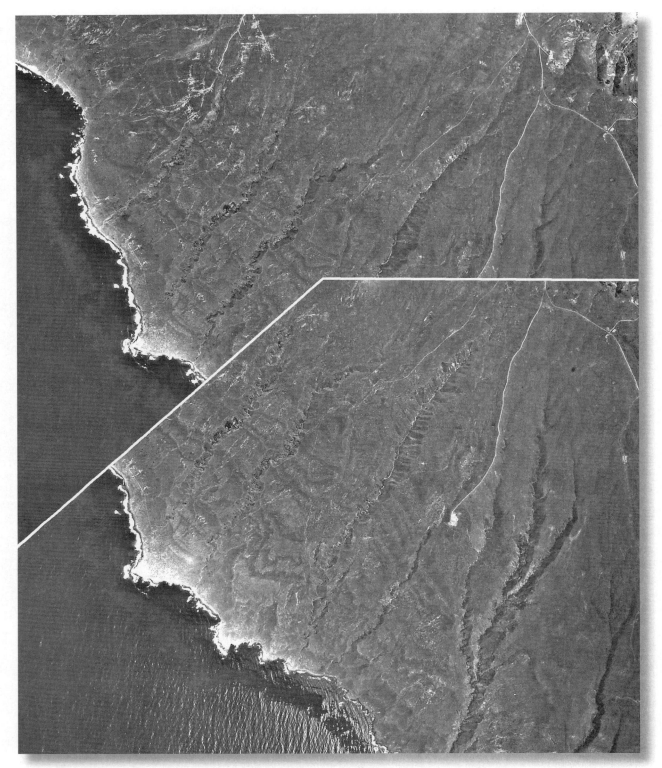

Figure 49-7: Stereogram of marine terraces on San Clemente Island, California (scale 1:40,000; USGS photographs, 1990; ↑N).

Name _____   Section _____

EXERCISE 49 PROBLEMS—PART I

The following questions are based on Map T-15, the "Point Reyes, California" quadrangle (scale 1:62,500; contour interval 80 feet) showing Point Reyes and Drakes Bay in Point Reyes National Seashore (38°01′38″N, 122°57′51″W).

1.   Explain the formation of Drakes Estero. Does it appear to be the feature of a shoreline of emergence or submergence?

2.   What suggests that Drakes Estero is quite shallow?

3.   (a)   Explain the formation of the hook-shaped lake southeast of "D Ranch."

      (b)   Explain the formation of the small pond and swamp just northwest of Drakes Beach.

4.   What is the dominant direction of the longshore current that produced Limantour Spit?

                    From _____ to _____

5.   What evidence of wave erosion is visible along the shore of Drakes Bay between Drakes Beach and Point Reyes?

6.   Describe the general coastal topography along the south face of Point Reyes itself.

Name _____ Section _____

EXERCISE 49 PROBLEMS—PART II

The following questions are based on Map T-16, the "Corpus Christi, Texas" topographic map (scale 1:250,000; contour interval 10 meters) and Figure 49-9, a satellite image you can view by going to the Lab Manual website or by scanning the QR code for this exercise. Padre Island is a prominent barrier island south of Corpus Christi, Texas (27°17′22″N, 97°24′21″W).

1. What is the highest elevation shown on Padre Island? _____ meters

2. (a) What is the most likely origin of Baffin Bay?

    (b) What is the evidence for this?

3. (a) Does it appear that "Laguna Madre" (between Padre Island and the mainland) is filling with sediment?

    (b) What is the evidence of this?

    (c) What are the likely sources of this sediment?

EXERCISE 49 PROBLEMS—PART III—GOOGLE EARTH™

To answer the following questions, go to the Hess *Physical Geography Laboratory Manual*, 12th edition, website at **www.MasteringGeography.com**, then Exercise 49 and select "Exercise 49 Part III Google Earth™" to open a KMZ file in Google Earth, or scan the QR code for this exercise and view "Exercise 49 Part III Google Earth™ video."

1. (a) Fly to Point 1. What kind of depositional coastal landform is shown here?

    (b) How does this kind of landform develop?

2. Fly to Point 2. What kind of depositional coastal landform is shown here?

3. (a) Fly to Point 3. What kind of depositional coastal landform is shown here?

    (b) What is the dominant direction of the longshore current along the shore in this bay—clockwise or counterclockwise?

    (c) How do you know?

4. Fly to Point 4. What kind of erosional coastal landform is shown here?

Name _____   Section _____

EXERCISE 49 PROBLEMS—PART IV

The following questions are based on Figure 49-8, a portion of the "San Clemente Island Central, California" quadrangle shown on the following page (scale 1:24,000; contour interval 25 feet; index contours drawn every fourth line; to view this map in color, go to the Lab Manual website or scan the QR code for this exercise), and the stereogram of San Clemente Island (Figure 49-7). The map and stereogram show a series of marine terraces on San Clemente Island in southern California (32°51′08″N, 118°29′58″W). The terraces have been incised by streams in several places.

1.   Using the graph below, construct a topographic profile from Point A to Point B. Plot index contours, and any intermediate contours necessary to accurately show significant changes in topography. The vertical exaggeration of the profile is 2×. The stereogram may be helpful in recognizing the extent of some terrace levels.

2.   (a)   How many marine terraces are shown on the profile?   _____

      (b)   Number the terraces on your topographic profile (with the terrace closest to sea level as number "1").

      (c)   With a blue pencil, number the terraces on the map in the same way.

3.   Do all of the terraces shown on your topographic profile extend, without interruption, to the southeast and northwest along this side of the island?

4.   Using the terraces shown on your topographic profile as a starting point, denote the extent of each terrace level across the entire map. Using a green pencil, outline the seaward lip of each terrace level shown on the map.

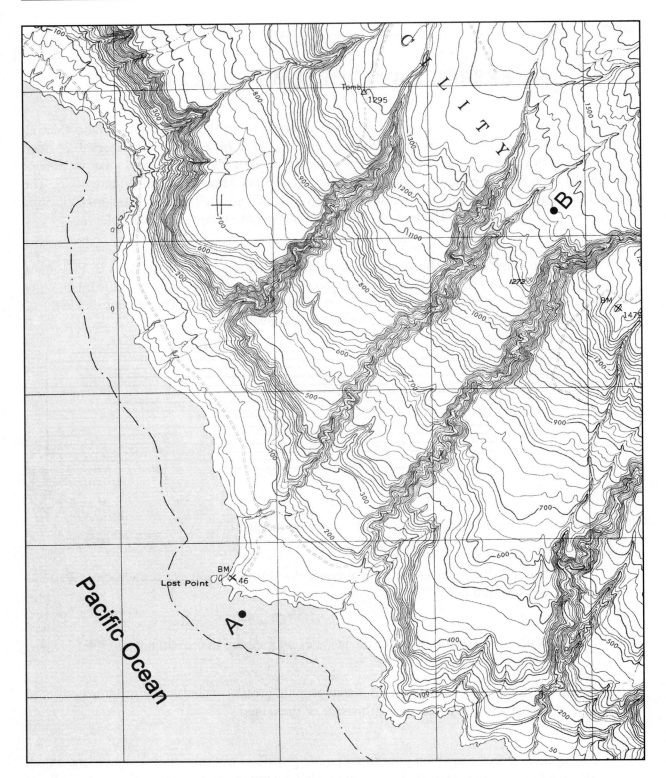

Figure 49-8: USGS "San Clemente Island Central, California" quadrangle (scale 1:24,000; contour interval 25 feet; note that the index contours on this map have been drawn every fourth line; ↑N).

APPENDIX I
Glossary

ablation Wastage of glacial ice through melting and sublimation.

absolute humidity A direct measure of the water vapor content of air, expressed as the mass of water vapor in a given volume of air, usually as grams of water per cubic meter of air.

accumulation (glacial ice accumulation) Addition of ice into a glacier by incorporation of snow.

adiabatic cooling Cooling by expansion, such as in rising air.

adiabatic warming Warming by compression, such as in descending air.

aerosols Solid or liquid particles suspended in the atmosphere; also called *particulates*.

air mass An extensive body of air that has relatively uniform properties in the horizontal dimension and moves as an entity.

albedo The reflectivity of a surface. The fraction of total solar radiation that is reflected back, unchanged, into space.

alluvial fan A fan-shaped depositional feature of alluvium laid down by a stream issuing from a mountain canyon.

alluvium Any stream-deposited sedimentary material; frequently alluvial deposits consist of rounded, sorted, and stratified rock debris.

alpine glacier Individual glacier that develops near a mountain crest line and normally moves downvalley for some distance.

analemma A "figure-8" chart showing the *declination of the Sun* (latitude of the vertical rays of the Sun) and the "equation of time" of the Sun throughout the year.

andesite A volcanic rock formed from lava intermediate in silica content between basalt and rhyolite; composed largely of a distinctive mineral association of plagioclase feldspar.

angle of incidence The angle at which the Sun's rays strike Earth's surface.

angle of repose The steepest angle that can be assumed by loose fragments on a slope without downslope movement.

annual snowline (firn limit) The elevation on a glacier above which winter snow is able to persist throughout the year; above the annual snowline, the accumulation of snow and ice exceeds ablation; also called the *equilibrium line.*

annular [drainage] pattern A network in which the major streams are arranged in a ringlike, concentric pattern in response to a structural dome or basin.

anticyclone A high-pressure center.

antitrade winds Tropical upper-air westerly winds near the top of the Hadley cells that blow toward the northeast in the Northern Hemisphere and toward the southeast in the Southern Hemisphere.

aphelion The point in Earth's elliptical orbit at which Earth is farthest from the Sun (about 152,171,500 kilometers or 94,555,000 miles).

aquifer A permeable subsurface rock layer than can store, transmit, and supply water.

arête A narrow, jagged, serrated spine of rock; remainder of a ridge crest after several glacial cirques have been cut back into an interfluve from opposite sides of a divide.

asthenosphere Plastic layer of the upper mantle that underlies the lithosphere. Its rock is very hot and therefore weak and easily deformed.

atmospheric window The range of wavelengths of infrared radiation to which the atmosphere is transparent (between approximately 8 and 12 micrometers).

average daily insolation Intensity of solar radiation received by a surface averaged over a 24-hour period.

average lapse rate The average rate of temperature decrease with height in the troposphere—about 6.5°C per 1000 meters (3.6°F per 1000 feet).

average monthly precipitation Average precipitation in each month of the year.

average monthly temperature Average temperature of each month of the year.

azimuth Description of direction in terms of the number of degrees clockwise from north.

backwash Water moving seaward after the momentum of the wave swash is overcome by gravity and friction.

bajada A continual alluvial surface that extends across the piedmont zone, slanting from the range toward the basin, in which it is difficult to distinguish between individual alluvial fans.

barchan dune A crescent-shaped sand dune with cusps of the crescent pointing downwind.

barrier island (barrier bar; offshore bar) Narrow offshore island composed of sediment; generally oriented parallel to shore.

basalt Fine-grained dark or black volcanic rock; formed from lava with low silica content.

base line Reference parallel from which townships are designated north or south in the Public Land Survey.

base reflectivity [radar image] Weather radar image showing the intensity of echoes very close to the horizon—usually about 0.5° above the horizon.

base velocity [radar image] Weather Doppler radar image showing the overall pattern of movement within a storm.

basin of interior drainage A topographic basin or valley that has no stream outlet leading to the ocean.

baymouth bar A spit that has become extended across the mouth of a bay to connect with a headland on the other side, transforming the bay into a lagoon.

beach drifting The zigzag movement of beach sediment in which the net result is a displacement parallel to the coast in a general downwind direction.

benchmark Metal disk embedded in rock or concrete that serves as reference marker for topographic maps; locations often labeled "BM" on topographic maps.

bergschrund The crevice at the head of an alpine glacier that forms in summer as the glacier pulls away from the headwall of a cirque.

biogeochemical cycles Planetwide cycles and flows of energy, chemicals, and nutrients into and out of ecosystems and biomes.

biome A large, recognizable assemblage of plants and animals in functional interaction with its environment.

bluff A relatively steep slope marking the outer edge of a floodplain.

broadleaf tree Tree with flat and expansive leaves.

calcium carbonate Chemical compound making up the mineral calcite; principal component of the rock limestone; $CaCO_3$.

caldera Large, steep-sided, roughly circular depression resulting from the explosion and/or collapse of a large volcano.

capacity (water vapor capacity) Maximum amount of water vapor that can be present in the air at a given temperature.

carbonic acid Mild acid formed when carbon dioxide dissolves in water; H_2CO_3.

cavern Large opening or cave, especially in limestone; often decorated with speleothems such as stalactites and stalagmites.

centripetal drainage pattern A basin structure in which the streams converge toward the center.

chaparral Shrubby vegetation of the mediterranean climatic region of North America.

circle of illumination The edge of the sunlit hemisphere that is a great circle separating Earth into a light half and a dark half.

cirque A broad amphitheater hollowed out at the head of a glacial valley by ice erosion.

climograph (climatic diagram) Chart showing the average monthly temperature and precipitation for a weather station.

col A pass or saddle through a ridge produced when two adjacent glacial cirques on opposite sides of a divide are cut back enough to remove part of the arête between them.

cold front The leading edge of a cool air mass actively displacing warm air.

collapse sinkhole (collapse doline) A sinkhole produced by the collapse of the roof of a subsurface cavern.

collapsed zone [of a glacier] Irregular topography resulting from the melting of ice blocks within the terminal moraine of a stagnant glacier.

composite reflectivity [radar image] Weather radar image displaying the strongest echo detected in each direction from the radar unit.

composite volcano (stratovolcano) Volcano with the classic symmetrical cone-shaped peak, produced by a combination of solidified lava flows and layers of pyroclastics.

condensation Process by which water vapor is converted to liquid water; a warming process because latent heat is released.

conditional instability A lapse rate somewhere between the dry and saturated adiabatic rates; rising air becomes unstable after condensation releases latent heat.

conic projection A family of maps in which one or more cones is set tangent to, or intersecting, a portion of the globe and the geographic grid is projected onto the cone(s).

conifer (gymnosperm) Seed-reproducing plants that carry their seeds in cones.

conformality The property of a map projection that maintains proper shapes of surface features.

conformal map [projection] A projection that maintains proper angular relationships over the entire map; over limited areas shows the correct shapes of features shown on a map.

continental ice sheet Large ice sheet covering a portion of a continental area.

continental rift valley Fault-produced valley resulting from spreading or rifting of continent.

contour interval Difference in elevation between two elevation contour lines.

contour line (elevation contour line) A line joining points of equal elevation.

Coriolis effect (Coriolis force) The apparent deflection of free-moving objects to the right in the Northern Hemisphere and to the left in the Southern Hemisphere, in response to the rotation of Earth.

creep (soil creep) The slowest and least perceptible form of mass wasting, which consists of a very gradual downhill movement of soil and regolith.

cumulonimbus [cloud] Cumuliform cloud of great vertical development, often associated with a thunderstorm.

cutoff meander A sweeping stream channel curve that is isolated from streamflow because the narrow meander neck has been cut through by stream erosion.

cyclone Low-pressure center.

cylindrical projection A family of maps derived from the concept of projection onto a paper cylinder that is tangential to, or intersecting with, a globe.

datum Elevation reference point for a map or stream gage.

Daylight-saving time Practice in many parts of the world to shift clocks ahead one hour for the spring, summer, and early fall, largely to conserve energy.

debris flow Streamlike flow of mud and water heavily laden with sediments of various sizes; a mudflow containing large boulders.

December solstice Day of the year when the vertical rays of the Sun strike the Tropic of Capricorn; on or about December 21.

deciduous tree A tree that experiences an annual period in which all leaves die and usually fall from the tree, due either to a cold season or a dry season.

declination arrow (declination diagram) Diagram on a map showing the relationships among true north, magnetic north, and grid north.

declination of the Sun Latitude receiving the vertical rays of the Sun.

dendritic [drainage] pattern A treelike, branching pattern that consists of a random merging of streams, with tributaries joining larger streams irregularly, but always at acute angles.

depression contour Elevation contour line with hachure marks pointing downslope into a depression.

deranged drainage pattern The chaotic stream drainage pattern resulting from the irregular topography found in recently glaciated landscapes.

desert varnish A dark shiny coating of iron and manganese oxides that forms on rock surfaces exposed to desert air for a long time.

dew point temperature (dew point) The critical air temperature at which water vapor saturation is reached.

disappearing stream Stream that abruptly disappears from the surface where it flows into an underground cavity; common in karst regions.

discharge (stream discharge) Volume of flow of a stream; typically expressed as cubic meters per second (cms) or cubic feet per second (cfs) passing a stream gage station.

Doppler effect Shift in the frequency of a sound wave or an electromagnetic wave due to either the movement of the observer or the source of the waves.

drainage basin (watershed) An area that contributes overland flow and groundwater to a specific stream (also called a catchment).

drainage divide The line of separation between runoff that descends into two different drainage basins.

drift (glacial drift) All material carried and deposited by glaciers.

drumlin A low, elongated hill formed by ice-sheet deposition and erosion. The long axis is aligned parallel with the direction of ice movements, and the end of the drumlin that faces the direction from which the ice came is blunter and slightly steeper than the other end.

dry adiabatic rate (dry adiabatic lapse rate) The rate at which a parcel of unsaturated air cools as it rises and warms as it descends (10°C per 1000 meters [5.5°F per 1000 feet]).

dynamic high High-pressure cell associated with prominently descending air.

dynamic low Low-pressure cell associated with prominently rising air.

earthflow Mass wasting process in which a portion of a water-saturated slope moves a short distance downhill.

earthquake Vibrations generated by abrupt movement of Earth's crust.

easterly wave A long but weak migratory low-pressure trough in the tropics.

ecological land unit An area with a distinct land cover, climate, lithology, and topography.

ecosystem The totality of interactions among organisms and the environment in the area of consideration.

El Niño Periodic atmospheric and oceanic phenomenon of the tropical Pacific that typically involves the weakening or reversal of the trade winds and the warming of surface water off the west coast of South America.

entrenched meanders A winding, sinuous stream valley with abrupt sides; often the outcome of the rejuvenation of a meandering stream.

environmental lapse rate The observed vertical temperature gradient of the troposphere.

epicenter Location on the surface directly above the center of fault rupture during an earthquake.

epiphytes Plants that live above ground level out of contact with soil, usually growing on trees or shrubs.

equal area map See *equivalence*.

equilibrium line See *annual snowline*.

equivalence The property of a map projection that maintains equal areal relationships in all parts of the map. Such maps are also called *equal area maps*.

equivalent map [projection] A projection that maintains constant area (size) relationships over the entire map; also called an "equal area projection."

esker Long, sinuous ridge of stratified glacial drift composed largely of glaciofluvial gravel and formed by the choking of subglacial streams during a time of glacial stagnation.

evaporation Process by which liquid water is converted to gaseous water vapor; a cooling process because latent heat is stored.

evergreen A tree or shrub that sheds its leaves on a sporadic or successive basis but at any given time appears to be fully leaved.

eye [of tropical cyclone] The nonstormy center of a tropical cyclone, which has a diameter of 16 to 40 kilometers (10 to 25 miles) and is a singular area of calmness in the storm that whirls around it.

fjord (fjorded coast) A glacial trough that has been partly drowned by the sea.

flash flood Sudden surge of flood water down a normally dry stream channel; commonly results from desert thunderstorms.

flood frequency curve Chart showing the relationship between a stream's peak discharge and its flood recurrence interval.

floodplain A flattish valley floor covered with stream-deposited sediments (alluvium) and subject to periodic or episodic inundation by overflow from the stream.

friction The force that impedes the relative motion of two objects in contact.

front A zone of discontinuity between unlike air masses.

frost wedging Fragmentation of rock due to expansion of water that freezes in rock openings.

gage height Height of stream above its local datum.

geographic information systems (GIS) Computerized systems for the capture, storage, retrieval, analysis, and display of spatial (geographic) data.

geostrophic wind A wind that moves parallel to the isobars as a result of the balance between the pressure gradient force and the Coriolis effect.

glacial abrasion The chipping and grinding effect of rock fragments embedded in the bottom of a glacier.

glacial plucking Action in which rock particles beneath the ice are grasped by the freezing of meltwater in joints and fractures and pried out and dragged along in the general flow of a glacier.

glacial steps Series of level or gently sloping bedrock benches alternating with steep drops in the down-valley profile of a glacial trough.

glacial trough A valley reshaped by an alpine glacier, usually U-shaped.

glaciofluvial deposition The action whereby much of the debris that is carried along by glaciers is eventually deposited or redeposited by glacial meltwater.

Global Positioning System (GPS) A satellite-based system for determining accurate positions on or near Earth's surface.

gnomonic map Maps on which a straight line represents a great circle.

GOES (Geostationary Operational Environmental Satellites) Weather satellites positioned over fixed locations on Earth to provide continuous multi-wavelength monitoring of the atmosphere and surface.

graben A block of land bounded by parallel faults in which the block has been downthrown, producing a structural valley with a straight, steep-sided fault scarp on both sides.

gradient Measure of slope steepness described as the elevation change over a given distance, such as feet per mile or meters per kilometer.

grassland Plant association dominated by grasses and forbs.

graticule The network of parallels and meridians on a map.

great circle Circle on a globe formed by the intersection of Earth's surface with any plane that passes through Earth's center. A path along a great circle represents the shortest distance between two points on a sphere.

Greenwich Mean Time (GMT) Time in the Greenwich time zone. Today more commonly called UTC or Universal Time Coordinated.

grid north (GN) North orientation on supplementary map grid system, such as the Universal Transverse Mercator grid.

groundwater Water found underground in the zone of saturation.

Hadley cells Two complete vertical convective circulation cells between the equator, where warm air rises in the ITCZ, and 25° to 30° north and south latitude, where much of the air subsides into the subtropical highs.

hanging valley (hanging trough) A tributary glacial trough, the bottom of which is considerably higher than the bottom of the principal trough that it joins.

headwall [of a cirque] The steep back wall of a glacial cirque.

highland ice field Largely unconfined ice sheet in high mountain area.

high pressure [cell] Area of relatively high atmospheric pressure.

horn A steep-sided, pyramidal mountain peak formed by expansive glacial plucking of the headwalls where three or more cirques intersect.

horst A relatively uplifted block of land between two parallel faults.

hot spot An area of volcanic activity within the interior of a lithospheric plate associated with magma rising up from the mantle below.

humidity Water vapor in the air.

hurricane A tropical cyclone with wind speeds of 64 knots or greater affecting North or Central America.

ice-contact deposit Depositional features that develop along the active margin of a glacier.

inches of mercury Measure of atmospheric pressure based on the height of a column of mercury in a liquid barometer.

index contour Every fourth or fifth elevation contour line on most contour line maps; thicker than regular contour lines; usually marked with elevation.

infrared [radiation] Electromagnetic radiation in the wavelength range of about 0.7 to 1000 micrometers; wavelengths just longer than visible light.

inselberg ("island mountain") Isolated summit rising abruptly from a low-relief surface.

insolation Incoming solar radiation.

interfluve The higher land or ridge above the valley sides that separates adjacent valleys.

International Date Line The line marking a time difference of an entire day from one side of the line to the other. Generally, this line falls on the 180th meridian except where it deviates to avoid separating an island group.

intertropical convergence zone (ITCZ) The region near or on the equator where the northeast trades and the southeast trades converge; dominated by air rising in thunderstorm updrafts; associated with high rainfall.

island arc (volcanic island arc) Chain of volcanic islands associated with an oceanic plate–oceanic plate subduction zone.

isobar A line on a map joining points of equal atmospheric pressure.

isoline A line on a map connecting points that have the same quality or intensity of a given phenomenon.

isotherm A line on a map joining points of equal temperature.

jet stream A rapidly moving current of air concentrated along a quasi-horizontal axis in the upper troposphere or in the stratosphere, characterized by strong vertical and lateral wind shears.

June solstice Day of the year when the vertical rays of the Sun strike the Tropic of Cancer on or about June 21.

kame A relatively steep-sided mound or conical hill composed of stratified drift found in areas of ice-sheet deposition and associated with meltwater deposition in close association with stagnant ice.

karst Topography developed as a consequence of subsurface solution.

kettle An irregular depression in a morainal surface created when blocks of stagnant ice eventually melt.

knot A unit of speed equal to 1 nautical mile, or 1.15 statute miles, per hour; 1.85 kilometers per hour.

Köppen climate classification system A climatic classification of the world devised by Wladimir Köppen.

lagoon A body of quiet salt or brackish water in an area between a barrier island or a barrier reef and the mainland.

landslide An abrupt and often catastrophic event in which a large mass of rock and earth slides bodily downslope in only a few seconds or minutes. An instantaneous collapse of a slope.

large-scale map A map with a scale that is a relatively large representative fraction and therefore portrays only a small portion of Earth's surface, but in considerable detail.

latent heat Energy stored or released when a substance changes state. For example, evaporation is a cooling process because latent heat is stored and condensation is a warming process because latent heat is released.

lateral moraine Well-defined ridge of unsorted debris (till) built up along the sides of valley glaciers, parallel to the valley walls.

latitude Location described as an angle measured north and south of the equator.

lava Molten magma that is extruded onto the surface of Earth, where it cools and solidifies.

lava dome (plug dome) Dome or bulge formed by the pushing up of viscous magma in a volcanic vent.

lens stereoscope Optical device for viewing stereograms.

lifting condensation level (LCL) The altitude at which rising air cools sufficiently to reach 100 percent relative humidity at the dew point temperature, and condensation begins.

linear fault trough Straight-line valley that marks the surface position of a fault, especially a strike-slip fault; formed by the erosion or settling of crushed rock along the trace of a fault.

lithospheric plates (lithosphere) Tectonic plates consisting of the crust and upper rigid mantle.

longitude Location described as an angle measured east and west from the prime meridian on Earth's surface.

longshore current A current in which water moves roughly parallel to the shoreline in a generally downwind direction; set up by the action of waves.

longwave radiation Wavelengths of thermal infrared radiation emitted by Earth and the atmosphere; also referred to as terrestrial radiation.

low pressure [cell] Area of relatively low atmospheric pressure.

magma Molten material below Earth's surface.

magnetic north (MN) For a given location, the compass direction toward the magnetic north pole.

mantle plume A location where molten mantle magma rises to, or almost to, Earth's surface; often found in the middle of a plate; associated with many *hot spots*.

March equinox One of two days of the year when the vertical rays of the Sun strike the equator; every location on Earth has equal day and night; occurs on or about March 20 each year.

marine terrace A platform of marine erosion that has been uplifted above sea level.

mass wasting Relatively short distance, downslope movement of broken rock material primarily under the direct influence of gravity.

meandering stream (meandering stream channel) Highly twisting or looped stream channel pattern.

meander scar A former stream meander or stream channel through which the stream no longer flows.

medial moraine A dark band of rocky debris down the middle of a glacier created by the union of the lateral moraines of two adjacent glaciers.

meltwater (glacial meltwater) Water resulting from the melting of glacial ice.

meridian An imaginary line of longitude extending from pole to pole, crossing all parallels at right angles, and being aligned in true north–south directions.

meteogram Chart plotting hourly changes in weather conditions for a location over a one-day period.

midlatitude cyclone Large migratory low-pressure system that occurs within the middle latitudes and moves generally with the westerlies. Also known as extratropical cyclones and wave cyclones.

midocean ridge A lengthy system of deep-sea mountain ranges, generally located at some distance from any continent; formed by divergent plate boundaries on the ocean floor.

millibar A measure of atmospheric pressure, consisting of one-thousandth part of a bar, or 1000 dynes per square centimeter.

mixing ratio Description of the water vapor content of the air expressed as the mass of water in a given mass of dry air; usually as grams of water vapor per kilogram of dry air.

moraine The largest and generally most conspicuous landform feature produced by glacial deposition of till, which consists of irregular rolling topography that rises somewhat above the level of the surrounding terrain.

mudflow Down-valley movement of a rapidly moving mixture of soil and water.

natural levee An embankment of slightly higher ground fringing a stream channel in a floodplain; formed by deposition during floods.

normal fault The result of tension (extension) producing a steeply inclined fault plane, with the block of land on one side being pushed up, or upthrown, in relation to the block on the other side, which is downthrown.

occluded front A complex front formed when a cold front overtakes a warm front.

occlusion Process of a cold front overtaking a warm front to form an occluded front.

oceanic trench (deep oceanic trench) Deep linear depression in the ocean floor where subduction is taking place.

offset stream A stream course displaced by lateral movement along a fault.

one-hour precipitation [radar image] Weather radar image showing estimate of precipitation that has fallen over the last hour.

orthographic projection Map projection in which Earth appears as it would from space.

outwash plain Extensive glaciofluvial feature that is a relatively smooth, flattish alluvial apron deposited beyond recessional or terminal moraines by streams issuing from ice.

overland flow The general movement of unchanneled surface water down the slope of the land surface.

oxbow lake A cutoff meander that initially holds water.

oxygen isotope analysis Using the ratio of ^{16}O ("oxygen 16") and ^{18}O ("oxygen 18") isotopes in compounds such as water and calcium carbonate to infer temperature and other conditions in the past.

Pacific Decadal Oscillation (PDO) Long-term pattern of sea surface temperature change between the northern/west tropical and eastern tropical Pacific Ocean.

paleomagnetism Past magnetic orientation.

parallel A circle resulting from a line connecting all points of equal latitude.

parallel [drainage] pattern A drainage pattern found in areas of pronounced regional slope, particularly if the gradient is gentle, with long streams flowing parallel to one another.

paternoster lakes A sequence of small lakes found in the shallow excavated depressions of glacial steps.

pediment A gently inclined bedrock platform that extends outward from a mountain front, usually in an arid region.

perihelion The point in its orbit at which a planet is nearest the Sun.

permeability The characteristic of soil or rock by which water can move through interconnected pore spaces.

piedmont (piedmont zone) Zone at the "foot of the mountains."

plane of the ecliptic The imaginary plane that passes through the Sun and through Earth at every position in its orbit around the Sun; the orbital plane of Earth.

planar projection A family of maps derived by the perspective extension of the geographic grid from a globe to a plane that is tangent to the globe at some point; also called *plane projection*.

plate tectonics A coherent theory of massive crustal rearrangement based on the movement of continent-sized lithospheric plates.

playa Dry lake bed in a basin of interior drainage.

Pleistocene Epoch An epoch of the Cenozoic era between the Pliocene and the Holocene; from about 2.59 million to 11,700 years ago.

plutonic rock Igneous rock formed below ground from the cooling and solidification of magma.

polar front The zone of contact between unlike air masses in the westerlies and the polar easterlies; associated with the subpolar lows; typically found at about 55°–65° N and S.

polarity (parallelism) [of rotation axis] A characteristic of Earth's axis wherein it always points toward Polaris (the North Star) at every position in Earth's orbit around the Sun.

porosity The amount of pore space between soil particles; a measure of soil's capacity to hold water and air.

potential evaporation The maximum amount of moisture that could be lost through evaporation if the water were available.

precipitation Drops of liquid or solid water falling from clouds.

pressure of air The force exerted by the atmosphere on a surface or walls of a container.

pressure gradient Change in atmospheric pressure over some horizontal distance.

pressure gradient force The propelling force exerted by a pressure gradient.

prime meridian The meridian passing through the Royal Observatory at Greenwich (England), just east of central London, and from which longitude is measured.

principal meridian Reference meridian from which ranges are designated east or west in the Public Land Survey.

pseudocylindrical projection (elliptical projection) A family of map projections in which the entire world is displayed in an oval shape.

Public Land Survey (township grid) Land survey system used throughout most of the United States west of the Mississippi River that uses a rectangular grid with townships and ranges.

pyroclastics Solid rock fragments ejected into the air by a volcanic eruption.

quadrangle Map (commonly a topographic map) delimited by lines of latitude and longitude on all four sides.

radar Radio detection and ranging.

radial [drainage] pattern Drainage pattern in which streams flow outward in all directions from a central dome or peak.

raster data GIS, image, or spatial data stored in arrays of pixels or data cells.

rating curve Chart showing the relationship between a stream's gage height and its discharge.

recessional moraine A glacial deposit of till formed during a pause in the retreat of the ice margin.

rectangular [drainage] pattern Pattern where streams follow sets of faults and/or joints, with prominent right-angled relationships.

recurrence interval (return period) [of a flood] Statistical estimate of the number of years between times when a given peak stream discharge will be reached or exceeded; for example, a "50-year flood" is the peak stream gage height likely to be reached once every 50 years.

relative humidity An expression of the amount of water vapor in the air in comparison with the total amount that could be there if the air were saturated. This is a ratio that is expressed as a percentage.

representative fraction (r.f.) The ratio that is an expression of a fractional map scale that compares map distance with ground distance.

reverse fault A fault produced from compression, with the upthrown block rising steeply above the downthrown block, so that the fault scarp would be severely oversteepened if erosion did not act to smooth the slope.

rhumb line (loxodrome) A true compass heading; a line of constant compass direction.

rhyolite Light-colored volcanic rock that forms from magmas with high silica content.

ria coastline An embayed coast with numerous estuaries; formed by the flooding of stream valleys by the sea.

ridge [of high pressure] Linear or elongated area of relatively high atmospheric pressure.

rockfall (fall) Mass wasting process in which weathered rock drops to the foot of a cliff or steep slope.

sag pond A pond caused by the collection of water from springs and/or runoff into sunken ground, resulting from the crushing of rock in an area of fault movement.

saturated adiabatic rate (saturated adiabatic lapse rate) The diminished rate of cooling, averaging about 6°C per 1000 meters (3.3°F per 1000 feet) of rising air above its lifting condensation level.

saturation [with water vapor] Circumstance in which the air contains the maximum amount of water vapor for a given temperature; condensation typically will begin when the air reaches saturation.

saturation mixing ratio Mixing ratio of a saturated parcel of air; one expression of water vapor capacity.

scale [of a map] The relationship between length measured on a map and the actual distance represented on Earth.

scarp (fault scarp) A steep escarpment or nearly vertical cliff formed by fault movement.

sea stack Pillars or towers that form where coastal cliffs wear back, leaving a remnant isolated from the mainland.

section Square tract of land, one mile to a side, in the Public Land Survey.

seif (longitudinal dune) Long, narrow desert dunes that usually occur in multiplicity and in parallel arrangement.

separates The size groups within the standard classification of soil particle sizes.

September equinox One of two days of the year when the vertical rays of the Sun strike the equator; every location on Earth has equal day and night; occurs on or about September 22 each year.

shield volcano Volcanoes built up in a lengthy outpouring of very fluid basaltic lava. Shield volcanoes are broad mountains with gentle slopes.

shrub Woody, low-growing perennial plant.

shutter ridge A small hill, moved by strike-slip faulting, that partially or completely closes off a stream valley.

silica Silicon dioxide (SiO_2) in any of several mineral forms.

sinkhole (doline) A small, rounded depression that is formed by the dissolution of surface limestone, typically at joint intersections.

sling psychrometer Instrument for measuring relative humidity consisting of a dry bulb and wet bulb thermometer mounted side-by-side; whirling promotes evaporation and cooling of the wet bulb.

slip face Steeper leeward side of a sand dune.

slump A slope collapse with a backward rotation; rotational slide.

small-scale map A map whose scale is a relatively small representative fraction and therefore shows a large portion of Earth's surface in limited detail.

soil horizon A recognizable vertical layer or zone within soil; distinguished from one another by differing characteristics.

soil profile A vertical cross section from Earth's surface down through the soil layers into the parent material beneath.

solar altitude Angle of the Sun above the horizon.

solar constant The fairly constant amount of solar radiation received at the top of the atmosphere, about 1372 watts per square meter (W/m^2; 1 watt = 1 joule per second) or slightly less than 2 calories per square centimeter per minute, or 2 langleys per minute.

specific humidity A direct measure of water-vapor content expressed as the mass of water vapor in a given mass of air (grams of vapor/kilograms of air).

spit A linear deposit of marine sediment that is attached to the land at one or both ends.

stable [air] Air that rises only if forced.

stalactite A pendant structure hanging downward from a cavern's roof.

stalagmite A projecting structure growing upward from a cavern's floor.

standard time zones Global system of 24 time zones, each 15° of longitude wide, within which it is the same time; boundaries of most standard time zones have been manipulated for the convenience of the local population.

star dune Pyramid-shaped sand dune with arms radiating out in three or more directions.

stationary front The common boundary between two air masses in a situation in which neither air mass is advancing or displacing the other.

station model (weather map station model) Standardized system of data presentation used on weather maps to convey conditions at a weather station.

stereogram (stereopair; stereo aerial photographs) Matched set of vertical aerial photographs for viewing in three dimensions with a stereoscope.

storm relative motion [radar image] Weather Doppler radar image showing wind directions as if the storm were stationary.

storm total precipitation [radar image] Weather radar image that estimates the total amount of precipitation that has fallen since the last one-hour pause in rainfall.

stream Water flowing within a channel, typically along a valley bottom.

streamflow Channeled movement of water along a valley bottom.

stream order Concept that describes the hierarchy of a drainage network.

stream rejuvenation When a stream gains downcutting ability, usually through regional tectonic uplift.

stream terrace Remnant of a previous valley floodplain of a rejuvenated stream.

strike-slip fault A fault produced by shearing, with adjacent blocks being displaced laterally with respect to one another. The movement is almost entirely horizontal.

subduction Descent of the edge of an oceanic lithospheric plate under the edge of an adjoining plate.

subtropical high (STH) Large, semipermanent high-pressure cells centered at about 30° latitude over the oceans, which have average diameters of 3200 kilometers (2000 miles) and are usually elongated east–west; characterized by persistent dry weather.

succulents Plants that have fleshy stems that store water.

Sun time Local time based on the position of the Sun in the sky; local Sun time noon is when the Sun reaches its highest point in the sky.

swallow hole The distinct opening at the bottom of some sinks through which surface drainage can pour directly into an underground channel.

swash The cascading forward motion of a breaking wave that rushes up the beach.

Système International (SI) The international system of measurement, popularly known as the "metric system" of measurement.

talus (scree) Pieces of angular weathered rock, of various sizes, that fall directly downslope.

talus cone Sloping, cone-shaped heaps of dislodged talus.

tangent rays [of the Sun] Sun's rays that are just skimming past Earth.

tarn Small lake in the shallow excavated depression of a cirque; sometimes used to describe any lake that forms on a glacial step.

temperature inversion A situation in which temperature increases with increasing altitude, the inverse of the normal condition.

terminal moraine A glacial deposit of till that builds up at the outermost extent of ice advance.

terminus [of a glacier] The end of a glacier, where ice is melting as quickly as it is flowing in.

thermal high High-pressure cell associated with cold surface conditions.

thermal low Low-pressure cell associated with warm surface conditions.

thrust fault (overthrust fault) A fault created by compression forcing the upthrown block to override the downthrown block at a relatively low angle.

till Rock debris that is deposited directly by moving or melting ice, with no meltwater flow or redeposition involved; typically consists of angular, unsorted, and unstratified rock debris.

tilted fault block mountain (fault-block mountain) An asymmetrical mountain formed by faulting on one side of a surface block without any faulting on the other side; results in a steep slope along the fault scarp and a relatively gentle slope on the other side.

tombolo A spit formed by sand deposition of waves converging in two directions on the landward side of a nearshore island, so that a spit connects the island to the land.

topographic map Large-scale map using elevation contour lines to depict the terrain.

topographic profile Side-view diagram showing elevation change along a line across a topographic map.

township 36-square-mile tract of land; township numbers refer to "tiers" north or south of a baseline in the Public Land Survey system.

trade winds The major easterly wind system of the tropics, issuing from the equatorward sides of the subtropical highs and diverging toward the west and toward the equator.

transverse dune A crescent-shaped sand dune that has convex sides facing the prevailing direction of wind and that occurs where the supply of sand is great. The crest is perpendicular to the wind and aligned in parallel waves across the land.

trellis drainage pattern A drainage pattern that is usually developed on alternating bands of hard and soft strata, with long parallel streams linked by short right-angled segments and joined by short tributaries.

tropical cyclone A migratory storm most significantly affecting the tropics and subtropics; consists of a prominent low-pressure center that is essentially circular in shape and has a steep pressure gradient outward from the center. When wind speed reaches 64 knots, it is called a hurricane in North America and the Caribbean.

troposphere The lowest thermal layer of the atmosphere, in which temperature decreases with height.

trough [of low pressure] Linear or elongated band of relatively low atmospheric pressure.

Universal Time Coordinated (UTC) or Coordinated Universal Time The world time standard reference; previously known as Greenwich Mean Time (GMT).

Universal Transverse Mercator grid (UTN) Supplementary rectangular grid system used on many topographic maps; marks off location in terms of the number of meters north or south of the equator, and east of a standard meridian.

unstable [air] Air that rises without being forced.

uvala A compound doline or chain of intersecting dolines.

valley That portion of the total terrain in which a drainage system is clearly established.

valley train A lengthy deposit of glaciofluvial alluvium confined to a valley bottom beyond the outwash plain.

vector data GIS data stored as points, lines, and polygons.

verbal map scale Map scale expressed with words.

vertical aerial photograph Aerial photograph taken with a camera pointing directly down toward the surface.

vertical exaggeration Exaggerated vertical scale of a topographic profile or raised-relief map; exaggerated steepness of slopes observed when viewing stereo aerial photographs.

vertical rays (of the Sun) Sun's rays that are striking perpendicular to the surface.

visible light Waves in the electromagnetic spectrum in the narrow band between about 0.4 and 0.7 micrometers in length; wavelengths of electromagnetic radiation to which the human eye is sensitive.

volcano A mountain or hill from which extrusive igneous material is ejected.

warm front The leading edge of an advancing warm air mass.

water table The top of the saturated zone of groundwater.

water vapor The gaseous state of water.

water vapor [satellite] images Weather satellite images showing the relative amount of water vapor in the mid-troposphere.

wave refraction Phenomenon whereby waves change their directional trend as they approach a shoreline.

westerlies The great wind system of the midlatitudes that flows basically from west to east around the world in the latitudinal zone between about 30° and 60° both north and south of the equator.

wind shear (vertical wind shear) Significant change in wind direction or speed in the vertical dimension.

yazoo stream A tributary unable to enter the main stream because of natural levees along the main stream.

zone of aeration The topmost hydrologic zone within the ground; contains a varying amount of water in the pore spaces.

zone of saturation The second hydrologic zone below the surface in which the pore spaces are filled with water; its upper boundary is the water table.

Zulu time Universal Time Coordinated (UTC; Greenwich Mean Time [GMT]) expressed on a 24-hour clock; abbreviated "Z."

APPENDIX II
Math Skills Practice Worksheet

This is not a test. The following practice problems reflect the kinds of math and charting skills you'll use in this Lab Manual. Hints for answering the questions are on the following page.

1. Describe the direction (north, southeast, etc.) of the arrows shown on the map of the United States to the right.

   (a) Arrow "a" is coming *from* the _____.

   (b) Arrow "b" is moving *toward* the _____.

2. The angle marked with an arrow in the circle to the right is _____° (degrees).

3. Express 35.25° in terms of whole degrees and minutes: 35.25° = _____° _____'

4. One meter equals 3.281 feet, so 4 meters = _____ feet

5. Express 0.375 as a percentage of 1: _____ percent.

6. In 3½ hours a car moving at 40 miles per hour will travel _____ miles.

7. If a mountain rises 25.0 centimeters (cm) in 125 years

   (a) The rate of uplift is _____ cm/year.

   (b) In 10,000 years, this mountain will be _____ cm taller.

8. Compute the average ("mean") of the following numbers: 7, 10, 10, 13, 20: _____

9. (a) Plot the data on the chart below and connect the points with a smooth curved line.

Height (meters)	Weight (kilograms)
4	9.0
3.5	5.5
3	4.0
2	2.3
1	1

   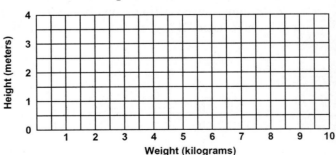

   (b) Based on the graph you've drawn above, estimate the height of an object with a weight of 3 kilograms: _____ meters

---

HINTS FOR MATH SKILLS PRACTICE WORKSHEET

General Suggestions: Before completing a lab calculation, to reduce the chance that you'll end up with an answer that is wildly wrong, it helps to estimate what the final answer *should* be. For example, if you are multiplying a very large number, such as 1,400,000, by a very small number, such as 1.2, you know that your final answer shouldn't be a lot larger than 1,400,000.

1. Envision a compass "rose" drawn next to this map ($\oplus$) with north at the top, south at the bottom, east to the right, and west to the left. Sometimes we describe direction based on the way we are heading (such as in navigation), and sometimes based on the direction from which something is coming (such as wind).

2. Remember, there are 360° in a full circle, so one-quarter of the way around is 90° and halfway around is 180°. Notice that the angle in the diagram is halfway between 180° and 270°.

3. When precise descriptions of angles are needed, portions or fractions of degrees are often indicated. Although this can be designated with "decimal" degrees (for example, one and-a-half degrees can be written 1.5°), it is traditional to indicate portions of degrees with *minutes* and *seconds*. One degree is divided into 60 minutes (60′; so 1° = 60′) and each minute can be further divided into 60 seconds (60″; so 1′ = 60″); this means that 1.5° would be written 1°30′. Remember, when describing angles, "minutes" and "seconds" are *not* referring to time.

4. This is an example of using a conversion formula. If 1 meter equals 3.281 feet, then 4 meters equals: 4 × 3.281 feet.

5. To express a decimal value as a percentage of one, simply multiply it by 100.

6. If a car travels 40 miles in one hour (a speed of 40 "miles per hour"), in 3.5 hours it will travel 3.5 × 40 miles.

7. (a) Here you are calculating a *rate of change*—in this case the number of centimeters per year (the change in distance over a period of time). The units of measure ("centimeters per year") can be written as a fraction (centimeters/year) and that tells you how to set up the calculation (in this case, centimeters divided by years): divide the total distance in centimeters by the total number of years to find the rate in cm/year. Now that you know the *rate* of change, use that value to calculate the answer for (b): For example, if a mountain rises 0.4 cm in one year, in 10,000 years it will rise 10,000 × 0.4 centimeters.

8. To compute an average, add up the total value of all the numbers (in this case 60) and divide by the number of values in the problem (in this case 5); so the average will be 60 ÷ 5.

9. (a) When plotting values on a graph, look first at the scales along both axis—in this case height and weight. Once you've added a smooth curve to your data points, you can interpolate (estimate) values between those in the original data set to find the answer for (b).

Weather Map Symbols

The charts and diagrams on the following pages show the symbols and codes used on standard weather maps.

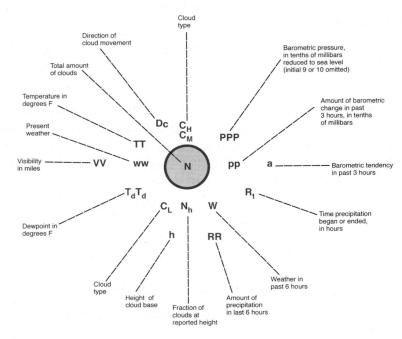

Figure III-1: Standard weather station model showing placement of codes. (From Hess, *McKnight's Physical Geography*, 12th ed.)

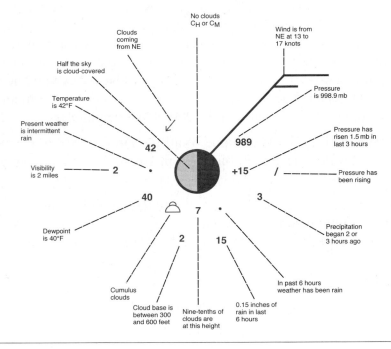

Figure III-2: Standard weather station model with sample data. (From Hess, *McKnight's Physical Geography*, 12th ed.)

W W
Present weather

00	Cloud development NOT observed or NOT observable during past hour.§
01	Clouds generally dissolving or becoming less developed during past hour.§
02	State of sky on the whole unchanged during past hour.§
03	Clouds generally forming or developing during past hour.§
04	Visibility reduced by smoke.
05	Dry haze.
06	Widespread dust in suspension in the air. Not raised by wind, at time of observation.
07	Dust or sand raised by the wind, at time of ob.
08	Well developed dust devil(s) within past hr.
09	Duststorm or sandstorm within sight of or at station during past hour.
10	Light fog.
11	Patches of shallow fog at station, NOT deeper than 6 feet on land.
12	More or less continuous shallow fog at station, NOT deeper than 6 feet on land.
13	Lightning visible, no thunder heard.
14	Precipitation within sight, but NOT reaching the ground at station.
15	Precipitation within sight, reaching the ground, but distant from station.
16	Precipitation within sight, reaching the ground, near to but Not at station.
17	Thunder heard, but no precipitation at the station.
18	Squall(s) within sight during past hour.
19	Funnel cloud(s) within sight during past hr.
20	Drizzle (NOT freezing and NOT falling as showers) during past hour, but NOT at time of ob.
21	Rain (NOT freezing and NOT falling as showers during past hr., but NOT at time of ob.
22	Snow (NOT falling as showers) during past hr., but NOT at time of ob.
23	Rain and snow (NOT falling as showers) during past hr., but NOT at time of observation.
24	Freezing drizzle or freezing rain (NOT falling as showers) during past hour, but Not at time of observation.
25	Showers or rain during past hour, but NOT at time of observation.
26	Showers of snow, or rain and snow, during past hour, but NOT at time of observation.
27	Showers of hail, or of hail and rain, during past hour, but NOT at time of observation.
28	Fog during past hour, but NOT at time of ob.
29	Thunderstorm (with or without precipitation) during past hour, but NOT at time of ob.
30	Slight or moderate duststorm or sandstorm, has decreased during past hour.
31	Slight or moderate duststorm or sandstorm, no appreciable change during past hour.
32	Slight or moderate duststorm or sandstorm, has increased during past hour.
33	Severe duststorm or sandstorm, has decreased during past hour.
34	Severe duststorm or sandstorm no appreciable change during past hour.
35	Severe duststorm or sandstorm, has increased during past hour.
36	Slight or moderate drifting snow, generally low.
37	Heavy drifting snow, generally low.
38	Slight or moderate drifting snow, generally high.
39	Heavy drifting snow, generally high.
40	Fog at distance at time of ob., but NOT at station during past hour.
41	Fog in patches.
42	Fog, sky discernible, has become thinner during past hour.
43	Fog, sky NOT discernible, has become thinner during past hour.
44	Fog, sky discernible, no appreciable change during past hour.
45	Fog, sky NOT discernible, no appreciable change during past hour.
46	Fog, sky discernible, has begun or become thicker during past hr.
47	Fog, sky NOT discernible, has begun or become thicker during past hour.
48	Fog, depositing rime, sky discernible.
49	Fog, depositing rime, sky NOT discernible.
50	Intermittent drizzle (NOT freezing) slight at time of observation.
51	Continuous drizzle (NOT freezing) slight at time of observation.
52	Intermittent drizzle (NOT freezing) moderate at time of ob.
53	Intermittent drizzle (NOT freezing), moderate at time of ob.
54	Intermittent drizzle (NOT freezing), thick at time of observation.
55	Continuous drizzle (NOT freezing), thick at time of observation.
56	Slight freezing drizzle.
57	Moderate or thick freezing drizzle.
58	Drizzle and rain slight.
59	Drizzle and rain, moderate or heavy.
60	Intermittent rain (NOT freezing), slight at time of observation.
61	Continuous rain (NOT freezing), slight at time of observation.
62	Intermittent rain (NOT freezing), moderate at time of ob.
63	Continuous rain (NOT freezing), moderate at time of observation.
64	Intermittent rain (NOT freezing), heavy at time of observation.
65	Continuous rain (NOT freezing), heavy at time of observation.
66	Slight freezing rain.
67	Moderate or heavy freezing rain.
68	Rain or drizzle and snow, slight.
69	Rain or drizzle and snow, mod, or heavy.
70	Intermittent fall of snow flakes, slight at time of observation.
71	Continuous fall of snowflakes, slight at time of observation.
72	Intermittent fall of snow flakes, moderate at time of observation.
73	Continuous fall of snowflakes, moderate at time of observation.
74	Intermittent fall of snow flakes, heavy at time of observation.
75	Continuous fall of snowflakes, heavy at time of observation.
76	Ice needles (with or without fog).
77	Granular snow (with or without fog).
78	Isolated starlike snow crystals (with or without fog).
79	Ice pellets (sleet, U.S. definition).
80	Slight rain shower(s).
81	Moderate or heavy rain shower(s).
82	Violent rain shower(s).
83	Slight shower(s) of rain and snow mixed.
84	Moderate or heavy shower(s) of rain and snow mixed.
85	Slight snow shower(s).
86	Moderate or heavy snow showers(s).
87	Slight showers(s) of soft or small hail with or without rain or rain and snow mixed.
88	Moderate or heavy shower(s) of soft or small hail with or without rain or rain and snow mixed.
89	Slight shower(s) of hail††, with or without rain or rain and snow mixed, not associated with thunder.
90	Moderate or heavy shower(s) of hail††, with or without rain or rain and snow mixed, not associated with thunder.
91	Slight rain at time of ob., thunderstorm during past hour, but NOT at time of observation.
92	Moderate or heavy rain at time of ob., thunderstorm during past hour, but NOT at time of observation.
93	Slight snow or rain and snow mixed or hail† at time of ob., thunderstorm during past hour, but not at time of ob.
94	Mod. or heavy snow, or rain and snow mixed or hail† at time of ob., thunderstorm during past hour, but NOT at time of observation.
95	Slight or mod. thunderstorm without hail†, but with rain and or snow at time of ob.
96	Slight or mod. thunderstorm with hail† at time of observation.
97	Heavy thunderstorm, without hail†, but with rain and or snow at time of observation.
98	Thunderstorm combined with duststorm or sandstorm at time of ob.
99	Heavy thunderstorm with hail† at time of ob.

Figure III-3: Standard weather station model symbols used to indicate "Present Weather" (ww). (From Hess, *McKnight's Physical Geography*, 12th ed.)

C_L Clouds of type C_L	C_M Clouds of type C_M	C_H Clouds of type C_H	W Past Weather	N_h*	a Barometer characteristics
0 No Sc, St, Cu, or Cb clouds.	**0** No Ac, As, Cu, or Ns clouds.	**0** No Ci, Cc, or $\overline{Cs}$ clouds.	**0** Clear or few clouds.	**0** No clouds.	**0** Rising then falling. Now higher than 3 hours ago.
1 Cu with little vertical development and seemingly flattened.	**1** Thin As (entire cloud layer semitransparent).	**1** Filaments of Ci, scattered and not increasing.	**1** Partly cloudy (scattered) or variable sky.	**1** Less than one-tenth or one-tenth.	**1** Rising, then steady; or rising, then rising more slowly. Now higher than, as, 3 hours ago.
2 Cu of considerable development, generally towering, with or without other Cu or Sc; bases all at same level.	**2** Thick As, or Ns.	**2** Dense Ci in patches or twisted sheaves, usually not increasing.	**2** Cloudy (broken or overcast).	**2** Two-or three-tenths.	**2** Rising steadily, or unsteady. Now higher than, 3 hours ago.
3 Cb with tops lacking clear-cut outlines, but distinctly not cirriform or anvil-shaped; with or without Cu, Sc or St.	**3** Thin Ac; cloud elements not changing much and at a single level.	**3** Ci, often anvil-shaped, derived from or associated with Cb.	**3** Sandstorm, or dust-storm, or drifting or blowing snow.	**3** Four-tenths.	**3** Falling or steady, then rising; or rising, then rising more quickly. Now higher than, 3 hours ago.
4 Sc formed by spreading out of Cu; Cu often present also.	**4** Thin Ac in patches; cloud elements and/ or occurring at more than one level.	**4** Ci, often hook-shaped, gradually spreading over the sky and usually thickening as a whole.	**4** Fog, or smoke, or thick dust haze.	**4** Five-tenths.	**4** Steady, Same as 3 hours ago.§
5 Sc not formed by spreading out of Cu.	**5** Thin Ac in bands or in a layer gradually spreading over sky and usually thickening as a whole.	**5** Ci and Cs, often in converging bands, or Cs alone; the continuous layer not reaching 45° altitude.	**5** Drizzle.	**5** Six-tenths.	**5** Falling, then rising. Same or lower than 3 hours ago.
6 St or Fs or both, but not Fs of bad weather	**6** Ac formed by the spreading out of Cu.	**6** Ci. and Cs, often in converging bands, or Cs alone; the continuous layer exceeding 45° altitude.	**6** Rain.	**6** Seven- or eight-tenths.	**6** Falling, then steady; or falling, then falling more slowly. Now lower than 3 hours ago.
7 Fs and /or Fc of bad weather (scud) usually under As and Ns.	**7** Doubel-layered Ac or a thick layer of Ac, not increasing; or As and Ac both present at same or different levels.	**7** Cs covering the entire sky.	**7** Snow, or rain and snow mixed, or ice pellets (sleet)	**7** Nine-tenths or overcast with openings.	**7** Falling, steady, or unsteady. Now lower than 3 hours ago.
8 Cu and Sc (not formed by spreading out of Cu) with bases at different levels.	**8** Ac in the form of Cu-shaped tufts or Ac with turrets.	**8** Cs not increasing and not covering entire sky; Ci and Cc may be present.	**8** Shower(s).	**8** Completley overcast.	**8** Steady or rising, then falling; or falling, then falling more quickly. Now lower than 3 hours ago.
9 Cb having a clearly fibrous (cirriform) top, often anvil-shaped, with or without Cu, Sc, St, or scud.	**9** Ac of a chaotic sky, usually at different levels; patches of dense Ci are usually present also.	**9** Cc alone or CC with some Ci or Cs, but the Cc being the main cirriform cloud present.	**9** Thunderstorm, with or without precipitation.	**9** Sky obscured.	

*Fraction representing how much of the total cloud cover is at the reported base height.

Figure III-4: Standard weather station model symbols for "Low Clouds" (C_L); "Middle Clouds" (C_M); "High Clouds" (C_H); "Past Weather" (W); "Total Cloud Cover" (N_h); "Barometric Tendency" (a). (From Hess, *McKnight's Physical Geography*, 12th ed.)

h (height of cloud base)	Approximate Cloud Height	
	Feet	Meters
0	0–149	0–49
1	150–299	50–99
2	300–599	100–199
3	600–999	200–299
4	1000–1999	300–599
5	2000–3499	600–999
6	3500–4999	1000–1499
7	5000–4699	1500–1999
8	6500–7999	2000–2499
9	>8000 or no clouds	>2500 or no clouds

Figure III-5: Standard weather station model codes for "Height of Cloud Base" (h). (Adapted from Hess, *McKnight's Physical Geography*, 12th ed.)

Symbol	Wind Speed (knots)
	Calm
	1–2
	3–7
	8–12
	13–17
	18–22
	23–27
	28–32
	33–37
	38–42
	43–47
	48–52
	53–57
	58–62
	63–67
	68–72
	73–77

Figure III-7: Standard weather station model symbols for "Wind Speed." (Adapted from Hess, *McKnight's Physical Geography*, 12th ed.)

R_t Code	Time of Precipitation
0	No precipitation
1	Less than one hour ago
2	1 to 2 hours ago
3	2 to 3 hours ago
4	3 to 4 hours ago
5	4 to 5 hours ago
6	5 to 6 hours ago
7	6 to 12 hours ago
8	More than 12 hours ago
9	Unknown

Figure III-6: Standard weather station model codes for "Time Precipitation Began or Ended" (R_t). (Adapted from Hess, *McKnight's Physical Geography*, 12th ed.)

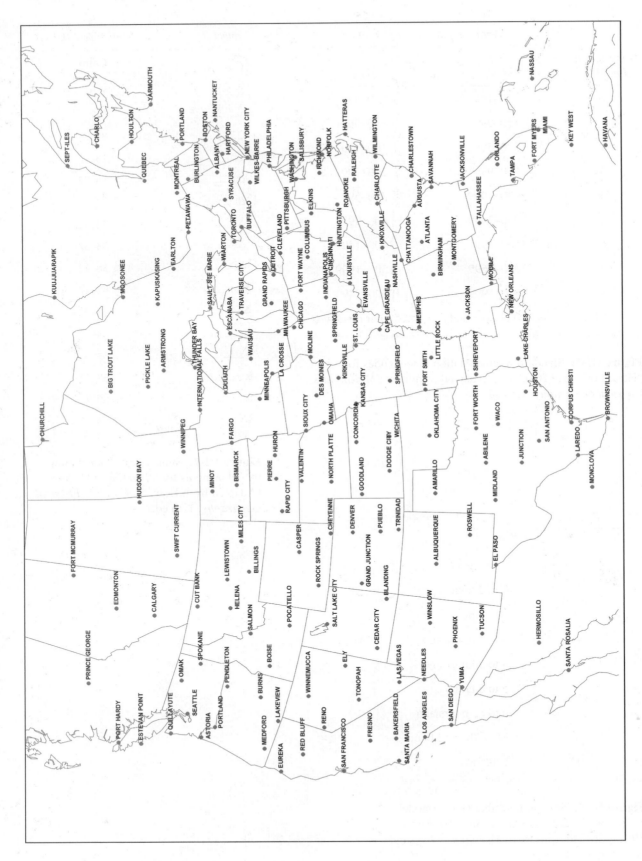

Figure III-8: Daily Weather Map station names and locations. (National Weather Service)

APPENDIX **IV**
Color Maps & Images

Thirty-two pages of color maps and images are reproduced in the back of the Lab Manual. On maps, aerial photographs, and satellite images, the scale, north arrow, and contour interval (for topographic maps) are provided. Graphic map scales are included for all full page topographic maps. Additional graphic scales for 1:24,000, 1:62,500, 1:63,360 and 1:250,000 are found inside the front cover of the Lab Manual.

In the table below, the map or image source, date of publication, and the approximate latitude and longitude of the southeast corner of the portion of the map or image shown are provided. Maps designated *USGS Quadrangle* or *USGS Topographic Map* are portions of U.S. Geological Survey topographic maps that were originally available in printed form. Maps and imagery designated *US Topo* are portions of U.S. Geological Survey maps issued in electronic form with layers that include both contour lines and aerial imagery.

Map	Name	Source	Scale	Date	Location (Southeast corner)	
					Latitude	Longitude
T-1	Hawaii, Hawaii	USGS Quadrangle	1:250,000	1975	19°10′N	155°10′W
T-2	Umnak, Alaska	USGS Quadrangle	1:250,000	1983	53°00′N	168°00′W
T-3	Deer Peak, Montana	US Topo Quadrangle	1:24,000	2011	46°53′N	114°30′W
T-4	Voltaire, North Dakota	USGS Quadrangle	1:24,000	1948	48°04′N	100°46′W
T-5	Jackson, Mississippi-Louisiana	USGS Quadrangle	1:250,000	1973	32°20′N	90°50′W
T-6	Johnson City, Tennessee-Virginia-Kentucky-North Carolina	USGS Quadrangle	1:250,000	1966	36°20′N	82°52′W
T-7	Canyonlands National Park	USGS Topographic Map	1:62,500	1968	38°30′N	109°56′W
T-8	Putman Hall, Florida	USGS Quadrangle	1:24,000	1993	29°38′N	81°53′W
T-9	Park City, Kentucky	US Topo Quadrangle	1:24,000	2016	37°03′N	86°03′W
T-10	Furnace Creek, California	USGS Quadrangle	1:62,500	1952	36°16′N	116°50′W
T-11	Whitewater, Wisconsin	USGS Quadrangle	1:62,500	1960	42°47′N	88°32′W
T-12	Sodus, New York	USGS Quadrangle	1:24,000	1952	43°08′N	77°03′W
T-13	Mono Craters, California	USGS Quadrangle	1:62,500	1953	37°48′N	119°06′W
T-14	Sumdum (D-4), Alaska	USGS Quadrangle	1:63,360	1984	57°45′N	133°05′W
T-15	Point Reyes, California	USGS Quadrangle	1:62,500	1954	37°58′N	122°48′W
T-16	Corpus Christi, Texas	USGS Quadrangle	1:250,000	1989	27°01′N	97°15′W
T-17	Greasewood Spring, Arizona (bottom center)	US Topo Quadrangle	1:24,000	2014	35°22′30″N	109°54′W
T-18	Greasewood Spring, Arizona (bottom right)	US Topo Quadrangle	1:24,000	2014	35°22′30″N	109°52′30″W

Map	Name	Source	Scale	Date	Location (Southeast corner)	
					Latitude	Longitude
T-19	Antelope Peak, Arizona	USGS Quadrangle	1:62,500	1963	32°54′N	112°09′W
T-20a	Bowknot Bend, Utah	US Topo Quadrangle	1:24,000	2014	38°33′N	110°01′W
T-20b	Bowknot Bend, Utah	US Topo Imagery	1:24,000	2011	38°33′N	110°01′W
T-21a	Mauna Loa, Hawaii	USGS Landsat 8	1:387,000	2014	18°52′N	155°33′W
T-21b	Umnak Island, Alaska	USGS Landsat 8	1:457,000	2014	53°04′N	167°49′W
T-22a	SP Mountain, Arizona	US Topo Imagery	1:24,000	2013	35°34′N	111°37′30″W
T-22b	SP Mountain, Arizona	USGS Quadrangle	1:24,000	1989	35°34′N	111°37′30″W
T-23a	Mono Lake, California	Landsat 7	1:140,000	1999	37°46′N	118°58′W
T-23b	Panum Crater, California	USGS Imagery Stereogram	1:40,000	1998	37°54′N	119°00′W
T-23c	Mono Craters, California	USGS Quadrangle	1:62,500	1953	37°46′N	119°00′W
T-24	Mount Dome, California	USGS Quadrangle	1:48,000	1950	41°46′N	121°32′W
T-25a	McKittrick Summit, California	USGS Quadrangle	1:24,000	1978	35°15′N	119°48′W
T-25b	McKittrick Summit, California	US Topo Imagery	1:24,000	2010	35°15′N	119°48′W
T-26a	Spring Hill, Idaho	US Topo Quadrangle	1:24,000	2011	44°16′N	113°41′W
T-26b	Mount Higgins, Washington	US Topo Quadrangle	1:24,000	2011	48°16′N	121°48′W
T-27a	Grotto Canyon, California	US Topo Imagery	1:24,000	2010	36°35′N	117°04′W
T-27b	Rome, Wisconsin	US Topo Quadrangle	1:43,000	2015	42°56′N	88°38′W
T-28a	January Sea-Level Temperature Map	Hess, *McKnight's Physical Geography*, 12th ed.				
T-28b	July Sea-Level Temperature Map	Hess, *McKnight's Physical Geography*, 12th ed.				
T-29	Climate Regions Map	Hess, *McKnight's Physical Geography*, 12th ed.				
T-30	Major Biomes Map	Hess, *McKnight's Physical Geography*, 12th ed.				
T-31a	Vegetation Map	NASA Imagery (2007)				
T-31b	McMurry, Canada	NASA Imagery (2016)				
T-31c	Los Angeles Traffic Map	Google MapsTM				
T-32a	Uinta Moutains, Utah	Esri ArcGIS Online/U.S. Forest Service				
T-32b	Uinta Moutains, Utah	Esri ArcGIS Online/U.S. Forest Service				
T-32c	Uinta Moutains, Utah	Esri ArcGIS Online/U.S. Forest Service				

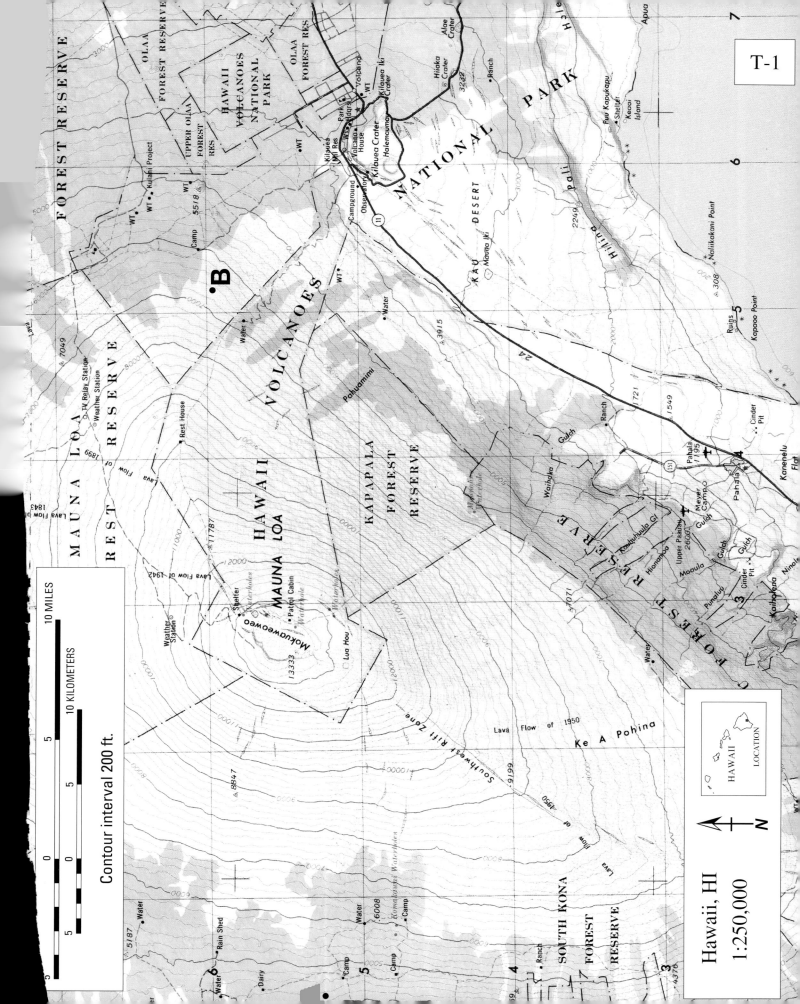

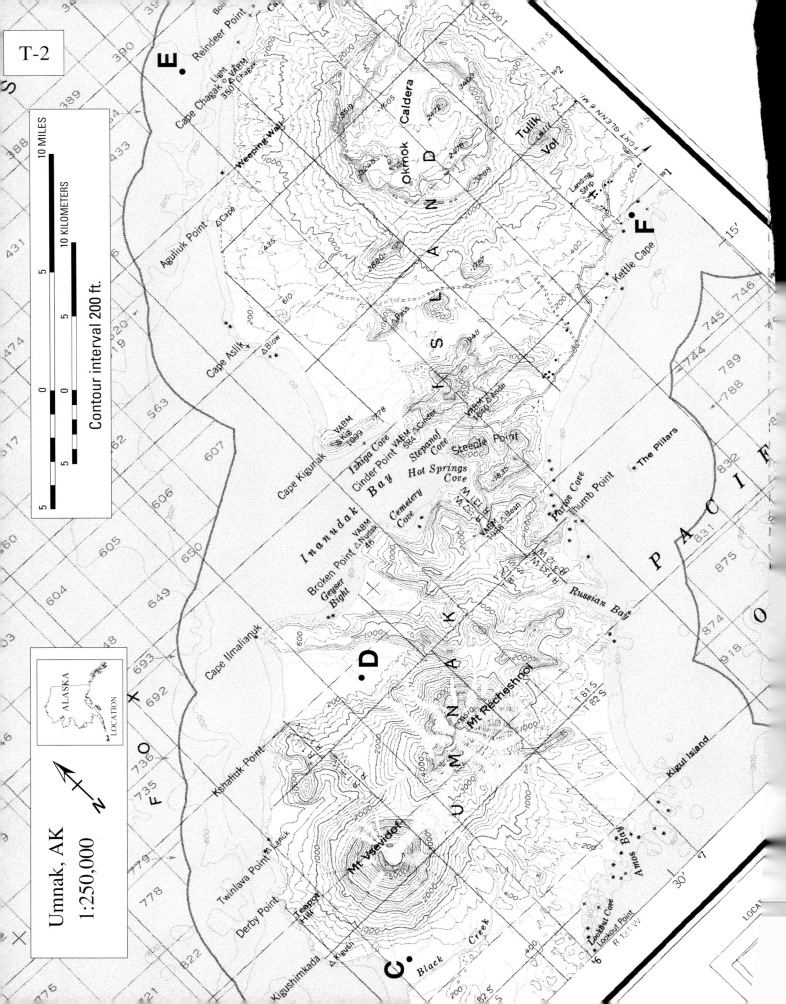

T-2

Umnak, AK
1:250,000

ALASKA

LOCATION

N

Contour interval 200 ft.

10 MILES

10 KILOMETERS

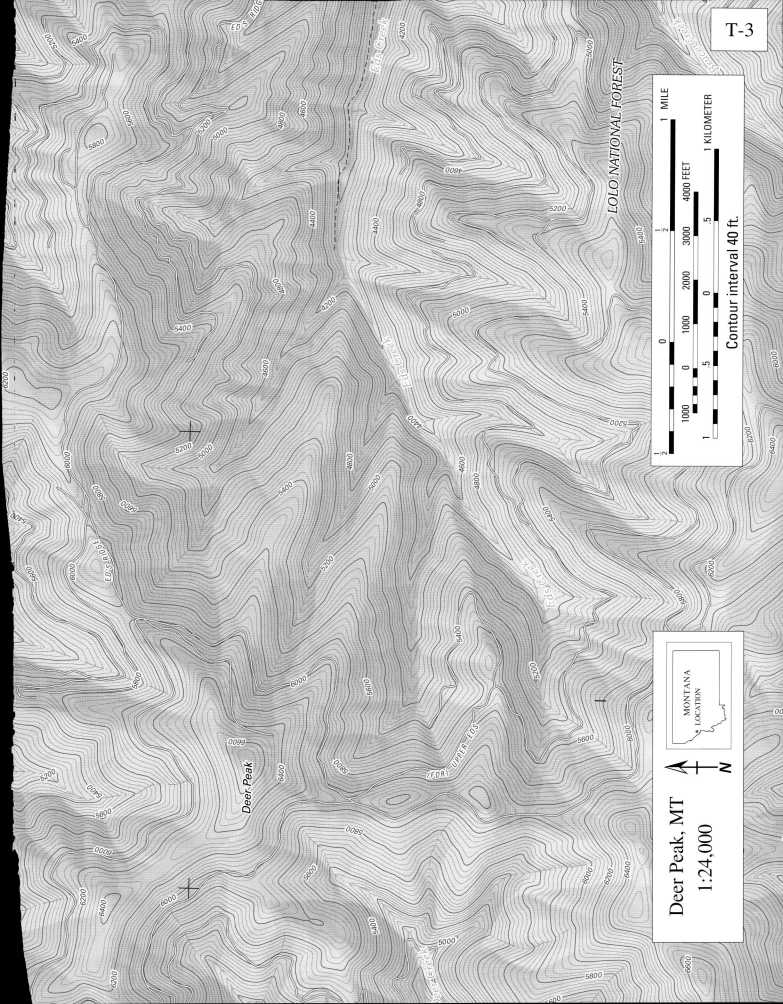

T-3

Deer Peak, MT
1:24,000

MONTANA
LOCATION

N

Contour interval 40 ft.

LOLO NATIONAL FOREST

Deer Peak

ED'S RIDGE

Ed's Creek

Eds Creek

Ed's Creek

(FDR) UPPER EDS

T-4

Voltaire, ND
1:24,000

LOCATION
NORTH DAKOTA

N

Contour interval 5 ft.

1 MILE

1 KILOMETER

Jackson, MS-LA
1:250,000

N

MISSISSIPPI
LOCATION

T-5

Contour interval 50 ft.

0   5   10 MILES

5   0   5   10 KILOMETERS

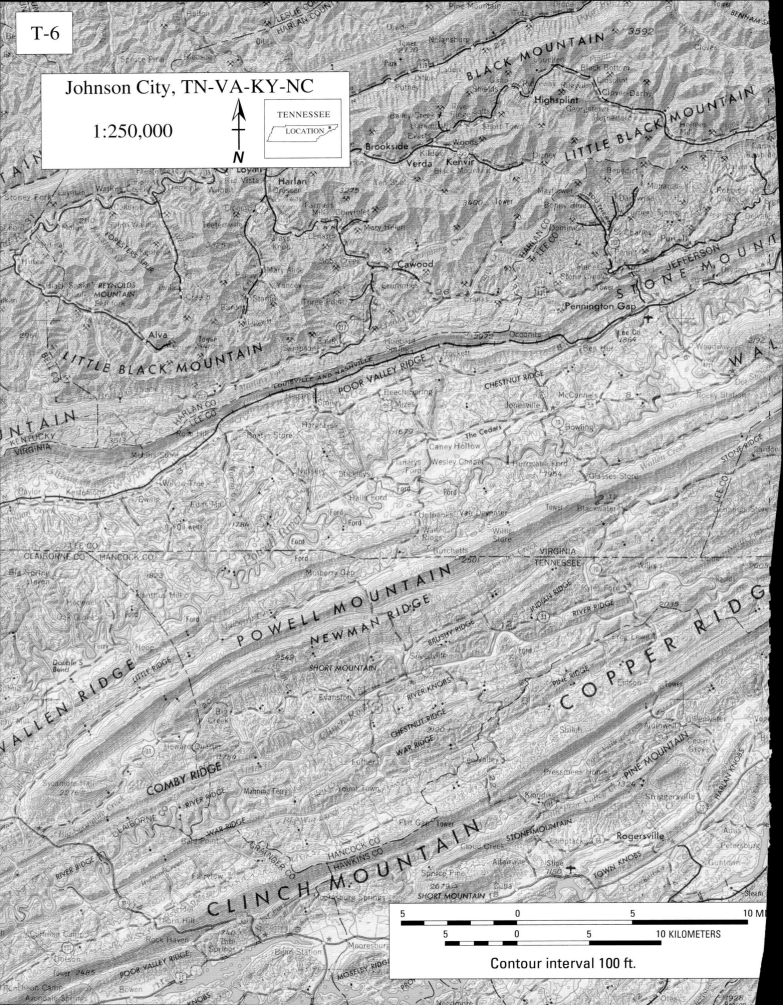

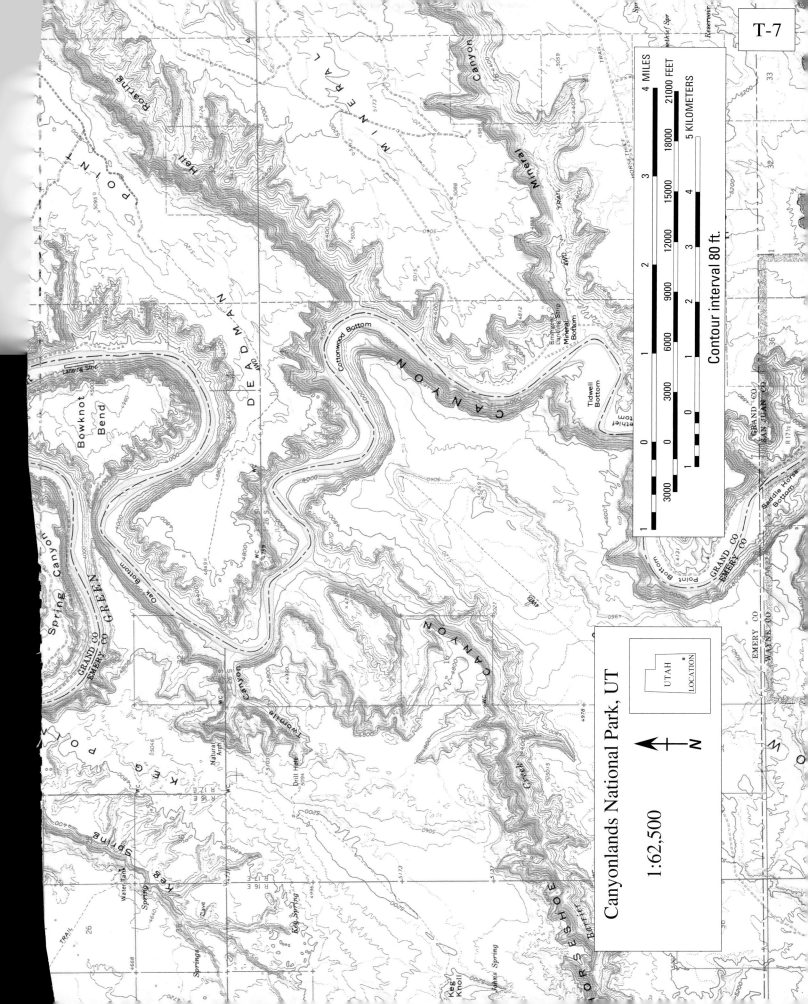

T-7

Canyonlands National Park, UT

1:62,500

N

Contour interval 80 ft.

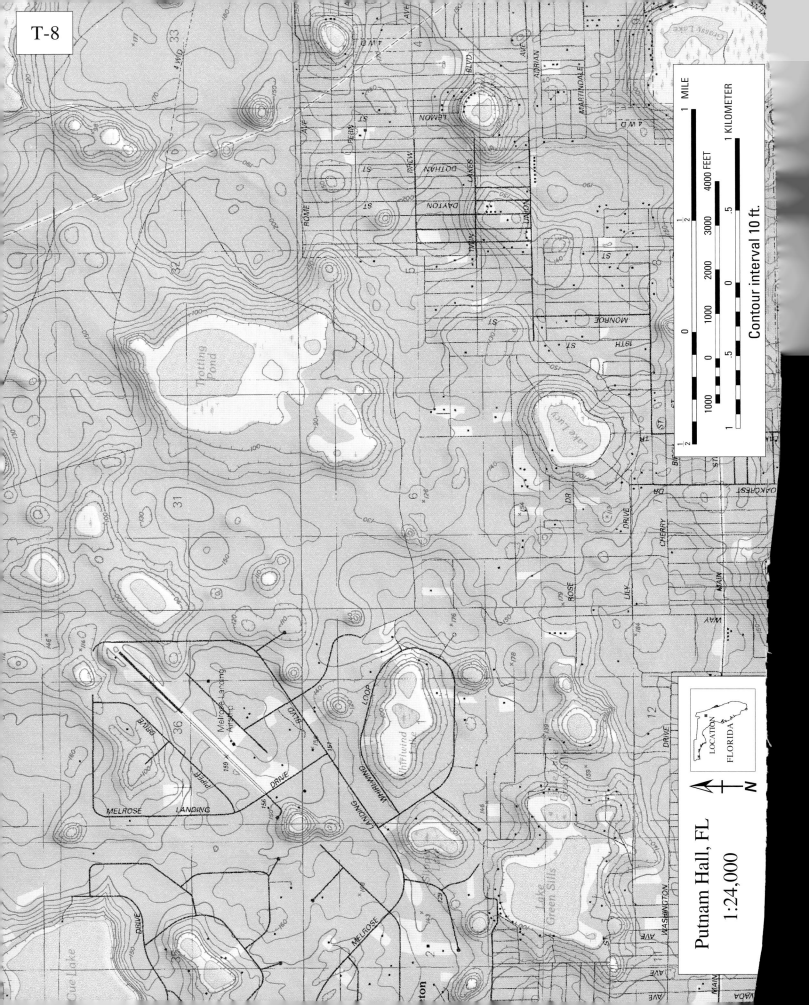

T-8

Putnam Hall, FL
1:24,000

LOCATION
FLORIDA

N

Contour interval 10 ft.

1 MILE

1 KILOMETER

T-9

The Knobs

DOYLE RD

255

SLAVE CAVE RD

MAMMOTH CAVE
NATIONAL PARK

Zion Cem

Bald
Knob

Opossum
Hollow

ED PARSLEY RD

BALD KNOB RD

PARK MAMMOTH RD

Oil
Well

RIHERD ESTATES RD

LOUISVILLE RD

LOUISVILLE RD

31

W FIRST ST

BM 635

650

STURGEON RD

BM
615

65

255

BM
585

Oil
Well

Oil
Well

Gardner Cr

BM
618

Oil
Well

Walnut Hill
Church

Oil Wells

RAY HOUCHIN RD

CRUMP RD

STINNET RD

IRON MOUNTAIN RD

MILLSTOWN RD

255

Oil
Well

BM 637

Fairview
Church

Oil Wells

Park City, KY
1:24,000

KENTUCKY

LOCATION

N

| 1/2 | | 0 | | 1/2 | | 1 MILE |

| 1000 | 0 | 1000 | 2000 | 3000 | 4000 FEET |

| 1 | | .5 | | 0 | | .5 | | 1 KILOMETER |

Contour interval 10 ft.

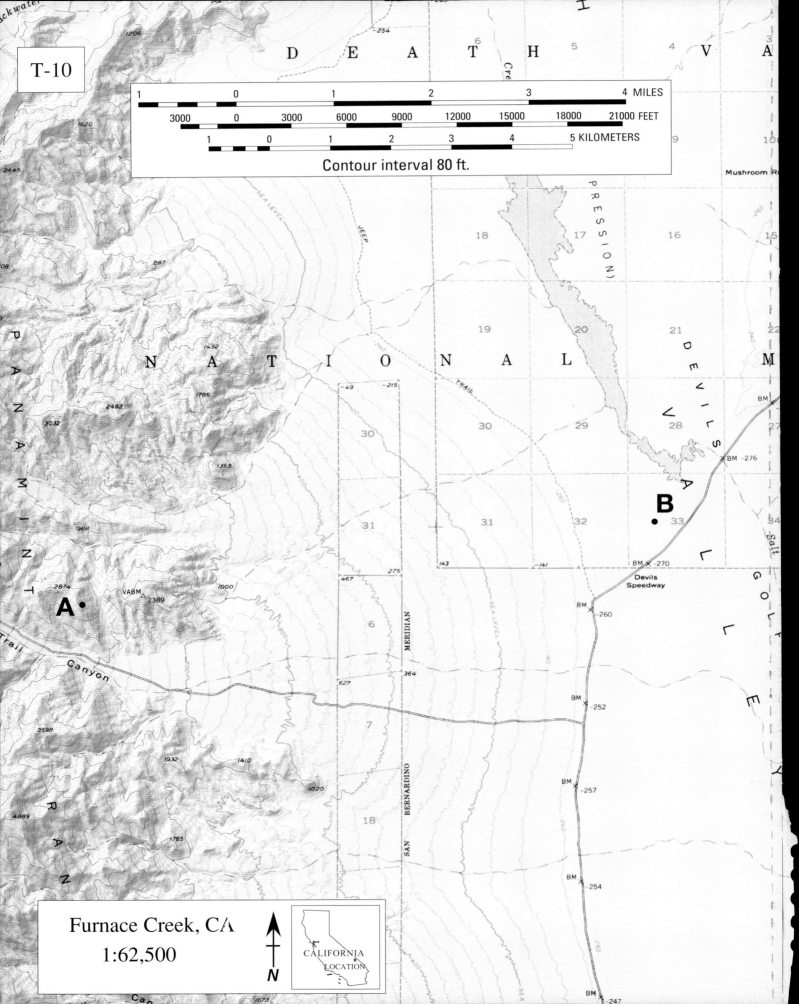

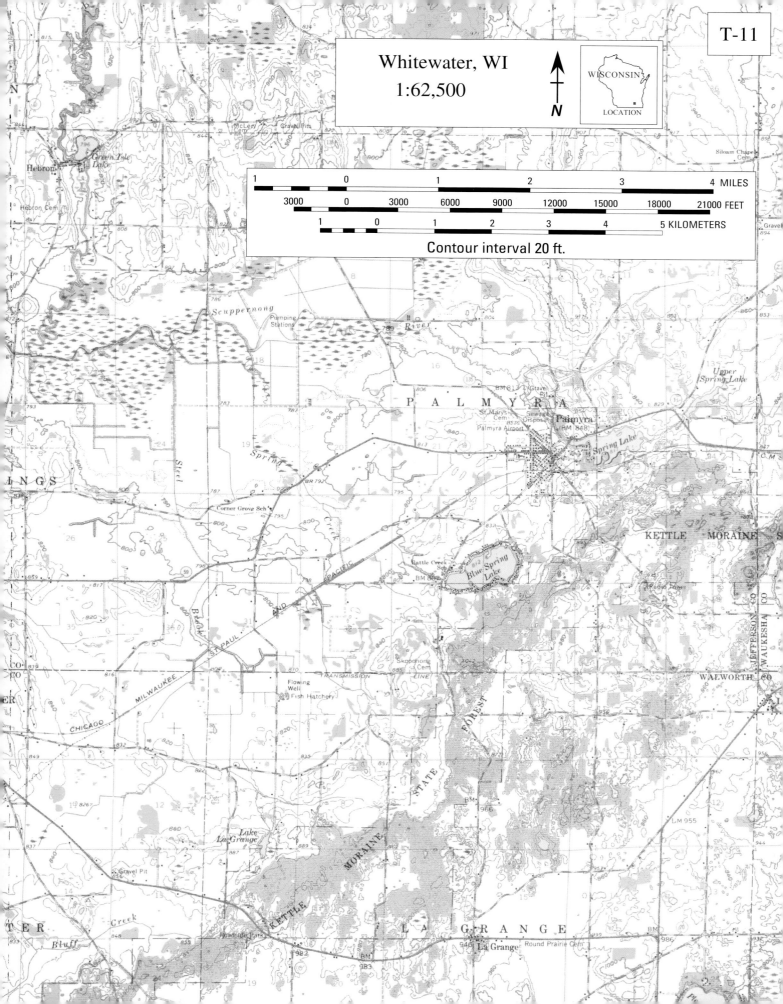

Whitewater, WI
1:62,500

WISCONSIN
LOCATION

| 1 | 0 | 1 | 2 | 3 | 4 MILES |

| 3000 | 0 | 3000 | 6000 | 9000 | 12000 | 15000 | 18000 | 21000 FEET |

| 1 | 0 | 1 | 2 | 3 | 4 | 5 KILOMETERS |

Contour interval 20 ft.

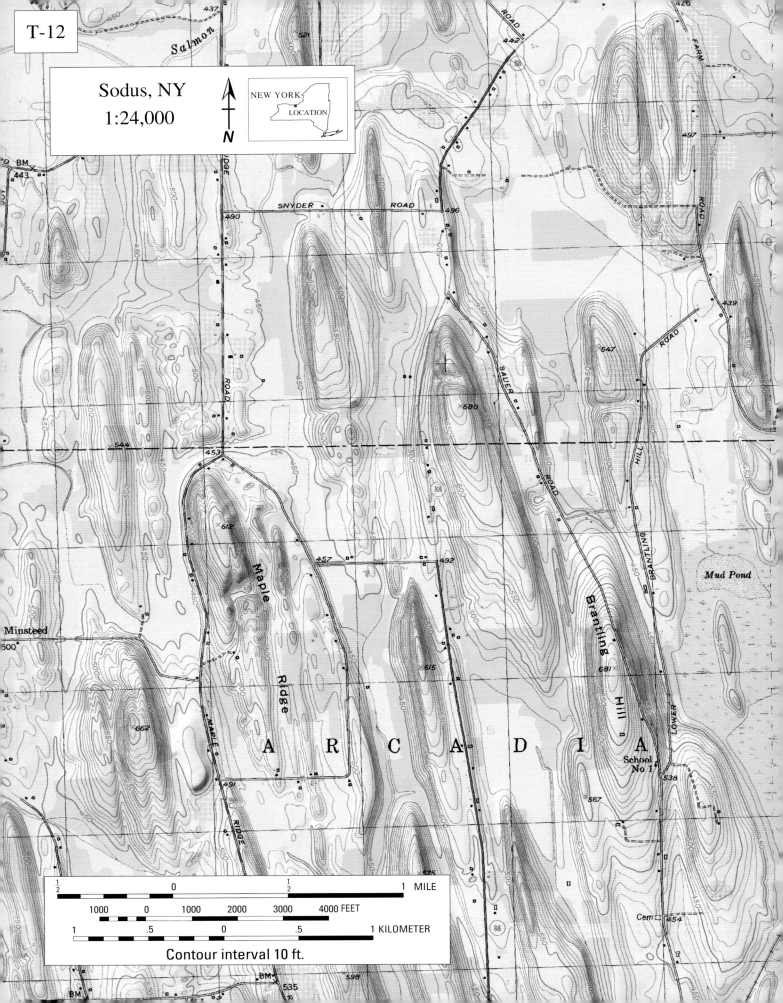

Mono Craters, CA
1:62,500

Contour interval 80 ft.

T-13

LOCATION
CALIFORNIA

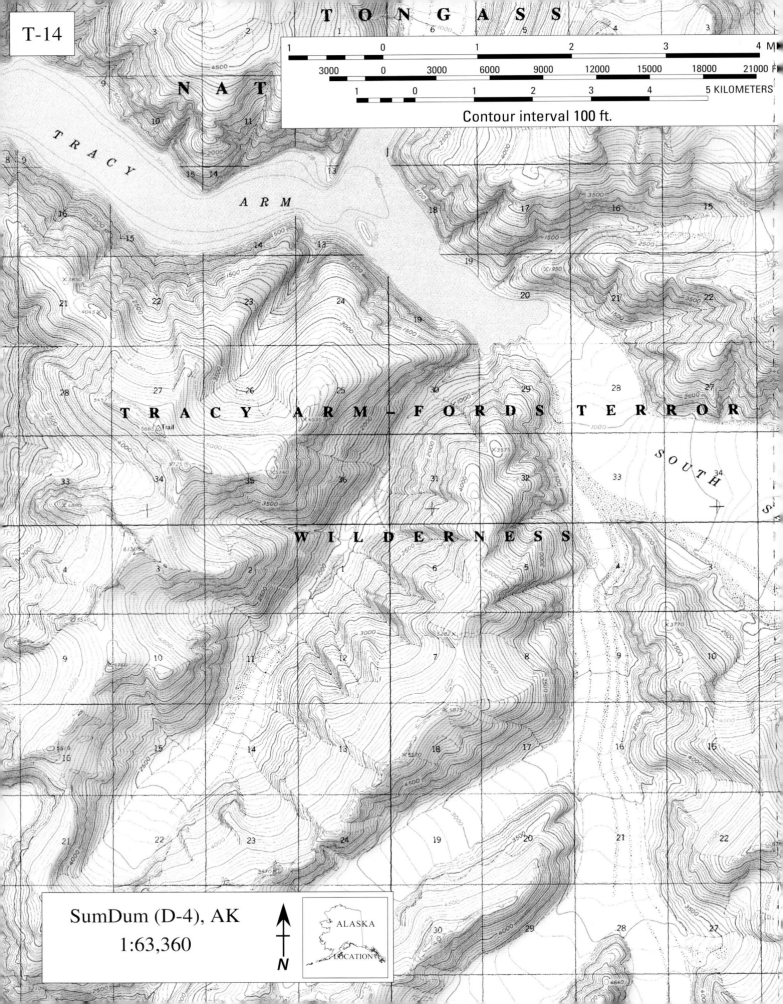

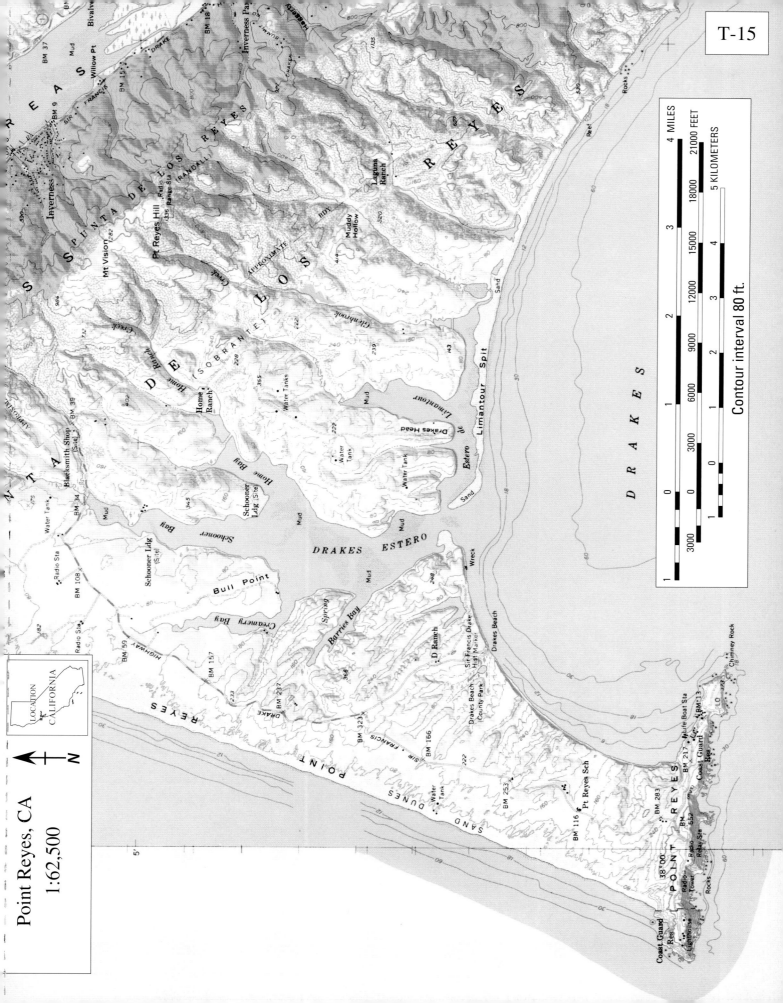

Point Reyes, CA
1:62,500

LOCATION
CALIFORNIA

N

T-15

Contour interval 80 ft.

T-16

Corpus Christi, TX
1:250,000

TEXAS
LOCATION

N

5   0   5   10 MILES

5   0   5   10 KILOMETERS

Contour interval 10 meters

Antelope Peak, AZ
1:62,500

ARIZONA
LOCATION

3550 III
(MOBILE)

UNITED STATES
DEPARTMENT OF THE INTERIOR
GEOLOGICAL SURVEY

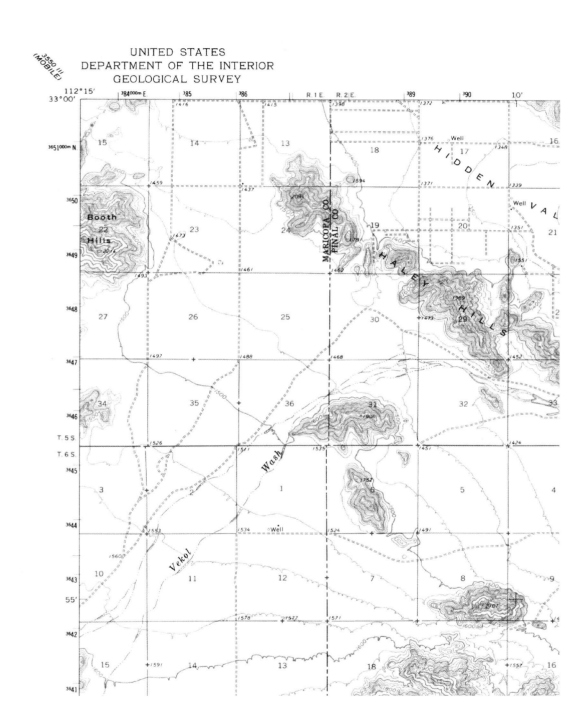

Contour interval 25 ft.

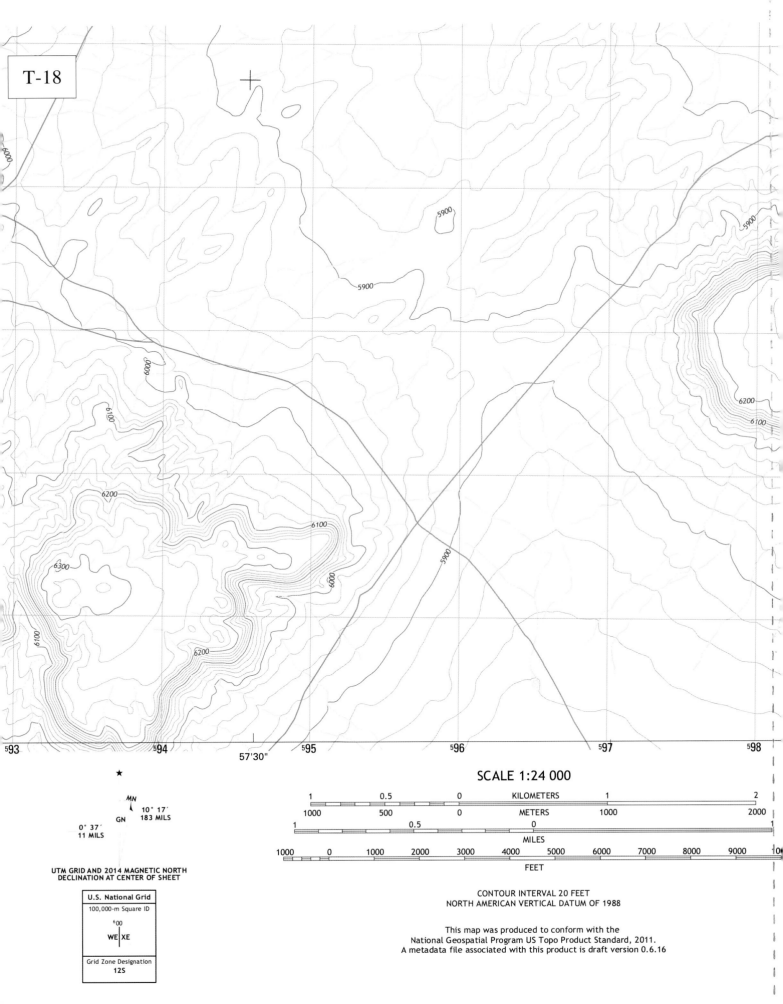

T-18

★

MN

10° 17'
183 MILS

GN

0° 37'
11 MILS

UTM GRID AND 2014 MAGNETIC NORTH
DECLINATION AT CENTER OF SHEET

U.S. National Grid

100,000-m Square ID

⁶00

WE | XE

Grid Zone Designation
12S

57'30"

SCALE 1:24 000

| 1 | 0.5 | 0 | KILOMETERS | 1 | 2 |

| 1000 | 500 | 0 | METERS | 1000 | 2000 |

| 1 | 0.5 | 0 | | | 1 |

MILES

| 1000 | 0 | 1000 | 2000 | 3000 | 4000 | 5000 | 6000 | 7000 | 8000 | 9000 | 10 |

FEET

CONTOUR INTERVAL 20 FEET
NORTH AMERICAN VERTICAL DATUM OF 1988

This map was produced to conform with the
National Geospatial Program US Topo Product Standard, 2011.
A metadata file associated with this product is draft version 0.6.16

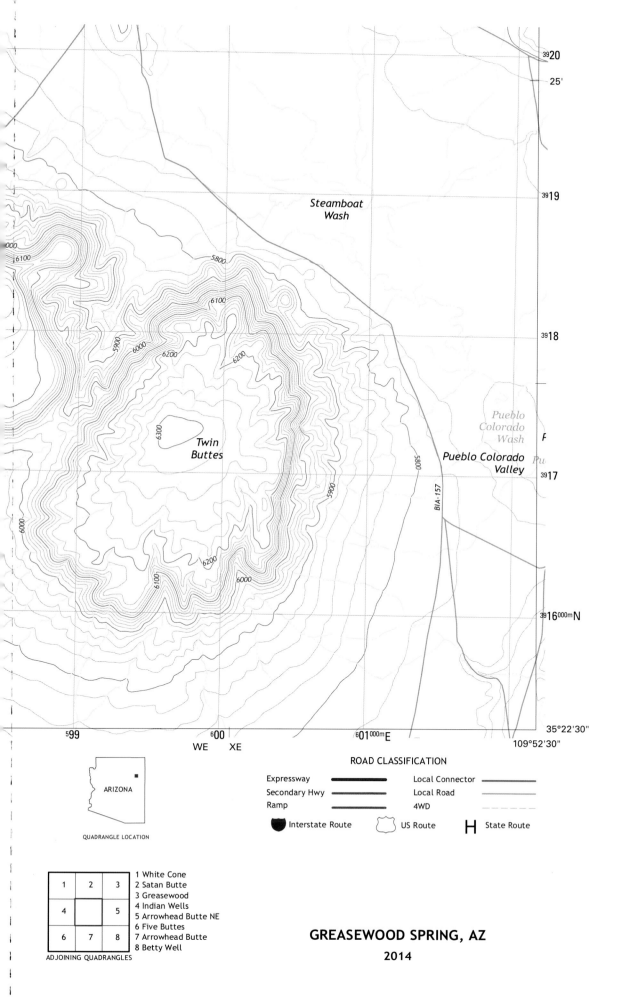

T-19

39**20**

25'

39**19**

Steamboat
Wash

6100

5800

6100

5900

6000

6200

6200

6000

6300

Twin
Buttes

39**18**

Pueblo
Colorado
Wash

F

5800

Pueblo Colorado
Valley

Pu

39**17**

BIA-157

5900

6200

6000

6100

6000

6000

39**16**000mN

35°22'30"

5**99**

6**00**

WE   XE

6**01**000mE

109°52'30"

ROAD CLASSIFICATION

ARIZONA

Expressway		Local Connector	
Secondary Hwy		Local Road	
Ramp		4WD	

Interstate Route   US Route   **H** State Route

QUADRANGLE LOCATION

1	2	3
4		5
6	7	8

ADJOINING QUADRANGLES

1 White Cone
2 Satan Butte
3 Greasewood
4 Indian Wells
5 Arrowhead Butte NE
6 Five Buttes
7 Arrowhead Butte
8 Betty Well

GREASEWOOD SPRING, AZ

2014

T-20

Twomile Canyon

Horseshoe Canyon

Green River

Half-circle-shaped hill

T-20a
Bowknot Bend, UT
1:24,000; contour interval 40 ft.

T-20b
Bowknot Bend, UT
US Topo imagery (2011)
1:24,000

N

T-21a
USGS Landsat 8 image of Mauna Loa
volcano on the Big Island of Hawai'i (2014)
1:387,000

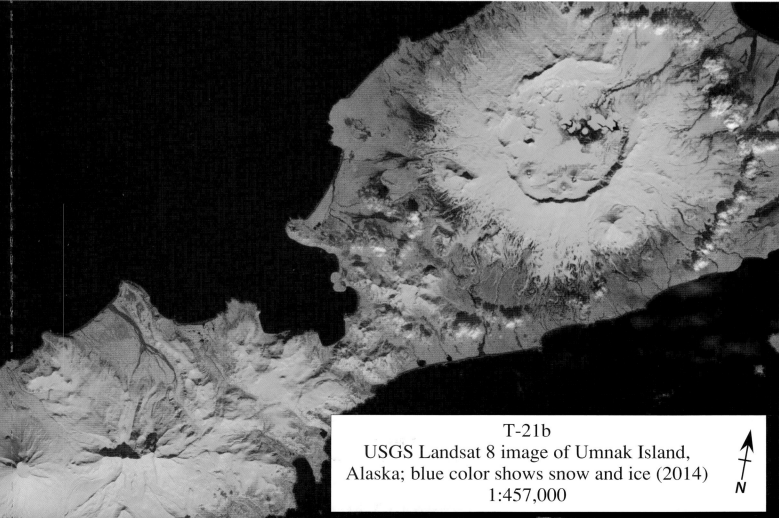

T-21b
USGS Landsat 8 image of Umnak Island,
Alaska; blue color shows snow and ice (2014)
1:457,000

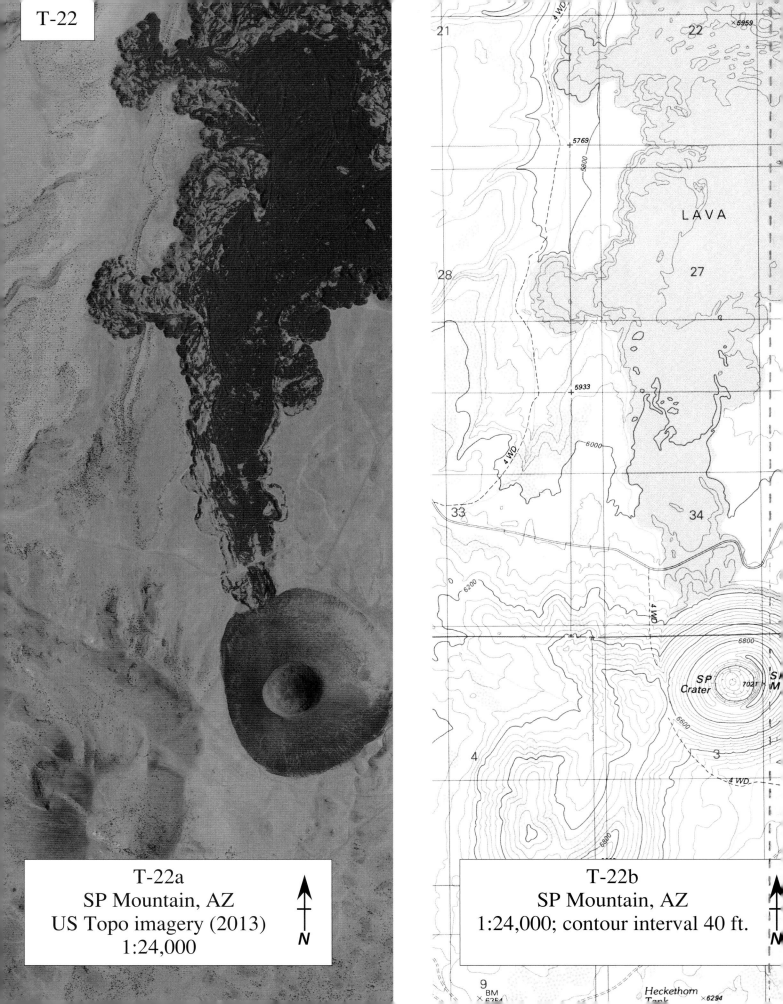

T-22

T-22a
SP Mountain, AZ
US Topo imagery (2013)
1:24,000

N

T-22b
SP Mountain, AZ
1:24,000; contour interval 40 ft.

N

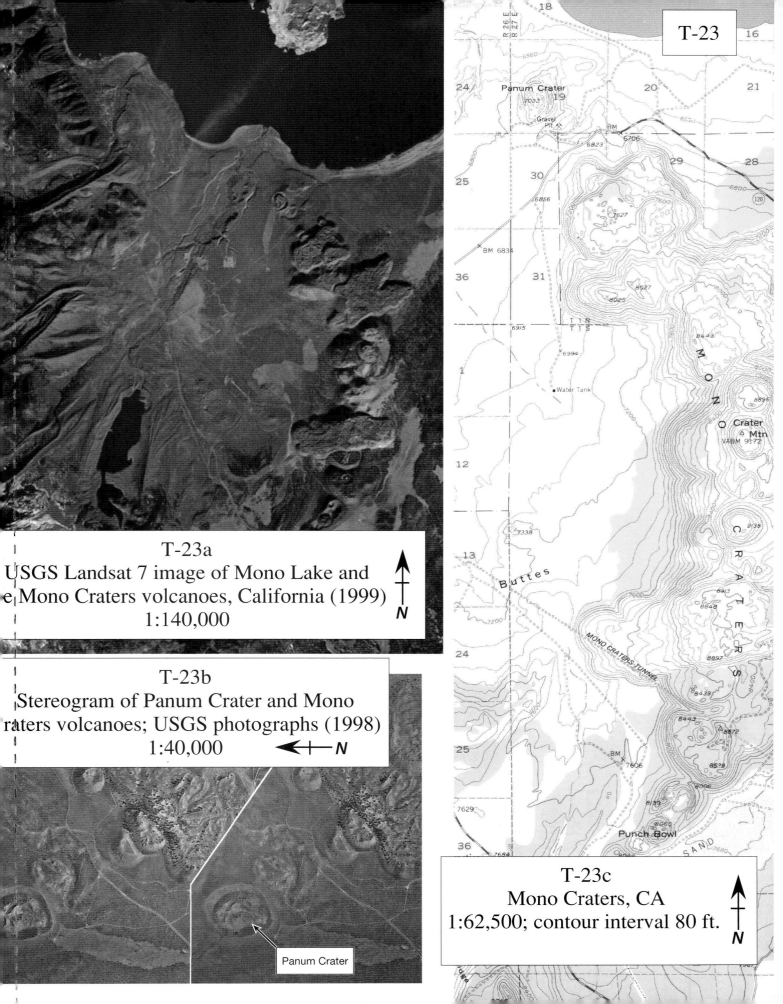

T-23

T-23a
USGS Landsat 7 image of Mono Lake and
the Mono Craters volcanoes, California (1999)
1:140,000

N

T-23b
Stereogram of Panum Crater and Mono
Craters volcanoes; USGS photographs (1998)
1:40,000

←—N

Panum Crater

T-23c
Mono Craters, CA
1:62,500; contour interval 80 ft.

N

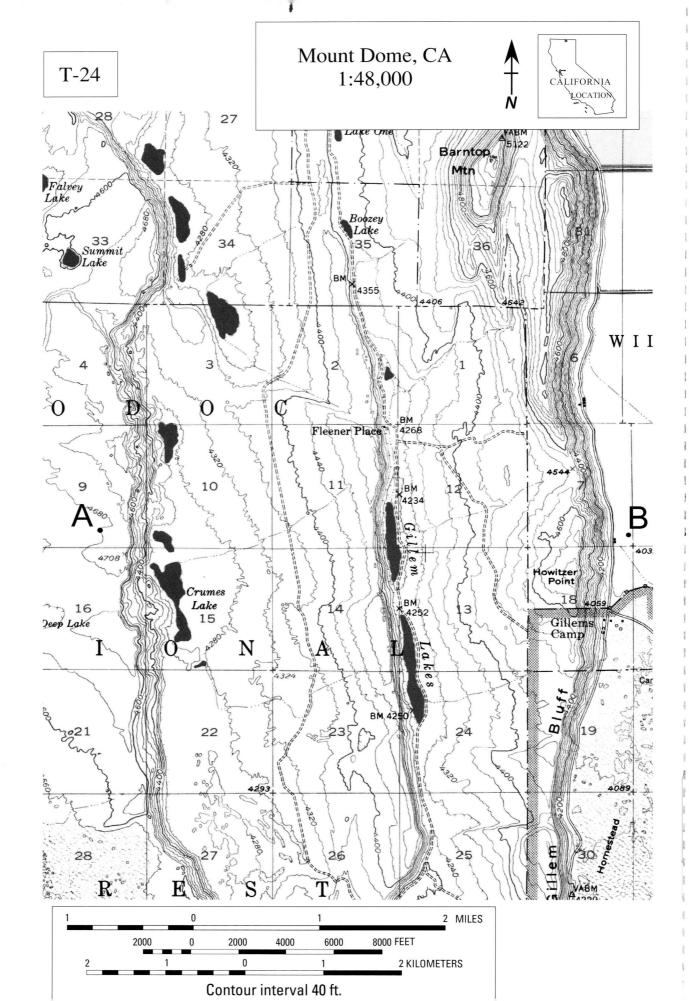

Mount Dome, CA
1:48,000

T-24

CALIFORNIA LOCATION

| 1 | 0 | 1 | 2 MILES |

| 2000 | 0 | 2000 | 4000 | 6000 | 8000 FEET |

| 2 | 1 | 0 | 1 | 2 KILOMETERS |

Contour interval 40 ft.

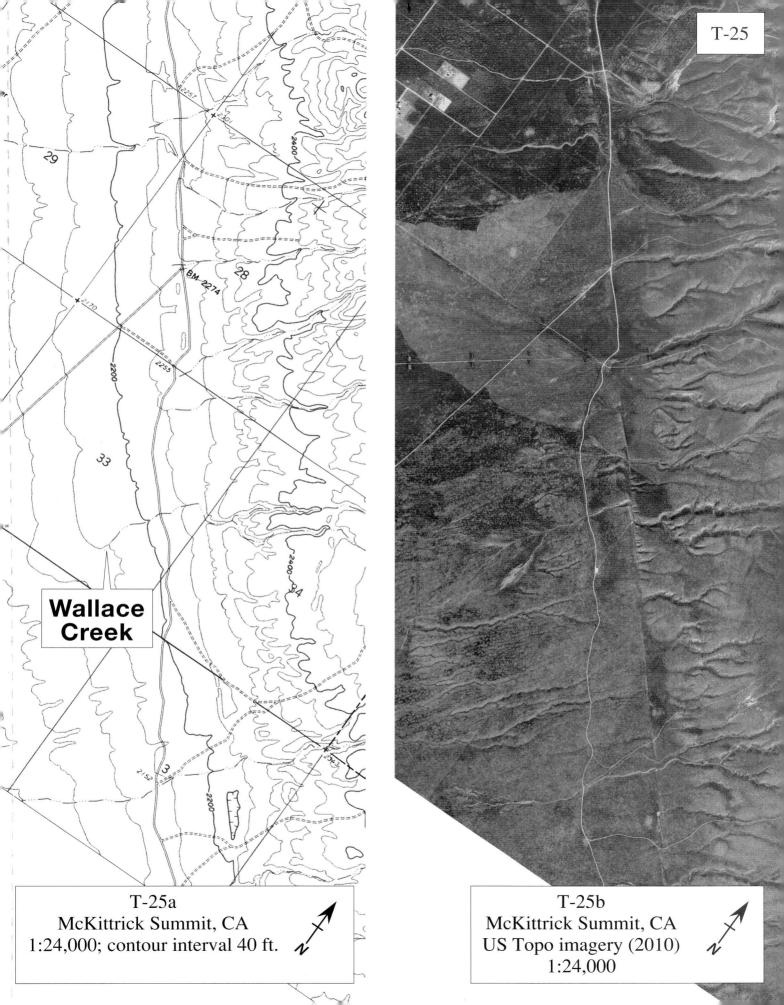

Wallace Creek

T-25

T-25a
McKittrick Summit, CA
1:24,000; contour interval 40 ft.

T-25b
McKittrick Summit, CA
US Topo imagery (2010)
1:24,000

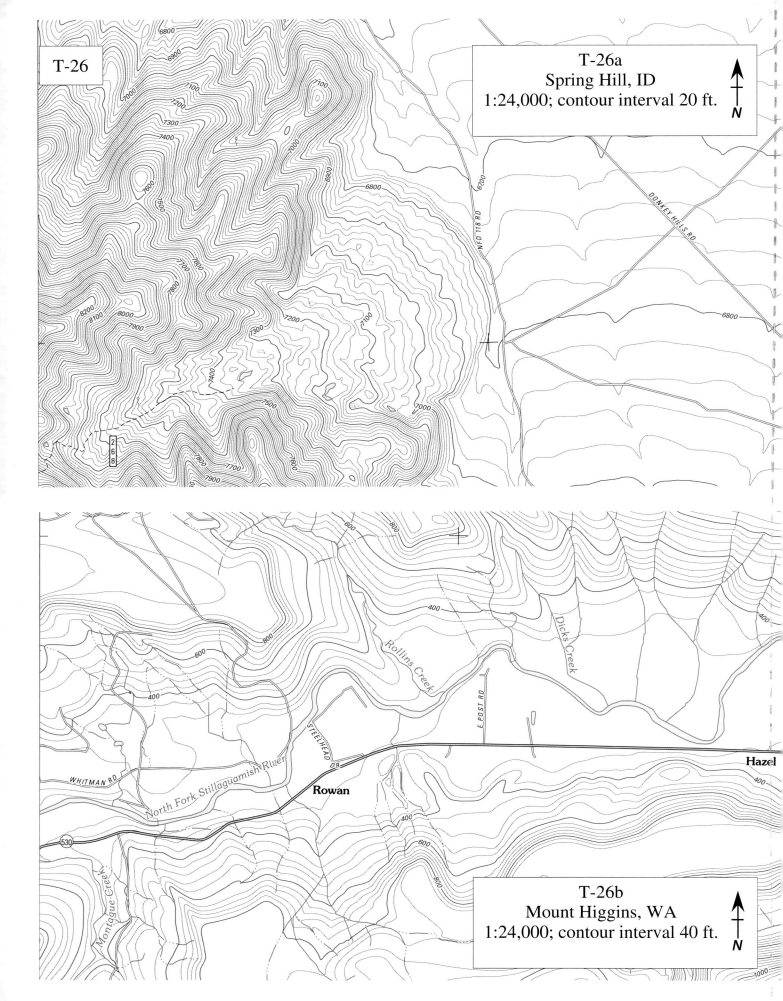

T-26

T-26a
Spring Hill, ID
1:24,000; contour interval 20 ft.

N

T-26b
Mount Higgins, WA
1:24,000; contour interval 40 ft.

N

T-27

T-27a
Grotto Canyon, CA
US Topo imagery (2010)
1:24,000
N

T-27b
Rome, WI
:43,000; contour interval 10 ft.
N

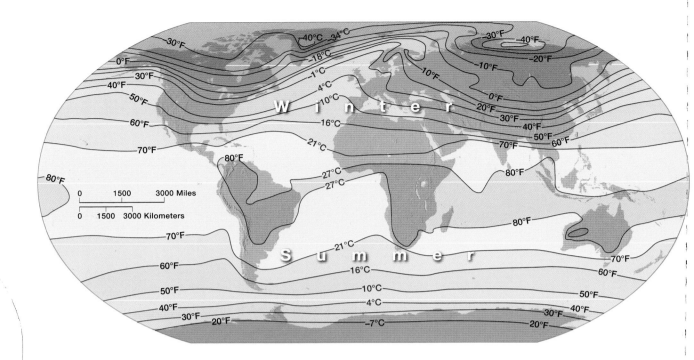

T-28a
Average January sea-level temperatures. (From Hess,
McKnight's Physical Geography, 12th ed.).

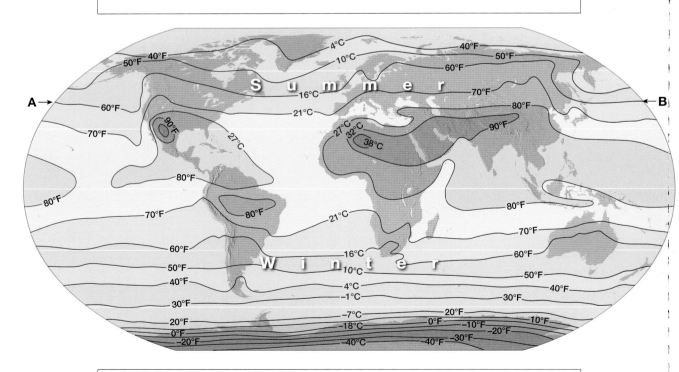

T-28b
Average July sea-level temperatures. (From Hess,
McKnight's Physical Geography, 12th ed.).

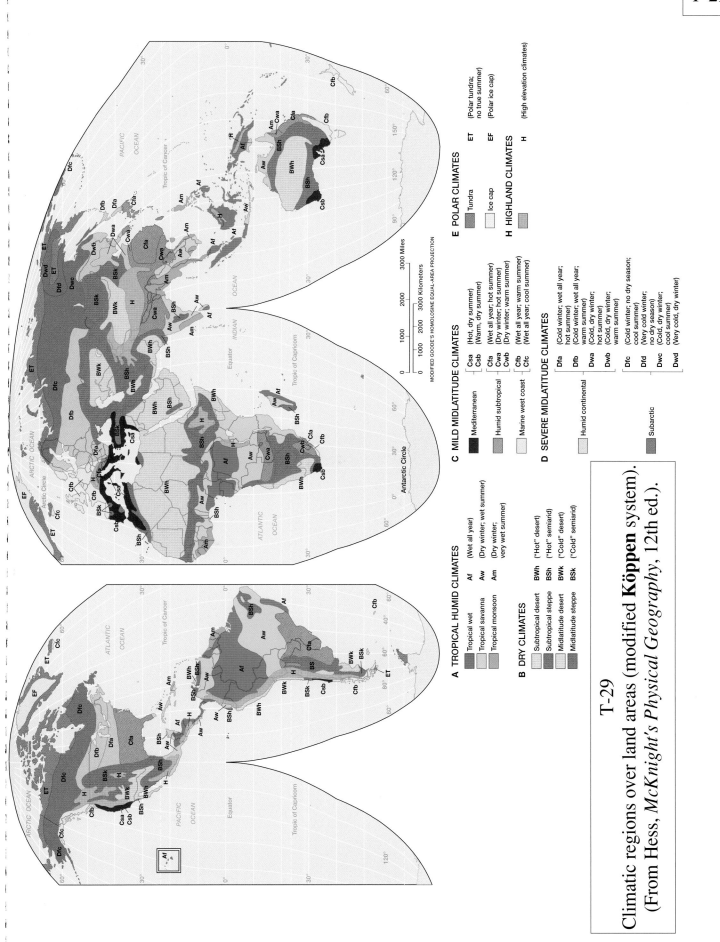

T-29

Climatic regions over land areas (modified **Köppen** system). (From Hess, *McKnight's Physical Geography*, 12th ed.).

A TROPICAL HUMID CLIMATES

Tropical wet	Af	(Wet all year)
Tropical savanna	Aw	(Dry winter; wet summer)
Tropical monsoon	Am	(Dry winter; very wet summer)

B DRY CLIMATES

Subtropical desert	BWh	("Hot" desert)
Subtropical steppe	BSh	("Hot" semiarid)
Midlatitude desert	BWk	("Cold" desert)
Midlatitude steppe	BSk	("Cold" semiarid)

C MILD MIDLATITUDE CLIMATES

Mediterranean	Csa	(Hot, dry summer)
	Csb	(Warm, dry summer)
Humid subtropical	Cfa	(Wet all year; hot summer)
	Cwa	(Dry winter; hot summer)
	Cwb	(Dry winter; warm summer)
Marine west coast	Cfb	(Wet all year; warm summer)
	Cfc	(Wet all year; cool summer)

D SEVERE MIDLATITUDE CLIMATES

Humid continental	Dfa	(Cold winter; wet all year; hot summer)
	Dfb	(Cold winter; wet all year; warm summer)
	Dwa	(Cold, dry winter; hot summer)
	Dwb	(Cold, dry winter; warm summer)
Subarctic	Dfc	(Cold winter; no dry season; cool summer)
	Dfd	(Very cold winter; no dry season)
	Dwc	(Cold, dry winter; cool summer)
	Dwd	(Very cold, dry winter)

E POLAR CLIMATES

| Tundra | ET | (Polar tundra; no true summer) |
| Ice cap | EF | (Polar ice cap) |

H HIGHLAND CLIMATES

| | H | (High elevation climates) |

MODIFIED GOODE'S HOMOLOSINE EQUAL-AREA PROJECTION

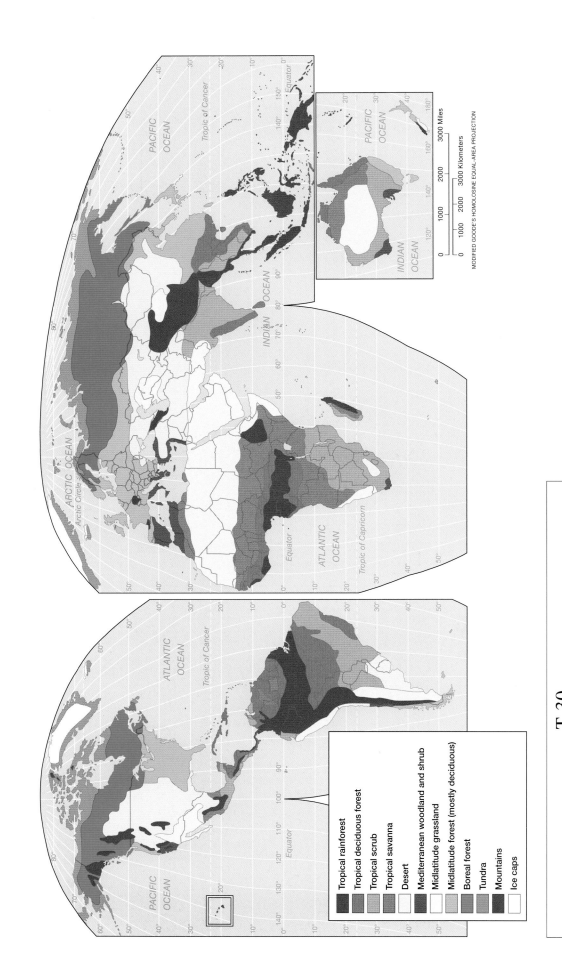

T-30
Major Biomes of the world.
(From Hess, *McKnight's Physical Geography*, 12th ed.).

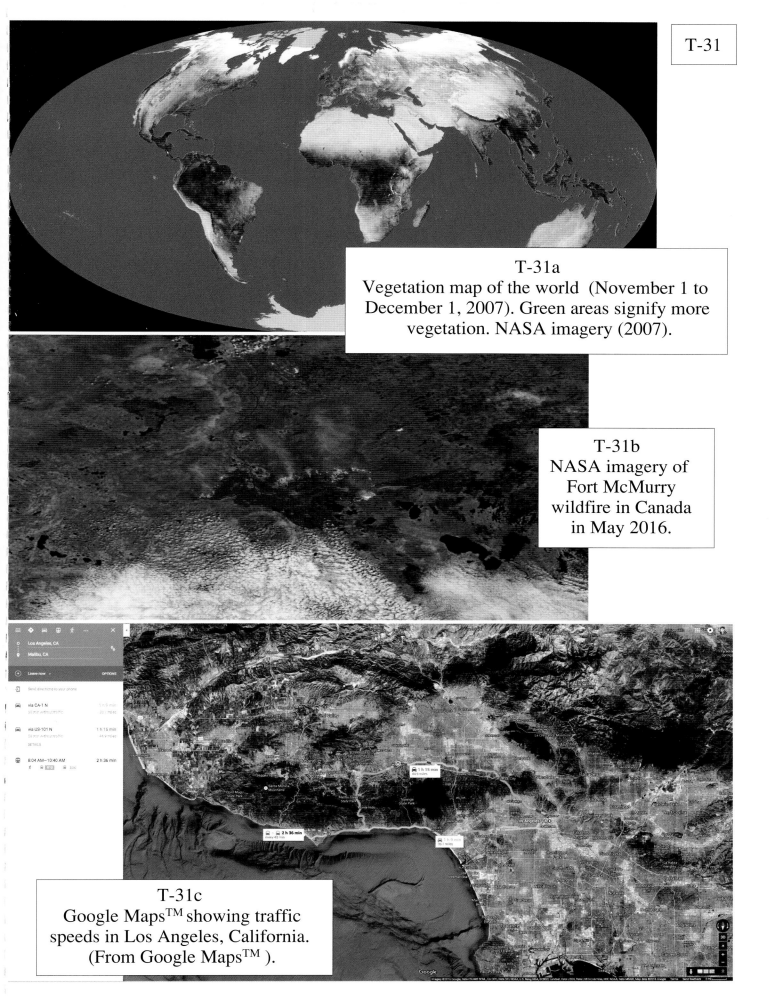

T-31a
Vegetation map of the world (November 1 to December 1, 2007). Green areas signify more vegetation. NASA imagery (2007).

T-31b
NASA imagery of Fort McMurry wildfire in Canada in May 2016.

T-31c
Google Maps™ showing traffic speeds in Los Angeles, California. (From Google Maps™).

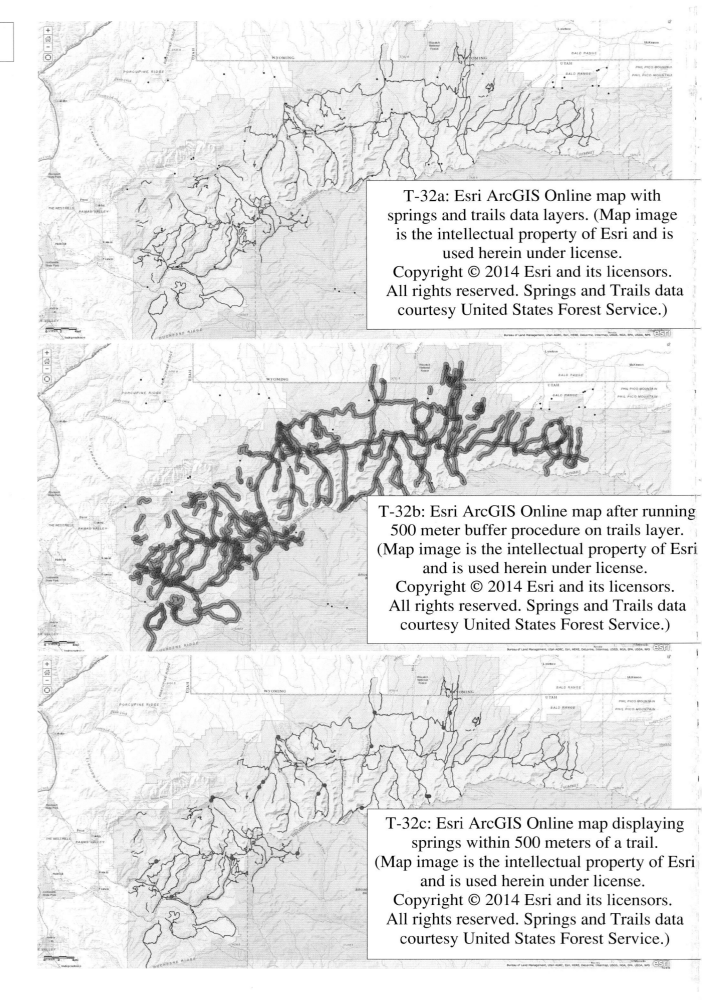

T-32

T-32a: Esri ArcGIS Online map with springs and trails data layers. (Map image is the intellectual property of Esri and is used herein under license.
Copyright © 2014 Esri and its licensors. All rights reserved. Springs and Trails data courtesy United States Forest Service.)

T-32b: Esri ArcGIS Online map after running 500 meter buffer procedure on trails layer. (Map image is the intellectual property of Esri and is used herein under license.
Copyright © 2014 Esri and its licensors. All rights reserved. Springs and Trails data courtesy United States Forest Service.)

T-32c: Esri ArcGIS Online map displaying springs within 500 meters of a trail. (Map image is the intellectual property of Esri and is used herein under license.
Copyright © 2014 Esri and its licensors. All rights reserved. Springs and Trails data courtesy United States Forest Service.)